Nanowire Field Effect Transistors: Principles and Applications

Dae Mann Kim · Yoon-Ha Jeong
Editors

Nanowire Field Effect Transistors: Principles and Applications

Springer

Editors
Dae Mann Kim
Korea Institute for Advanced Study
Seoul
South Korea (Republic of Korea)

Yoon-Ha Jeong
Department of Creative IT Excellence Engineering
POSTECH
Gyeongbuk
South Korea (Republic of Korea)

ISBN 978-1-4939-4572-6 ISBN 978-1-4614-8124-9 (eBook)
DOI 10.1007/978-1-4614-8124-9
Springer New York Heidelberg Dordrecht London

Softcover reprint of the hardcover 1st edition 2014

Springer is part of Springer Science+Business Media (www.springer.com)

Contents

Chapter 1
Quantum Wire and Sub-Bands

Dae Mann Kim, Bong Koo Kang and Yoon-Ha Jeong

Abstract The quantum mechanics is the basic science supporting the nanotechnology and most of the key concepts operative in the nano electronic devices are derived from it. In terms of the impact made by the theory, the quantum mechanics is onc of thc grcatcst thcorics. In this chapter, the essential features of the quantum mechanics are compactly highlighted, using a few examples. The examples chosen for discussion are directly related to the design and operation of the silicon nano wire field-effect transistor. Specifically, the sub-bands formed in the nano wire are considered, together with the density of states.

1.1 Schrödinger Equation

The classical mechanics is based upon Newton's equation of motion. Similarly, the quantum mechanics is based upon the wave equation, called the Schrödinger equation. These two equations represent the basic postulates, the validity of which is to be proven by the agreement between the theoretical result derived from it and experimental data. The Schrödinger equation reads as

$$ih\frac{\partial\psi(\underline{r},t)}{\partial t} = \hat{H}\psi(\underline{r},t), \hbar \equiv h/2\pi \tag{1.1}$$

D. M. Kim (✉)
Korea Institute for Advanced Study, Seoul, Korea
e-mail: dmkim@kias.re.kr

B. K. Kang · Y.-H. Jeong
POSTECH, Pohang, Korea
e-mail: bkkang@postech.ac.kr

Y.-H. Jeong
e-mail: yhjeong@postech.ac.kr

D. M. Kim and Y.-H. Jeong (eds.), *Nanowire Field Effect Transistors: Principles and Applications*, DOI: 10.1007/978-1-4614-8124-9_1,

Here, the Hamiltonian $\hat{H}$ represents the total energy of a system under consideration and naturally consists of the kinetic and potential energies, i.e.,

$$\hat{H} = \frac{p^2}{2m} + V \tag{1.2}$$

where p is the linear momentum, m the mass of the particle, and V the potential energy. The constant h is the universal Planck constant with the value $6.626 \times 10^{-34}\,Js$ and $\Psi(\underline{r}, t)$ is the wavefunction.

A basic postulate of the quantum mechanics is to represent physical quantities by operators. For instance, the momentum and energy are associated with the operators given by

$$p_x \rightarrow -i\hbar\frac{\partial}{\partial x} \tag{1.3}$$

$$E \rightarrow i\hbar\frac{\partial}{\partial t} \tag{1.4}$$

Hence, the Hamiltonian is represented by the operator,

$$\hat{H} = -\frac{\hbar^2}{2m}\left(\frac{\partial^2}{\partial x^2} + \frac{\partial^2}{\partial y^2} + \frac{\partial^2}{\partial z^2}\right) + V \tag{1.5}$$

Upon inserting (1.5) into (1.1), the Schrödinger equation of a particle is given by

$$i\hbar\frac{\partial\psi(\underline{r}, t)}{\partial t} = \left[-\frac{\hbar^2}{2m}\left(\frac{\partial^2}{\partial x^2} + \frac{\partial^2}{\partial y^2} + \frac{\partial^2}{\partial z^2}\right) + V(\underline{r})\right]\psi(\underline{r}, t) \tag{1.6}$$

The equation is the linear, second-order partial differential equation. The bulk of the quantum mechanics is to solve the wave equation and extract the dynamical information of the system under consideration. In so doing, a few additional postulates of quantum mechanics are needed.

Postulates of Quantum Mechanics: The essential contents of the postulates are as follows: (1) a dynamical system, e.g., an atom, molecule, is associated with a wavefunction, $\psi(\underline{r}, t)$ which contains all possible information of the system, (2) the wavefunction evolves in time according the Schrödinger equation, (3) the quantity, $\psi^*(r, t)\psi(r, t)d\underline{r}$ represents the probability of finding the system between the spatial element, $\underline{r}$ and $\underline{r} + d\underline{r}$ at time t, with $\psi^*(r, t)\psi(r, t)$ representing the probability density. Next, the Schrödinger equation is applied to a few simple examples which are intimately related to the nanoelectronic devices.

1.2 Energy Eigenequation

The time-dependent wave Eq. (1.6) can be cast into the time-independent equation by applying the separation of variable technique and looking for the solution in the form,

$$\psi(\underline{r},t) = T(t)u(\underline{r}) = e^{-i(Et/\hbar)}u(\underline{r}) \tag{1.7}$$

where E denotes the energy of the system. Upon inserting (1.7) into (1.6) and dividing both sides by (1.7), one finds

$$E = \frac{\hat{H}u(\underline{r})}{u(\underline{r})} \tag{1.8}$$

That is,

$$\hat{H}u(\underline{r}) = Eu(\underline{r}) \tag{1.9}$$

The resulting time-independent Schrödinger Eq. (1.9) represents the eigenequation, in which an operator, in this case the Hamiltonian $\hat{H}$, acting on u reproduces the same function, $u(\underline{r})$ multiplied by a constant E. Here, u and E are called the energy eigenfunction and eigenvalue, respectively. Quantum mechanics consists essentially of solving the energy eigenequation and obtain the dynamical information out of $u(\underline{r})$.

Momentum Eigenfunction: Consider a free particle moving along the x-axis. The 1-dimensional (1D) eigenequation of the momentum reads as

$$-i\hbar\frac{\partial}{\partial x}u(x) = p_x u(x), \ \hat{p}_x \equiv -i\hbar\frac{\partial}{\partial x} \tag{1.10}$$

where $u(x)$ and p_x are the eigenfunction and eigenvalue of the momentum, respectively.

Equation (1.10) is to be rearranged as

$$\frac{\partial u(x)}{u(x)} = \frac{ip_x}{\hbar}\partial x = ik_x\partial x, \ k_x \equiv p_x/\hbar \tag{1.11}$$

and one can carry out the simple integrations on both sides, obtaining

$$u(x) = N \exp ik_x x \tag{1.12}$$

Here, N is the constant of integration and k_x is called the wave vector and plays an identical role as the optical wave vector.

The wave vector k_x is determined by the boundary conditions imposed. For example, when a periodic boundary condition is imposed in the interval, $0 \leq x \leq L$, i.e., $u(x = 0) = u(x = L)$, then it follows from (1.12) that k_x takes up a discrete set of eigenvalues, satisfying the condition,

$$k_{xn}L = 2\pi n, \ n = 0, \pm 1, \pm 2, \ldots \tag{1.13}$$

The constant of integration, N at our disposal can be used for normalizing the eigenfunction, i.e.,

$$1 = \int_0^L dx u^* u = N^2 L \tag{1.14}$$

Hence, one can write the normalized eigenfunction of the momentum as

$$u_n(x) = \frac{1}{\sqrt{L}} e^{in(2\pi/L)x} \tag{1.15}$$

1.3 Bound States in Quantum Well

A potential well is often called the quantum well. The problem of a particle in a potential well is interesting and useful, providing valuable insights of the bound states. It also offers an example, in which Schrödinger equation can be solved analytically and the general features of bound states are brought out. Moreover, the results obtained are directly applicable to nanoelectronic devices.

Infinite Square-Well Potential: Thus, consider a particle confined in the 1D infinite square-well potential with width L. The potential is then specified as

$$V(x) = \begin{cases} 0 & |x| \leq L \\ \infty & \text{otherwise} \end{cases} \tag{1.16}$$

Inside the well, $V = 0$ and one can write the wave equation as

$$i\hbar \frac{\partial}{\partial t} \psi(x,t) = -\frac{\hbar^2}{2m} \frac{\partial^2}{\partial x^2} \psi(x,t) \tag{1.17}$$

(see 1.6), and look for the solution in the form

$$\psi(x,t) = e^{-iEt/\hbar} u(x) \tag{1.18}$$

and insert (1.18) into (1.17), obtaining

$$\frac{\partial^2 u}{\partial x^2} + k^2 u = 0,\ k^2 \equiv p^2/\hbar^2 = 2mE/\hbar^2 \tag{1.19}$$

where k is the wave vector.

The Eq. (1.19) is identical to the harmonic oscillator equation, with t replaced by x. Hence, cos kx, sin kx are the possible solutions, even and odd in x. Moreover, the infinite barrier height requires that $u(x)$ vanish at the edges of the well, so that the probability of finding the particle outside the well is kept zero. This boundary condition is readily satisfied by choosing the solution, sin kx. The eigenfunction satisfying the boundary condition $u_n(x = 0) = u_n(x = L) = 0$ is then given by

$$u_n(x) \propto \sin k_n x,\ k_n = n\pi/L,\ n = 1, 2, \ldots \tag{1.20}$$

Since u^*u is the probability density, its integration over the potential well should be equal to unity, i.e.,

$$N^2 \int_{-0}^{L} dx\, u^*(x)u(x) = 1,\ u(x) = N \sin kx \tag{1.21}$$

One can find N by performing the integration and write the normalized eigenfunction as

$$u_n(x) = (2/L)^{1/2} \sin k_n x,\ k_n = 2n\pi/L,\ n = 1,2,3,\ldots \tag{1.22}$$

The energy levels resulting from (1.22) are given by

$$E_n = \frac{\hbar^2 \pi^2}{2mL^2} n^2;\quad n = 1,\ 2,\ 3,\ldots \tag{1.23}$$

The set of discrete energy levels as given in (1.23) is a typical example of the energy quantization. Evidently, the quantization arises from the boundary conditions imposed on the eigenfunctions. The integer, n appearing (1.23) is called the quantum number and each eigenfunction in (1.22) represents a single quantum state.

Exclusion principle: A class of particles called fermions, such as electrons, holes, and protons, etc., obey a fundamental principle, namely the Pauli exclusion principle, according to which a quantum state can accommodate only a single fermion. This principle plays a key role for specifying concentrations of electrons and holes.

It is also important to notice from (1.23) that the lowest ground state energy is not zero but finite. In fact E_1 increases with decreasing width of the potential well. This is in contrast with the classical theory. Classically, a particle can be completely at rest in the potential well, that is $E_1 = 0$. Quantum mechanically, however, it is impossible for a particle to be at complete rest. This again is due to the fundamental principle of quantum mechanics, namely the uncertainty principle. The typical wavefunctions, probability densities are sketched in Fig. 1.1.

1.4 Particle in 3D Box

An electron in solids is modeled as a free particle in 3D box, and the box is in turn often represented by 3D infinite square-well potential. It is therefore important to consider the motion of a particle in such a potential. Obviously, the 3D cubic box is represented by

$$V(\underline{r}) = \begin{cases} 0 & 0 \le x,\ y,\ z \le L \\ \infty & \text{otherwise} \end{cases} \tag{1.24}$$

and the energy eigenequation of a particle therein reads as

$$-\frac{\hbar^2}{2m}\left(\frac{\partial^2}{\partial x^2} + \frac{\partial^2}{\partial y^2} + \frac{\partial^2}{\partial z^2}\right)u(x,\ y,\ z) = Eu(x,\ y,\ z) \tag{1.25}$$

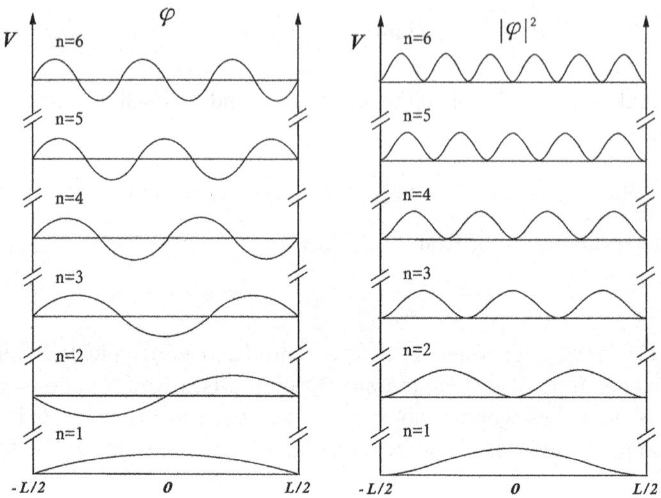

Fig. 1.1 The energy eigenfunctions and corresponding probability densities in the infinite square-well potential

where the kinetic energy is composed of the 3D motion in x, y, z directions.

The Eq. (1.25) can again be solved by using the separation of variable technique and by seeking the solution in the form,

$$u(x, y, z) = f(x)g(y)h(z) \tag{1.26}$$

Upon inserting (1.26) into (1.25), performing the differentiations and dividing both sides by (1.26), there results

$$-\frac{\hbar^2}{2m}\left[\frac{f(x)''}{f(x)} + \frac{g(y)''}{g(y)} + \frac{h(z)''}{h(z)}\right] = E \tag{1.27}$$

with the double primes denoting the second derivatives with respect to x, y, and z, respectively. Here, each term in (1.27) depends only on x, y, and z, respectively, and therefore, the solution can be readily obtained by putting each term to constant energy values, E_x, E_y, and E_z. The resulting three separate equations involving x, y, and z, respectively, are identical to the 1D eigenequation (1.19) and one can therefore write the normalized eigenfunction as

$$u_{\underline{n}}(x, y, z) = \left(\frac{2}{L}\right)^{3/2} \sin\left(\frac{n_x\pi}{L}x\right) \sin\left(\frac{n_y\pi}{L}y\right) \sin\left(\frac{n_z\pi}{L}z\right) \tag{1.28}$$

and the energy eigenvalue is given by

$$E_{\underline{n}} = \frac{\hbar^2\pi^2}{2mL^2}\left(n_x^2 + n_y^2 + n_z^2\right) \tag{1.29}$$

Hence, the total wavefunction is given from (1.18) by

$$\psi(\underline{r},t) = \exp - [i(E_n t/\hbar)]u_n(x, y, z) \tag{1.30}$$

Naturally, the quantum numbers n_x, n_y, and n_z are positive to circumvent the redundancy of the solution. The ground state corresponds to $n_x = n_y = n_z = 1$, while the first excited state is characterized by three different combinations of the quantum numbers, $n_x = 2$, $n_y = n_z = 1$, $n_y = 2$, $n_x = n_z = 1$ and $n_z = 2$, $n_x = n_y = 1$. This indicates the threefold degeneracy in the first excited state, that is, three quantum states have identical energy levels. The degree of degeneracy increases in higher lying energy levels.

1.5 Density of States

As mentioned, an electron in solids can be taken as a free particle, confined therein by the potential barrier at the surface. The boundary condition used requires that the wavefunction vanish at the edge of the potential well, ensuring thereby that electrons are well confined in solids in one of those bound states in the box. On the other hand, the periodic boundary condition is often used to focus on the motion of the electron freely propagating in the bulk solid. The periodic boundary condition states in essence that a particle exiting at $x + L$ reenters the box at x as indicated in Fig. 1.2.

Now, the eigenfunction of an electron propagating freely can be expressed in the traveling wave form as

$$u(\underline{r}) = \frac{1}{\sqrt{V}}e^{+i\underline{k}\cdot\underline{r}}, \quad \underline{k}\cdot\underline{r} = k_x x + k_y y + k_z z \tag{1.31}$$

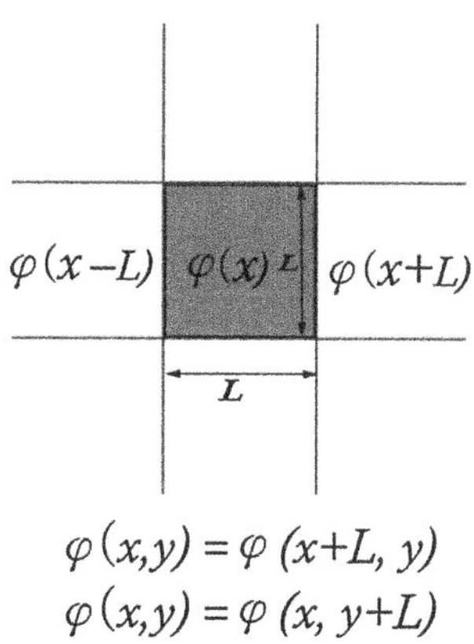

Fig. 1.2 The illustration of the periodic boundary condition

where $V = L^3$ is the volume of the box. Note that $\exp ix = \cos x + i\sin x$ and a linear combination of eigenfunctions, $\cos x$ and $\sin x$ is also the eigenfunction. Hence, (1.31) is the solution of the eigenequation (1.25). Upon inserting (1.31) into (1.18), there results

$$\Psi(\underline{r}, t) = \frac{1}{\sqrt{V}} \exp - i(\omega t \mp i\underline{k} \cdot \underline{r}), \; \omega = E/\hbar \tag{1.32}$$

and represents explicitly the electron moving freely in the conduction band in solids. It is also clear that the positive branch in the exponent in (1.32) represents the forward component traveling in the $\underline{k}$ direction while the negative component represents the component traveling in the opposite direction.

The periodic boundary conditions are specified as

$$u(x, y, z) = u(x + L, y, z) = u(x, y + L, z) = u(x, y, z + L) \tag{1.33}$$

This boundary condition can be satisfied by constraining the wave vector $\underline{k}$ by

$$k_x L = 2\pi n_x, \; k_y L = 2\pi n_y, \; k_z L = 2\pi n_z \tag{1.34}$$

Then, the total energy of the electron is given by

$$E_{\underline{n}} = \frac{\hbar^2 k^2}{2m} = \frac{\hbar^2}{2m}\left(\frac{2\pi}{L}\right)^2 \left(n_x^2 + n_y^2 + n_z^2\right) = \frac{\hbar^2}{2m}\left(\frac{2\pi}{L}\right)^2 \underline{n} \cdot \underline{n} \tag{1.35}$$

The $\underline{k}$ vector specified by (1.34) or the corresponding momentum, $\underline{p} = \hbar\underline{k}$ represents a single quantum state with the eigenfunction (1.32) and eigenenergy (1.35). Therefore, there is one to one correspondence between $\underline{k}$ (k_x, k_y, k_z) and the integer $\underline{n}$ (n_x, n_y, n_z).

An important quantity entering in the analysis of semiconductor devices is the number of quantum states in the interval k and $k + \mathrm{d}k$ or equivalently in the interval E and $E + \mathrm{d}E$ of a free particle in 1D, 2D, or 3D environments. The number of states can be found by introducing 1D, 2D, and 3D $\underline{k}$ spaces, as shown in Fig. 1.3. Here, the k space is scaled with $2\pi/L$, and the basic volume elements or cells containing a single dot, that is, one single quantum state in 3, 2, and 1 dimensions are given by

$$(2\pi/L)^3; \; (2\pi/L)^2, \; (2\pi/L)^1 \tag{1.36}$$

It is also clear from Fig. 1.3 that the corresponding differential volume elements between k and $k + \mathrm{d}k$ are specified, respectively, by

$$4\pi k^2 \mathrm{d}k, \; 2\pi k \mathrm{d}k, \; 2\mathrm{d}k \tag{1.37}$$

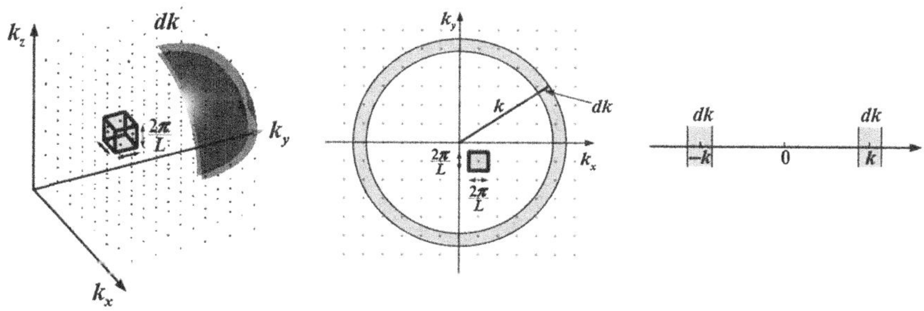

Fig. 1.3 The 3D (*left*), 2D (*middle*), and 1D (*right*) volume elements in the range from k and $k + dk$. Each dot represents one quantum state

1.5.1 3D Density of States

For a particle in 3D space, the number of quantum states, that is, the number of points in the differential volume element in 3D k space is obtained by dividing the differential volume in (1.37) by the corresponding unit cell in (1.36):

$$\frac{4\pi k^2 \mathrm{d}k}{(2\pi/L)^3} \tag{1.38}$$

Now, it is important to point out that for each quantum state for given $\underline{k}$, there are two independent quantum states, corresponding to the spin states of the electron, i.e., spin up and spin down states. Therefore, the number of the quantum states between k and $k + \mathrm{d}k$ per unit volume is given by

$$g_{3D}(k)\mathrm{d}k = 2\frac{4\pi k^2 \mathrm{d}k}{(2\pi/L)^3}\frac{1}{L^3} = \frac{k^2 \mathrm{d}k}{\pi^2} \tag{1.39}$$

That is,

$$g_{3D}(k) = \frac{k^2}{\pi^2} \tag{1.40}$$

The quantity $g_{3D}(k)$ derived in (1.40) is called the 3D density of states in k space. One can express the same number of quantum states in energy space by transcribing k into E, using the dispersion relation of the free particle,

$$E = \frac{\hbar^2 k^2}{2m} \tag{1.41}$$

By using (1.41), one can express k and dk by E and dE, obtaining

$$g_{3D}(k)\mathrm{d}k \equiv g_{3D}(E)\mathrm{d}E = \frac{\sqrt{2}m^{3/2}E^{1/2}}{\pi^2\hbar^3}\mathrm{d}E \tag{1.42}$$

That is,

$$g_{3D}(E) = \frac{\sqrt{2}m^{3/2}E^{1/2}}{\pi^2\hbar^3} \tag{1.43}$$

The quantity $g_{3D}(E)$ is called the 3D density of states in E space and is a key parameter used extensively for modeling the semiconductor devices. Note that $g_{3D}(E)$ follows the power law $g(E) \propto E^{1/2}$.

1.5.2 2D and 1D Densities of States

For 2D electron motion, one can obtain the number of states between k and $k + dk$ by following exactly the same procedure as was used for 3D case. Here, one can divide 2D differential volume element by 2D unit cell given in (1.36) and (1.37), obtaining

$$g_{2D}(k)dk = 2\frac{2\pi k dk}{(2\pi/L)^2}\frac{1}{L^2} = \frac{kdk}{\pi} \tag{1.44}$$

That is,

$$g_{2D}(k) = \frac{k}{\pi} \tag{1.45}$$

In terms of E, the 2D density of states can likewise be expressed as

$$g_{2D}(E) = \frac{m}{\pi\hbar^2} \tag{1.46}$$

and is shown to be independent of E.

One can carry out similar analysis for the 1D density of states, obtaining

$$g_{1D}(k) = \frac{2}{\pi} \tag{1.47}$$

or in terms of E

$$g_{1D}(E) = \frac{\sqrt{2}m^{1/2}}{\pi\hbar}\frac{1}{E^{1/2}} \tag{1.48}$$

The 3D density of states is a key factor for modeling the bulk semiconductor devices such as MOSFET. The 2D and 1D densities of states also play key roles for modeling devices in quantum well and nanowire, respectively. Figure 1.4 shows the density of states versus energy in 3D, 2D, and 1D systems, respectively.

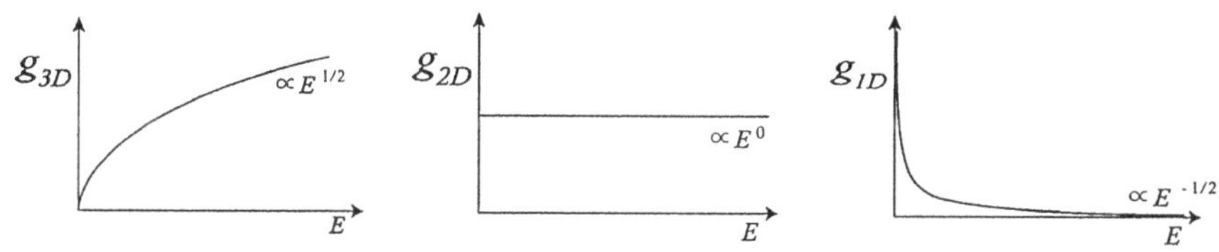

Fig. 1.4 The 3D (*left*), 2D (*middle*), and 1D (*right*) densities of states versus energy

1.6 Particle in Quantum Well

A potential well with a finite depth is called the quantum well and it is an essential element in various semiconductor and optoelectronic device structures, for example, laser diode. The analysis of a particle in the quantum well is therefore important and is considered. Figure 1.5 shows a quantum well with depth V and width W and the potential is given by

$$V(x) = \begin{cases} 0 & |x| \le W/2 \\ V & |x| \ge W/2 \end{cases} \tag{1.49}$$

Inside the well, $V = 0$ and the energy eigenequation is the same as (1.19) and one can write

$$\frac{\partial^2 u}{\partial x^2} + k^2 u = 0, \ k^2 \equiv 2mE/\hbar^2, \ |x| \le W/2 \tag{1.50a}$$

For the bound state, $E \le V$ and the eigenequation outside the well is given by

$$\frac{\partial^2 u}{\partial x^2} - \kappa^2 u = 0; \ \kappa^2 \equiv 2m(V - E)/h^2, \ |x| > W/2 \tag{1.50b}$$

Clearly one can find the solutions of (1.50a) in terms of sinusoidal functions inside the well, and exponential functions $\exp \pm \kappa x$ outside the well. Hence, the even and odd solutions are given by

$$u_e(x) = N \begin{cases} Ae^{\kappa x}; & x < -W/2 \\ \cos kx & |x| \le W/2 \\ Ae^{-\kappa x} & x > W/2 \end{cases} \tag{1.51a}$$

$$u_o(x) = N \begin{cases} -Ae^{\kappa x}; & x < -W/2 \\ \sin kx & |x| \le W/2 \\ Ae^{-\kappa x} & x > W/2 \end{cases} \tag{1.51b}$$

Here, the exponential functions have been chosen such that $u(x)$ vanishes at infinity, as it should. Also, the two constants, A and N can be used to satisfy the boundary conditions and to normalize $u(x)$.

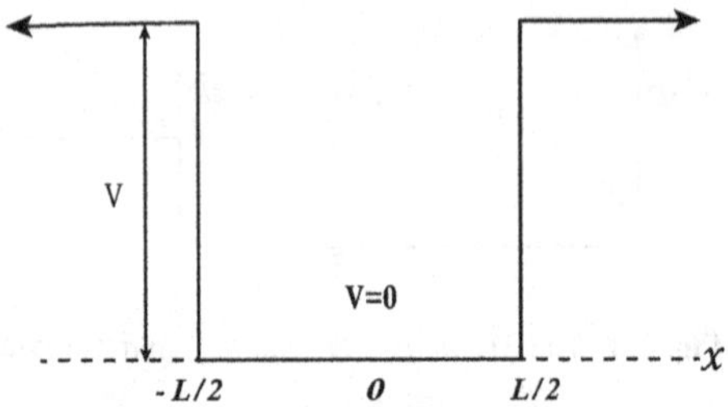

Fig. 1.5 A quantum well with the potential depth *V* and width, *L*

There are two boundary conditions to be met, namely that $u(x)$ and its derivative $\partial u(x)/\partial x$ be continuous everywhere. The first requirement is needed since, $u^*(x)u(x)$ represents the probability density and should therefore vary continuously versus x. The second condition is required, since the derivative is proportional to the momentum of the particle and should also be continuous everywhere.

The two boundary conditions are automatically satisfied both within and outside the quantum well, since $u_e(x)$ and $u_o(x)$ are analytical functions. Thus, the conditions should be imposed on the two edges of the well. But because $u(x)$ is even or odd, once the boundary conditions are satisfied at $W/2$ the conditions are also met at $-W/2$. For the case of $u_e(x)$, the two boundary conditions are specified by

$$\cos\xi = Ae^{-\eta};\ \xi \equiv \frac{kW}{2},\ \eta \equiv \frac{\kappa W}{2} \tag{1.52a}$$

$$-k\sin\xi = -\kappa Ae^{-\eta} \tag{1.52b}$$

The two conditions can be combined into one by multiplying both sides of (1.52b) by W/2 and dividing it with (1.52a):

$$\xi\tan\xi = \eta \tag{1.53}$$

One can likewise specify the boundary conditions for $u_0(x)$ at $x = W/2$:

$$\sin\xi = Ae^{-\eta};\ \xi \equiv \frac{kW}{2},\ \eta \equiv \frac{\kappa W}{2} \tag{1.54a}$$

$$k\cos\xi = -\kappa Ae^{-\eta} \tag{1.54b}$$

and combine the two into one as

$$-\xi\cot\xi = \eta \tag{1.55}$$

Now, the parameters, ξ and η are related with each other from (1.50b, 1.52a) as

$$\xi^2 + \eta^2 \equiv \left(\frac{kW}{2}\right)^2 + \left(\frac{\kappa W}{2}\right)^2 = \frac{mVW^2}{2\hbar^2} \tag{1.56}$$

Therefore, the analysis has been reduced to determining k and κ from (1.53, 1.56) for even eigenfunction and from (1.55) and (1.56) for odd eigenfunction. Once k

and κ are determined, the eigenfunctions $u_e(x)$, $u_0(x)$ can satisfy the required boundary conditions and the energy eigenvalues can be found from (1.50a). The values of ξ, η satisfying conditions can be found by a graphical means as follows.

In Fig. 1.6, η is plotted versus ξ specified by (1.53) and (1.55). Also plotted is the family of circles given by (1.56) for various V or W. Thus, finding the values of ξ and η or equivalently k and κ has been reduced to finding the coordinates of the crossing points of the two curves (1.53) and (1.56) for $u_e(x)$ and (1.55) and (1.56) for $u_0(x)$, respectively.

Typical eigenfunctions found in this manner are plotted in Fig. 1.7. The general features of the bound states emerge from the analysis. With increased V or W, and the radius of the circle there are more crossing points, that is, more bound states in the well. Also, the lowest ground state is associated with $u_e(x)$ and higher lying states alternate between $u_e(x)$ and $u_0(x)$. Moreover, there exists at least one bound state, regardless of the potential depth.

In the limit of the infinite well depth, the intersection points should occur at

$$\xi_n \equiv \frac{k_n W}{2} = \frac{\pi}{2}(2n + 1),\ n = 0, 1, 2, \ldots \tag{1.57a}$$

and

$$\xi_n \equiv \frac{k_n W}{2} = n\pi;\ n = 1, 2, \ldots \tag{1.57b}$$

for $u_e(x)$ and $u_0(x)$, respectively.

Hence, when combined (1.57a) and (1.57b) yield the energy eigenvalues given by

$$E_n = \frac{\hbar^2 \pi^2}{2mW^2} n^2;\ n = 1, 2, 3, \ldots \tag{1.58}$$

in agreement with the analytical result of the infinite potential well, as it should.

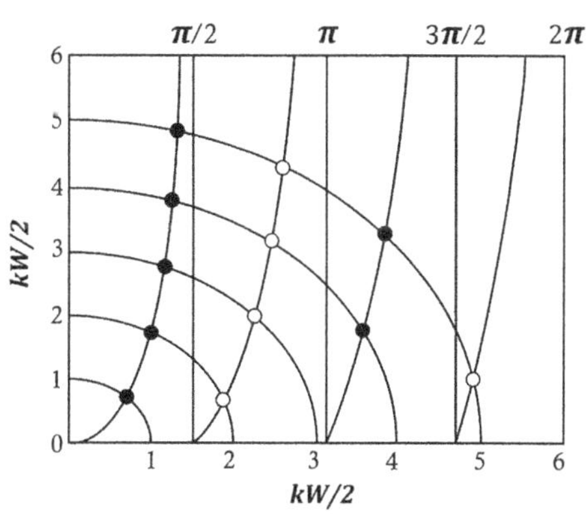

Fig. 1.6 The graphical scheme for finding the energy eigenvalues in a quantum well. Plotted are $\eta = \xi\tan\xi$ (*thick lines*), $\eta = -\xi\cot\xi$ (*thin lines*) and a family of *circles* representing $\xi^2 + \eta^2 = mVW^2/2\hbar^2$. The intersection points between two *curves* determine the energy levels of the bound states

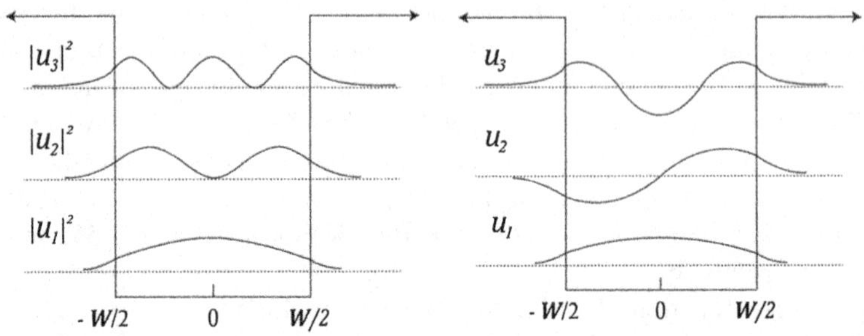

Fig. 1.7 The even and odd eigenfunctions (*left*) and probability densities (*right*) in the quantum well

1.7 Quantum Well and Wire

As mentioned, the quantum well and wire are essential elements in nanoelectronic devices and can be processed by using various techniques such as molecular beam epitaxy and metal-organic chemical vapor deposition techniques. For instance, with the use of these epitaxial deposition techniques, the atomic layers of varying thicknesses and band gaps can be grown and interfaced, forming a quantum well for the electrons in the conduction band and that of holes in the valence band as sketched in Fig. 1.8. The electrons and holes move as free particles in conduction and valence bands in bulk solids. But once the quantum wells are formed, the dynamics of these charge carriers become different and discussed next.

Quantum Well: Consider electrons in a quantum well. These electrons are confined in one direction, say in the z-direction, while propagating freely in x, y directions, forming thereby the 2D electron gas. The potential energy is then given by

$$V(x, y, z) = \begin{cases} 0 & |z| \leq W/2 \\ V & |z| \geq W/2 \end{cases} \tag{1.59}$$

The 2D electron gas system can be readily analyzed by generalizing the results obtained from the simple potential wells considered in previous sections. The energy eigenequation of the 2D electron gas system reads as

$$\left[-\frac{\hbar^2}{2m_x}\frac{\partial^2}{\partial x^2} - \frac{\hbar^2}{2m_y}\frac{\partial^2}{\partial y^2} - \frac{\hbar^2}{2m_z}\frac{\partial^2}{\partial z^2} + V(x, y, z)\right]u(x, y, z) = Eu(x, y.z) \tag{1.60}$$

Here, m_x, m_y, and m_z denote the effective electron masses in x, y, and z directions, respectively. In solids, the mass of electrons differs in general from the rest mass and depends on crystallographic directions. This is due to the wave nature of electrons. The eigenequation can again be reduced to three independent 1D equations by putting

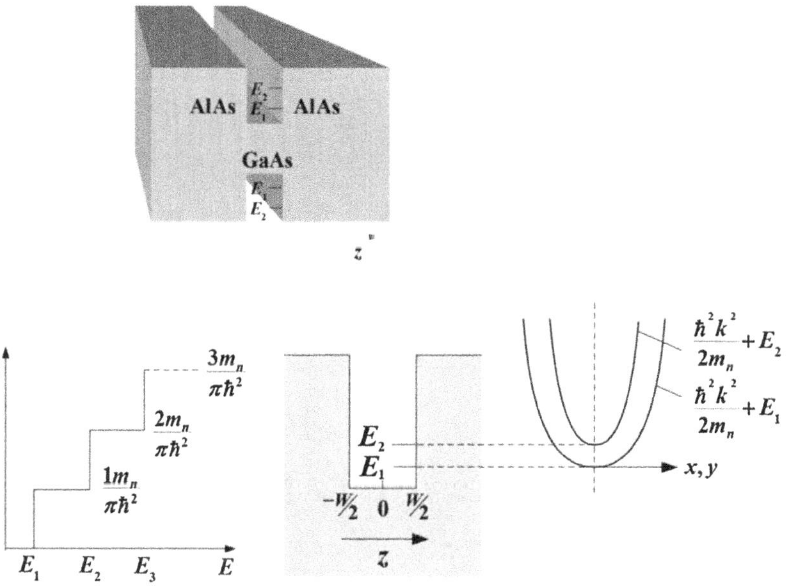

Fig. 1.8 The quantum well in the conduction band for electrons and in the valence band for holes (*top*), 2D density of states versus energy (*bottom left*), sub-bands, *E*–*k* dispersion *curves* (*bottom right*)

$$u(x,y,z) = u(x)u(y)u(z) \tag{1.61}$$

and performing the usual procedure for separating the variables. In x, y directions, one obtains two 1D eigenequation of a free particle, while in the z-direction, there ensues the energy eigenequation in a quantum well, as discussed in the previous section. Hence, the total electron energy consists of the kinetic energies in x, y directions, and the quantized energy formed along the z-direction, that is,

$$E_n = \frac{\hbar^2 k_x^2}{2m_x} + \frac{\hbar^2 k_y^2}{2m_y} + E_{zn} \tag{1.62}$$

Sketched in Fig. 1.8 are the energy levels of the 2D electron gas system, together with the density of states. The discrete energy levels originating from the spatial confinement in the z-direction are called the sub-bands. Since the 2D density of states is constant, independent of energy, the number of the quantum states increases stepwise whenever E crosses those discrete sub-levels. Also, the sub-band states with E_{zn} are inseparably associated with quantum states corresponding to the free motion of electrons in x, y directions, respectively (see Fig. 1.8).

Quantum Wire: A particle in a quantum well has two degrees of freedom. In quantum wire, however, as sketched in Fig. 1.9, the electron is confined in, for example, y, z directions while free to move along the x direction. The energy

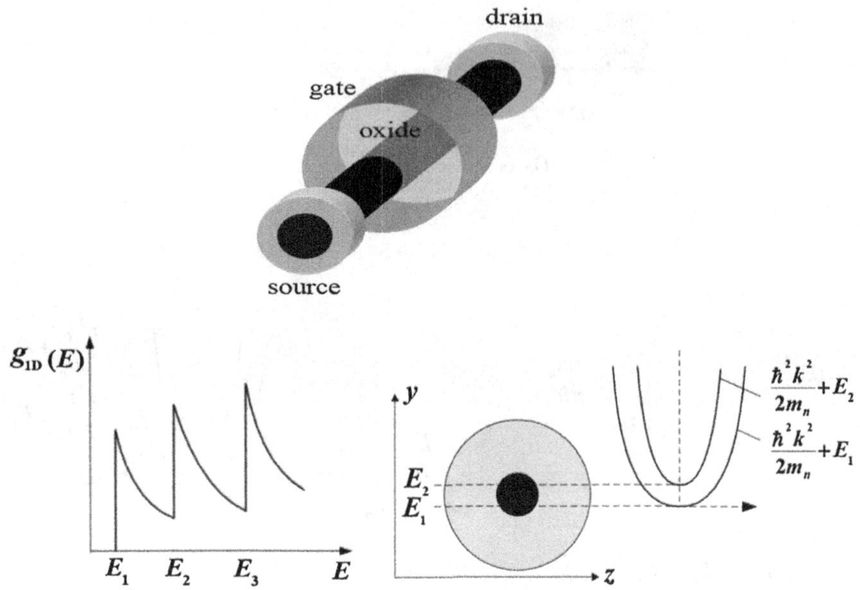

Fig. 1.9 The quantum wire FET (*top*), 1D density of states versus energy (*bottom left*) and sub-bands, *E*–*k* dispersion curves (*bottom right*)

spectrum then consists of two sets of discrete energy levels corresponding to the spatial confinements in *y*, *z* directions and the energy level associated with free propagation in the *x* direction. The total energy is thus given by

$$E_{n,m} = \frac{\hbar^2 k_x^2}{2m_x} + E_{ym} + E_{zn} \tag{1.63}$$

where the quantum numbers, *m*, *n*, denote the discrete sub-levels. Figure 1.9 shows the energy spectrum of the nanowire, together with the corresponding density of states. Since the 1D density of states varies as a function of energy according to the power law, $E^{-1/2}$, the number of states exhibits the saw-tooth-like shape as a function of *E*.

The quantum wells have become an important element of various semiconductor devices. For example, in high efficiency laser diodes, electrons and holes are injected into the respective quantum wells in the junction region. Hence, those electrons and holes are allowed to have longer interaction time, while confined in the well for radiative recombination. Additionally, the operation of MOSFETs in bulk silicon is based upon the 2D electrons or holes in the gate voltage induced quantum well. In addition, FETs in the quantum wires have emerged as a promising device in nano electronics. A central theme of this book is focused on discussing the operation, characteristics, and applications of the nano wire field-effect transistors.

1.8 Problems

1.1 Consider a particle moving freely in 3D space.

a. Write down the eigenequation of the 3D momentum.
b. Find the eigenfunction satisfying the periodic boundary conditions in the interval $0 \leq x \leq L_x, 0 \leq y \leq L_y, 0 \leq z \leq L_z$.

1.2 Verify that (1.28) is the solution of the eigenequation (1.25).

1.3 Using the relation $E = \hbar^2 k^2/2m$, derive (1.43, 1.46, and 1.48).

1.4 Consider a quantum well with the potential depth $V = 3.1\,\text{eV}$ and the width ranging from 1 to 10 nm.

a. Find numerically the lowest four eigenstates and eigenenergies of an electron.
b. Find the magnitudes of the electron velocity residing in each eigenstate and compare with the thermal speed of at electron at room temperature. The thermal speed is determined from

$$mv_x^2/2m = k_B T/2,$$

where k_B is the Boltzmann constant.

1.5 Consider a quantum wire with a fixed potential $V = 3.1\,\text{eV}$ and the widths,

i. $L_x = 2$ nm, $L_y = 8$ nm, (ii) $L_x = L_y = 4$ nm.
a. Find by numerical means the lowest four eigenstates and eigenenergies of an electron.
b. Compare and interpret the two results obtained.
c. Find the degeneracy of four eigenstates.

Suggested Reading

1. Kim, D. M. (2010) *Introductory quantum mechanics for semiconductor nanotechnology*, John Wiley-VCH, New Jersey.
2. Robinett, R. W. (2006). *Quantum mechanics, classical results, modern systems and visualized examples*. Oxford: Oxford University Press.
3. Liboff, R. L. (2002). *Introductory quantum mechanics* (4th Ed.). Reading, MA: Addison Wesley Publishing Company.
4. Singh, J. (1996). *Quantum mechanics, fundamentals & applications to technology*, John Wiley & Sons, New Jersey.
5. Kroemer, H. (1994). *Quantum mechanics for engineering, materials science, and applied physics, International edition*. New Jersey: Prentice Hall.
6. Yariv, A. (1982). *An introduction to theory and applications of quantum mechanics*, John Wiley & Sons, New Jersey.

Chapter 2
Carrier Concentration and Transport

Dae Mann Kim, Bong Koo Kang and Yoon-Ha Jeong

Abstract The semiconductor devices operate based upon the charge control, and two factors are involved in the control, namely the concentration and transport of charge carriers. This chapter is addressed to these two basic quantities. The concentration depends upon the doping level, temperature, and other electronic properties of the semiconductor and will be discussed, starting from the Fermi and Boltzmann distribution functions. The carrier transport in semiconductor devices has been attributed primarily to the drift and diffusion. With the downsizing of the semiconductor devices, however, the ballistic transport becomes important as well as tunneling. The transport processes are compactly discussed.

2.1 Boltzmann and Fermi Distribution Functions

The microscopic dynamics of electrons, atoms, and molecules are connected to the macroscopic world via the cumulative effects of a large number of such particles. The statistical description of the ensemble of such microscopic objects is therefore important, and the quantum statistics is an essential part of the description. Highlighted in this section are the main results of the quantum statistics.

Fermions and Bosons: The statistical description of the particles differs depending on the kinds of particles. There are in general three different kinds:

D. M. Kim (✉)
Korea Institute for Advanced Study, Seoul, Korea
e-mail: dmkim@kias.re.kr

B. K. Kang · Y.-H. Jeong
POSTECH, Pohang, Korea
e-mail: bkkang@postech.ac.kr

Y.-H. Jeong
e-mail: yhjeong@postech.ac.kr

D. M. Kim and Y.-H. Jeong (eds.), *Nanowire Field Effect Transistors: Principles and Applications*, DOI: 10.1007/978-1-4614-8124-9_2,

(1) identical but distinguishable, e.g., atoms and molecules, (2) fermions such as electrons, holes, protons, neutrons, which obey the Pauli exclusion principle and Fermi-Dirac statistics, and (3) bosons such as photons, which obey the Bose-Einstein statistics. In this section, the distinguishable particles and fermions will be chosen for discussion.

Distinguishable Particles: Given an ensemble of distinguishable particles at thermodynamic equilibrium, the probability of a particle having a total energy E at the position r is given by

$$f_0(\underline{r}, \underline{v}) = Ne^{-E(\underline{r})/k_BT} \tag{2.1a}$$

where the total energy consists of the kinetic and potential energies,

$$E(\underline{r}) = \frac{1}{2}mv^2 + V(\underline{r}) \tag{2.1b}$$

and N is the normalization constant, k_B the Boltzmann constant, and T the absolute temperature in Kelvin. The f_0 function introduced in (2.1a) is known as the Boltzmann distribution function in equilibrium, and the exponential factor appearing therein is called the Boltzmann probability factor. For a system of free particles in which $V = 0$, $f_0(\underline{r}, \underline{v})$ simplifies as

$$f_0(\underline{v}) = \left(\frac{m}{2\pi k_BT}\right)^{3/2} e^{-mv^2/2k_BT}, \; v^2 = v_x^2 + v_y^2 + v_z^2 \tag{2.2}$$

where the normalization constant is found by performing the integration of (2.1a) over the entire velocity space and equating the result to unity (problem 2.1).

The distribution function (2.2) well represents the properties of the thermodynamic equilibrium. First, $f_0(\underline{v})$ is symmetric in velocity v, which represents the fact that there is no preferred direction in $\underline{v}$. A particle moving from left to right is balanced by another particle moving from right to left. A process balanced by its inverse process in equilibrium is known as the detailed balancing. Since $f_0(\underline{v})$ is even in $\underline{v}$, the average velocity in equilibrium is zero:

$$\langle \underline{v} \rangle = \int_{-\infty}^{\infty} dv_x \int_{-\infty}^{\infty} dv_y \int_{-\infty}^{\infty} dv_z \underline{v} f_0(\underline{v}) = 0 \tag{2.3}$$

However, $\langle v_x^2 \rangle$ is not zero and can be readily evaluated as,

$$\langle v_x^2 \rangle = \int_{-\infty}^{\infty} dv_x v_x^2 \int_{-\infty}^{\infty} dv_y \int_{-\infty}^{\infty} dv_z f_0(v) = k_BT/m \tag{2.4}$$

(problem 2.1). By the same token, one can write by inspection,

$$\langle v_x^2 \rangle = \langle v_y^2 \rangle = \langle v_z^2 \rangle = k_BT/m \tag{2.5}$$

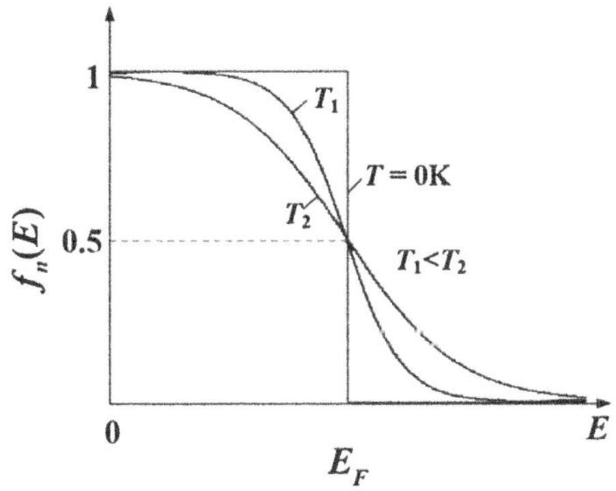

Fig. 2.1 The Fermi distribution function versus energy for different temperatures

and obtain the celebrated equipartition theorem, namely

$$\frac{1}{2}m\langle v_x^2\rangle = \frac{1}{2}m\langle v_y^2\rangle = \frac{1}{2}m\langle v_z^2\rangle = \frac{k_BT}{2} \tag{2.6}$$

Fermions and Fermi Level: Given an ensemble of electrons, the probability of an electron having the energy E in equilibrium is given by

$$f_n(E) \equiv \frac{1}{1 + e^{(E - E_F)/k_BT}} \tag{2.7}$$

The function $f(E)$ is known as the Fermi distribution function or occupation factor and can be derived beginning from the quantum statistics of fermions. The detailed derivation is relegated to the reference books listed. The parameter E_F is called the Fermi energy or level, a key quantity which is utilized extensively in the semiconductor statistics as well as the device modeling.

Figure 2.1 plots the Fermi distribution function or the occupation factor versus energy. At $T = 0$, $f(\varepsilon)$ is simply a step function. Thus, for $E < E_F$, it is equal to unity, representing 100 % probability of occupation. For $E > E_F$ on the other hand, the probability of occupation is zero. For $T \neq 0$, the overall shape of $f(E)$ is generally preserved except for the fact that the occupation curve is rounded off near E_F. Specifically, $f(E)$ is less than unity in the range of a few k_BT below E_F, while it tails out a few k_BT beyond E_F. This represents the fact that the occupation probability is transferred from below E_F to above E_F. This tailing of the occupation probability beyond E_F is called the Boltzmann tail. With increasing T, the Boltzmann tail becomes progressively pronounced. The difference between the Boltzmann and the Fermi distribution functions is rooted in the fact that electrons as fermions are indistinguishable and obey the Pauli exclusion principle.

2.2 Carrier Densities in Intrinsic Semiconductors

A solid is generally characterized by a series of energy bands, which are separated by forbidden gaps. In allowed bands, electrons and holes move around as free particles with the effective masses, m_n and m_p as determined by the properties of the solid. Of particular interest are the valence and conduction bands, as illustrated in Fig. 2.2. Here, the valence band consists of the highest energy states occupied by the outermost lying electrons, called valence electrons of the host atoms constituting the solid. The conduction band is the next higher lying allowed energy band.

Conductors, Insulators, and Semiconductors: Solids are in general classified into metals, insulators, and semiconductors. These are in turn characterized by differing configurations of the valence and conduction bands. In metals, the valence and conduction bands overlap and there is no forbidden gap in between (Fig. 2.2). This means that the valence electrons can move up to the conduction band under bias and readily contribute to the current.

An insulator, e.g., the silicon dioxide is characterized by the conduction and valence bands with relatively narrow widths and separated by a large bandgap, typically 10 eV or more. In an insulator, the bonds between neighboring atoms are strong and are difficult to break, resulting in a large bandgap. Thus, very few electrons are thermally excited into the conduction band at room temperature to conduct the current.

The semiconductor has the conduction and valence bands with thicker widths and narrower bandgap, compared with insulator. Specifically, the bandgap ranges from about 0.5 eV to a few eV. The bonds between neighboring atoms are moderately strong and therefore are relatively easy to be broken at room temperature. Thus, an electron can be promoted into the conduction band via the band-

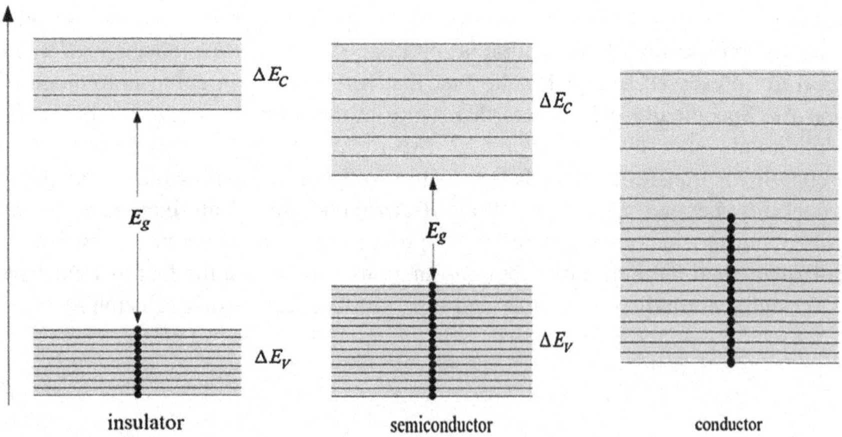

Fig. 2.2 The energy bands in conductors, insulators, and semiconductors

to-band thermal excitation, leaving behind a hole in the valence band at room temperature to conduct the current.

Carrier Concentration in Intrinsic Semiconductor: As noted, an electron in the valence band can be thermally excited from the valence band to the conduction band, leaving behind a hole. Hence, in intrinsic semiconductor, the electron and hole concentrations are identical, i.e., $n = p \equiv n_i$. Furthermore, the holes are to be taken as the positive charge carriers with the effective mass, m_p.

Equilibrium: The semiconductor statistics in equilibrium is considered next. In so doing, the characteristics of the equilibrium are pointed out. First, the physical quantities are time invariant, since every process is balanced by its inverse process (detailed balancing). Additionally, the electron and hole concentrations, n, p, are quantified by a single Fermi level, E_F. Finally, the law of mass action holds true, that is, $np = n_i^2$ with n_i denoting the intrinsic carrier concentration.

3D Electron Density: The electron density in the conduction band is specified by the electron density of states as discussed in Chap. 1 and the Fermi occupation factor, given in (2.7):

$$n = \int_{E_C}^{E_C + \Delta E_c} \mathrm{d}E g_{n3D}(E) f_n(E) \tag{2.8}$$

Here, ΔE_C is the conduction bandwidth, and the bottom of the conduction band E_C serves as the reference point from which to define the electron energy moving with the effective mass, m_n. Also, $g_{n3D}(E)\mathrm{d}E$ represents the total number of states per unit volume between E and $E + \mathrm{d}E$. When multiplied with the Fermi occupation factor, $f_n(E)$, the product represents the number of electrons in the same energy interval as illustrated in Fig. 2.3.

Upon inserting Eq. (1.43) for $g_{n3D}(E)$ and (2.7) for $F_n(E)$ into (2.8) and making an approximation, $\Delta E_C/k_B T \to \infty$, one can introduce a new variable of integration, $\eta = (E - E_C)/k_B T$, and cast the integration in (2.8) into a compact form given by

$$n = \frac{2}{\sqrt{\pi}} N_c F_{1/2}(\eta_{Fn}), \; N_C \equiv 2\left(\frac{2\pi m_n k_B T}{h^2}\right)^{3/2}, \; \eta_{Fn} \equiv (E_F - E_c)/k_B T \tag{2.9a}$$

where the integral

$$F_{1/2}(\eta_F) \equiv \int_0^\infty \frac{\eta^{1/2}\mathrm{d}\eta}{1 + e^{\eta - \eta_F}} \tag{2.9b}$$

is known as the Fermi 1/2 integral and the constant, N_C is the effective density of states at the conduction band (problem 2.2). The approximation, $\Delta E_C/k_B T \to \infty$, is well taken, since ΔE_C is typically few eV, while $k_B T \approx 25\,\mathrm{meV}$ at room temperature. Moreover, $f_n(E)$ falls off rapidly with the energy, $E - E_C$; hence, the contribution from those states a few $k_B T$ above E_C is negligible, as clear from Fig. 2.3.

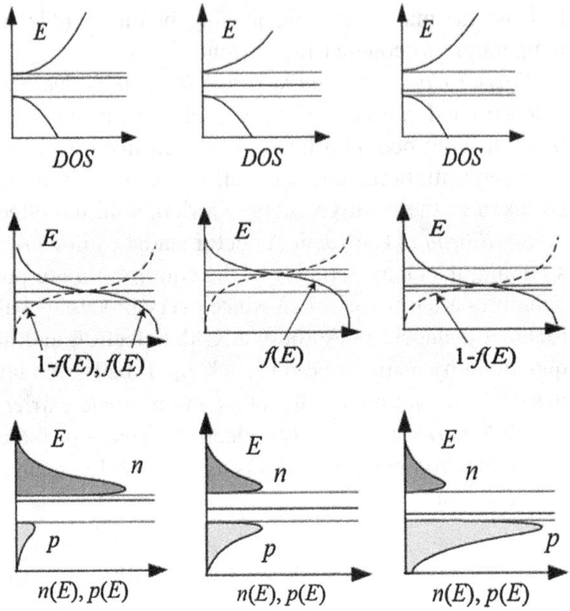

Fig. 2.3 The graphic description of electron and hole concentrations in terms of 3D density of states, Fermi occupation factor for electrons and holes, and the profiles of n and p versus energy

Non-Degenerate Statistics: The non-degenerate statistics corresponds to the case in which E_F remains below E_C by a few k_BT, so that $\eta_F < 0$ and $\exp -\eta_F \gg 1$. In this case, the Fermi 1/2 integral in (2.9b) can be readily integrated to yield,

$$F_{1/2}(\eta_{F_n}) \approx \exp(\eta_{F_n}) \int_0^\infty d\eta e^{-\eta}\eta^{1/2} = \exp(\eta_{F_n})\sqrt{\pi}/2 \tag{2.10}$$

Here again, a new variable of integration $\eta = \xi^2$ has been used (problem 2.2). Hence, by inserting (2.10) into (2.9a), the non-degenerate electron concentration is obtained in a simple analytic form as

$$n = N_C \exp - [(E_C - E_F)/k_BT] \tag{2.11a}$$

Or in terms of intrinsic concentration, n_i, one can write

$$\begin{aligned} n &= N_C \exp - [(E_C - E_F - E_{Fi} + E_{Fi})/k_BT] \\ &= n_i \exp[(E_F - E_{Fi})/k_BT] \end{aligned} \tag{2.11b}$$

where E_{Fi} is the intrinsic Fermi level. Figure 2.4 plots the Fermi 1/2 integral numerically evaluated together with the Boltzmann approximation given in (2.10) versus η_{Fn}. Clearly, these two curves practically coincide in the non-degenerate region, in which $\eta_{Fn} \equiv (E_F - E_C)/k_BT \leq -2$. For $\eta_{Fn} \geq -2$, however, the two curves progressively depart from each other. In this degenerate region, the increase

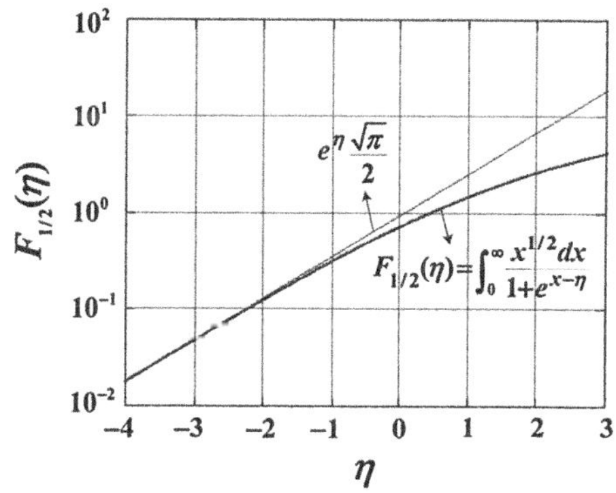

Fig. 2.4 The Fermi 1/2 integral and the Boltzmann approximation for electrons

in n with η_{Fn} slows down considerably, compared with the Boltzmann approximation.

Hole Concentration: The hole concentration is given by

$$p = \int_{E_V - \Delta E_V}^{E_V} dE g_{p3D}(E) f_p(E) \tag{2.12a}$$

where ΔE_V is the valence band width and

$$g_{p3D}(E) = \frac{1}{2\pi^2}\left(\frac{2m_p}{\hbar^2}\right)^{3/2}(E_V - E)^{1/2} \tag{2.12b}$$

is the hole density of states in the valence band with E_V serving as the reference level, and m_p is the hole effective mass. Also, $g_p(E)dE$ represents the number of hole states per unit volume between E and $E - dE$. By definition, the hole occupation factor is the probability that the state at E is not occupied by the electron, that is,

$$f_p(E) \equiv 1 - \frac{1}{1 + e^{(E - E_F)/k_BT}} = \frac{1}{1 + e^{(E_F - E)/k_BT}} \tag{2.12c}$$

Thus, by inserting (2.12b, c) into (2.12a), one can again recast the integral as

$$p = \frac{2}{\sqrt{\pi}} N_V F_{1/2}(\eta_{Fp}),\ N_V = 2\left(\frac{2\pi m_p k_B T}{h^2}\right)^{3/2},\ \eta_{Fp} \equiv (E_V - E_F)/k_BT \tag{2.13a}$$

Here, N_V is the effective density of states at the valence band, and the approximation, $\Delta E_V/k_BT \to \infty$, has been used for the same reasons as discussed (problem 2.2).

For the non-degenerate regime in which E_F stays above E_V by a few k_BT, the Fermi 1/2 integral again simplifies as

$$F_{1/2}\left(\eta_{F_p}\right) \approx \exp\left(\eta_{F_p}\right)\int_0^{\infty} d\eta e^{-\eta}\eta^{1/2} = \exp(\eta_{F_p})\sqrt{\pi}/2 \tag{2.14}$$

and p can be expressed as

$$p = N_V e^{-(E_F - E_V)/k_B T} \tag{2.15a}$$

Or equivalently

$$p = n_i \exp[(E_{Fi} - E_F)/k_B T] \tag{2.15b}$$

In an intrinsic semiconductor $n = p = n_i$, hence in the intrinsic concentration, n_i is specified as

$$n_i \equiv \sqrt{np} = \sqrt{N_C N_V} e^{-E_G/2k_B T}, \; E_G \equiv E_C - E_V \tag{2.16}$$

It is clear from (2.16) that n_i exponentially increases with decreasing bandgap, E_G, as more electron hole pairs are thermally excited across the smaller bandgap.

Intrinsic Fermi Level, E_{Fi}: The Fermi level is determined in general from the charge neutrality condition. In the intrinsic semiconductor, the condition is simply given by $n = p$. Therefore, one can write from (2.11a) and (2.15a)

$$N_C e^{-(E_C - E_{Fi})/k_B T} = N_V e^{-(E_{Fi} - E_V)/k_B T} \tag{2.17a}$$

and obtain

$$E_{Fi} = \frac{1}{2}(E_C + E_V) + \frac{3k_B T}{4} \ln \frac{m_p}{m_n}, \; \ln \frac{N_V}{N_C} = \frac{3}{2} \ln \frac{m_p}{m_n} \tag{2.17b}$$

Thus, E_{Fi} is located practically at the midgap. The slight departure from the midgap arises from the difference in effective masses, m_n, m_p, so that the difference amounts to a fraction of the thermal energy, $k_B T$ of 25 meV at room temperature.

2.3 Carrier Concentration in Extrinsic Semiconductors

As mentioned, the control of n and p is a key factor for device operation and is done by doping impurity atoms, i.e., donor and acceptor atoms. The statistics of n and p in the doped semiconductor, i.e., the extrinsic semiconductor is briefly summarized in this section. First, the physics of doping is discussed, taking silicon as an example. Silicon is located in Column IV in the periodic table, and there are four outermost valence electrons. Si atoms are therefore bonded together by the covalent bond, that is, by sharing one valence electron between two nearest atoms, as illustrated in Fig. 2.5. As shown in the figure, doping consists of incorporating donor and acceptor atoms at the substitutional sites.

Donors and Acceptors: The donor atoms are from Column *V* in periodic table, e.g., phosphorus or arsenic atoms and have five valence electrons, and four of these valence electrons replace the role of four valence electrons in the silicon atom. The remaining fifth electron forms a hydrogen-like atom with one excess proton in the

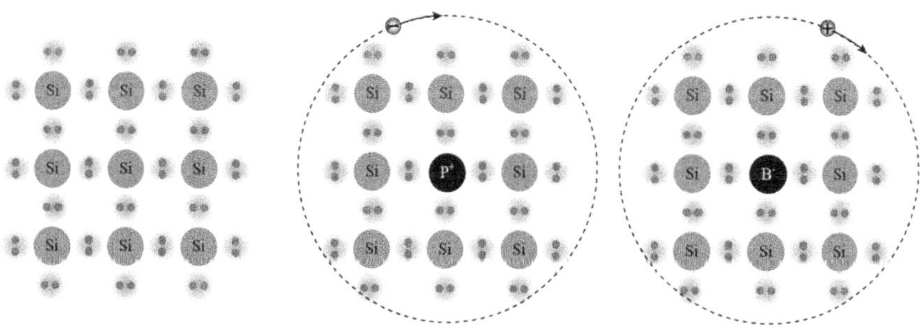

Fig. 2.5 The donor and acceptor atoms embedded in tetrahedrally bonded silicon. The donor and acceptor atoms form loosely bound hydrogenic atoms, consisting of positively charged nucleus and electron in the former and negatively charged nucleus and hole in the latter

nucleus, that is, P^+ or As^+ ion core (Fig. 2.5). The acceptor atoms on the other hand are from Column III, for instance, boron with three valence electrons. These acceptor atoms cannot complete the covalent bonding in silicon but readily accept instead an electron from other Si-Si bonds, thereby completing the bonding. In the process, a hole is created in the valence band and the resulting negative boron ion core and hole again form a hydrogenic atom.

The ionization energy of the donor atom is estimated by using the ionization potential of the hydrogen atom. The ionization energy of the H atom is the energy required to release the electron from the ground state to the vacuum level and is specified as

$$E_{\text{Hion}} = \frac{e^4 m_0}{2h^2(4\pi\varepsilon_0)^2} = 13.64\,\text{eV} \tag{2.18}$$

where m_0 is the rest mass of the electron and ε_0 the vacuum permittivity. The derivation of (2.18) can again be found in the books listed for suggested reading list. By the same token, the ionization energy, E_D, of the donor atom can be specified by using (2.18), but with m_0 replaced by the effective mass, m_n, and ε_0 by the silicon permittivity, ε_S:

$$E_D = \frac{e^4 m_n}{2h^2(4\pi\varepsilon_S)^2} = \frac{e^4 m_0}{2h^2(4\pi\varepsilon_0)^2}\left(\frac{m_n}{m_0}\right)\left(\frac{\varepsilon_0}{\varepsilon_S}\right)^2 = 13.64\left(\frac{m_n}{m_0}\right)\left(\frac{\varepsilon_0}{\varepsilon_S}\right)^2\,\text{eV} \tag{2.19}$$

Thus, with $m_n/m_0 \approx 0.98,\ 0.2$, depending on the crystallographic directions and $\varepsilon_S/\varepsilon_0 \approx 12$, the ionization energy of the donor atom is reduced to about 20–100 meV, a few thermal energies at room temperature, as sketched in Fig. 2.6. Therefore, the fifth valence electron in the donor atom is loosely bound and can be readily promoted to the conduction band, hence the name the donor. Here, the ionization consists of releasing the electron to the conduction band.

A similar estimation can be made with acceptor atoms, and the ionization energy of the hole can also be shown to be about the same as that of electrons in

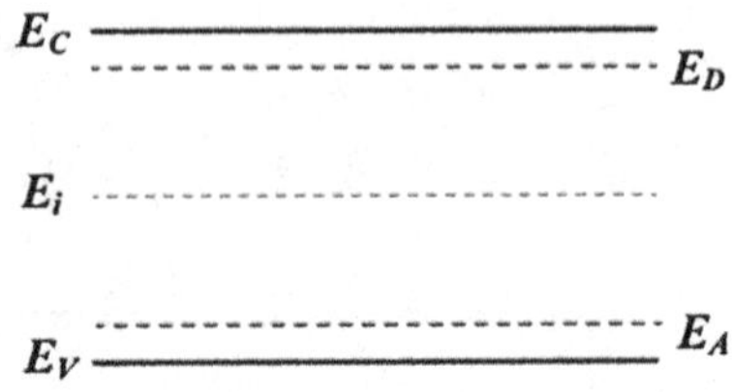

Fig. 2.6 Donor and acceptor levels in the energy gap. The *solid line* represents the extended nature of the electron states in conduction and valence bands, while the *broken lines* denote the localized states

donor atoms. Therefore, the acceptor atoms readily accept electrons from the valence band, thereby creating holes in the valence band, hence the name the acceptor (Fig. 2.6).

Fermi Level and Carrier Concentration: E_F in extrinsic semiconductors in equilibrium can again be found from the charge neutrality condition:

$$n + N_A^- = p + N_D^+ \tag{2.20}$$

where the ionized donor and acceptor concentrations, N_D^+, N_A^- can be specified in terms of the doping level, N_D, N_A, as

$$N_D^+ = \frac{N_D}{1 + g_D e^{(E_F - E_D)/k_B T}} \tag{2.21a}$$

and

$$N_A^- = \frac{N_A}{1 + g_A e^{(E_A - E_F)/k_B T}} \tag{2.21b}$$

Here, E_D, E_A are the donor and acceptor energy levels with respect to E_C, E_V, respectively (Fig. 2.6), and $g_D = 2$, $g_A = 4$ are the degeneracy factors, the discussion of which is again relegated to the suggested reading material. Upon inserting (2.9a, 2.13a) for n and p and (2.21a) for N_D^+ and N_A^- into (2.20) the Fermi level in extrinsic semiconductors can be found for given N_D, N_A, and temperature, T.

Figure 2.7 shows E_F thus found numerically versus temperature in n- and p-type silicon for different N_D, N_A. In n-type silicon, in which $N_A = 0$, E_F is raised above E_{Fi} at the midgap with increasing N_D, as it should. Also at fixed N_D, E_F is lowered with increasing temperature, T to merge with E_{Fi}. This is because the electron concentration at high T is primarily contributed by the thermal excitation, regardless of the doping level. In p-type silicon, where $N_D = 0$, E_F versus T curves essentially mirror the corresponding curves in n-type silicon.

The electron concentration, n, in the n-type silicon is plotted versus $1000/T$ in Fig. 2.8 for different doping levels. As clear from the figure, there are three regimes of the electron concentration. In the intrinsic regime for large T, n is determined primarily by temperature regardless of N_D, as more electrons holes pairs are thermally generated via the band-to-band excitation. With decreasing T, the intrinsic region is followed by the saturation regime, in which $n \approx N_D$. In this

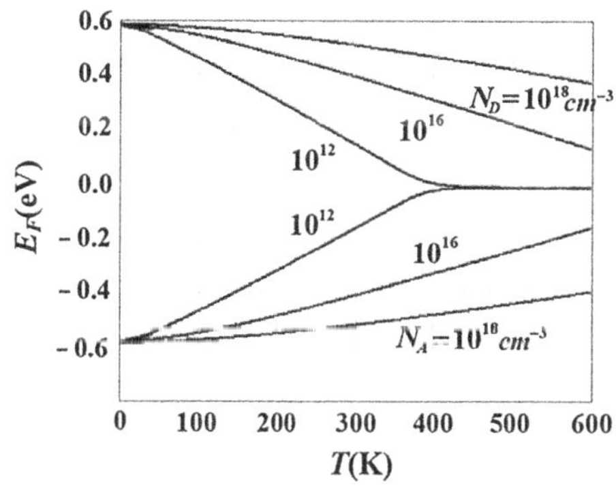

Fig. 2.7 The Fermi level versus T in silicon for different doping levels of donor and acceptor atoms. The bandgap is weakly dependent on T as $E_G(T) = 1.17 - \alpha T^2/(T + \beta)$ eV; $\alpha = 4.73 \times 10^{-4}$, $\beta = 636$ and $m_n = 1.1\, m_0$

Fig. 2.8 The electron concentration in silicon versus $1000/T$ for different doping levels

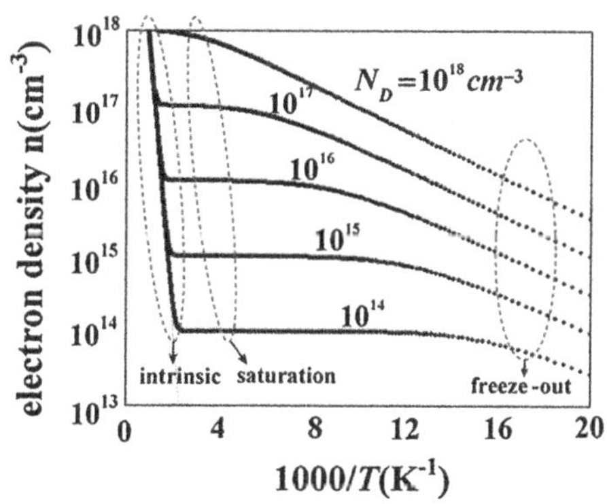

regime, nearly all of the donor atoms are ionized, as E_D is still above E_F (see Figs. 2.6 and 2.7). With T further decreasing, the freeze out regime ensues in which n decreases exponentially with decreasing T. In this regime, there is practically no thermal excitation of e–h pairs and the electrons in the conduction band are captured back by the donors. This is consistent with E_F approaching and surpassing E_D level (Fig. 2.7).

Fermi Potential: In the non-degenerate regime, n and p can be specified analytically in terms of n_i [see (2.11b) and (2.15b)]. Hence, in saturation region in which nearly all the impurity atoms are ionized, one can write

$$n \approx N_D = n_i e^{(E_F - E_{Fi})/k_B T} \tag{2.22}$$

so that the Fermi potential,

$$q\varphi_{Fn} \equiv E_F - E_{Fi} \approx E_F - E_i$$

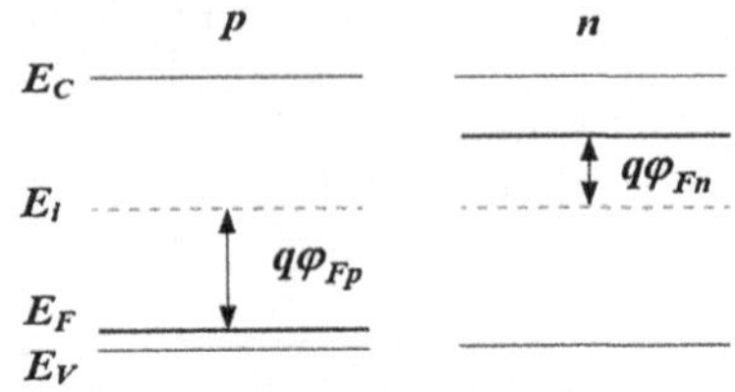

Fig. 2.9 The Fermi potentials in n- and p-type semiconductors

defined as the difference between E_F and the midgap is easily found from (2.22) as

$$\varphi_{\mathrm{F}n} = (k_{\mathrm{B}}T/q)\ln(N_D/n_i) \tag{2.23}$$

One can likewise specify the Fermi potential, $q\varphi_{\mathrm{F}p} \equiv E_i - E_{\mathrm{F}}$, in the p-type semiconductor as

$$\varphi_{\mathrm{F}p} = (k_{\mathrm{B}}T/q)\ln(N_A/n_i) \tag{2.24}$$

Thus, E_{F} in n-type semiconductors is raised above E_i, while it is lowered in p-type below E_i, as shown in Fig. 2.9.

2.4 Carrier Transport: Drift and Diffusion

The carrier transport is another key factor for the charge control. The transport has been driven mainly by drift and diffusion in semiconductor devices. With the downsizing of the device dimension, however, the ballistic transport has become important. The drift is based upon the average velocity the carriers attain in between collisions in the presence of an external electric field. The diffusion is driven, on the other hand, by the concentration gradient. These transport coefficients are briefly summarized as follows.

Mobility, μ: As mentioned, the drift velocity, $v_{\mathrm{d}n}$, of electrons is driven by the external electric field, E, in between collisions. Thus, given the mean collision time, τ_n $v_{\mathrm{d}n}$ can be specified by the product of the acceleration and the mean collision time:

$$v_{dn} \approx \frac{(-q)\mathrm{E}}{m_n}\tau_n \equiv -\mu_n\mathrm{E}, \mu_n \equiv \frac{q\tau_n}{m_n} \tag{2.25}$$

As clear from (2.25), the mobility, μ_n, thus defined is the response function connecting the input field, E, to the output drift velocity and is specified in terms of the mean collision time, τ_n, and the effective mass, m_n, of the electron. The total velocity of an electron then consists of two components, i.e., the random thermal velocity and the drift velocity. Obviously, only the latter component contributes to the current, as illustrated in Fig. 2.10. One can likewise express the hole drift velocity and mobility as

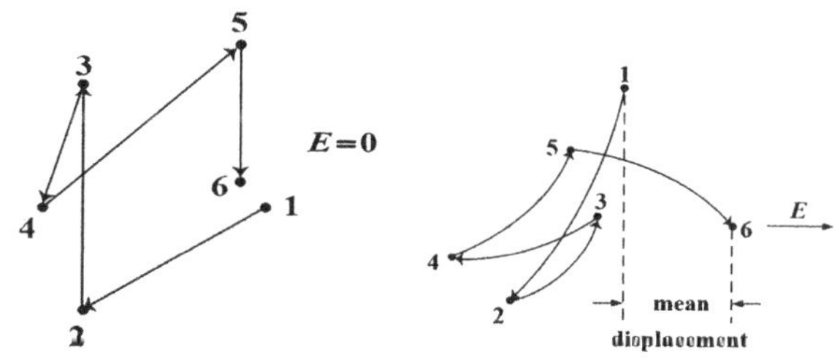

Fig. 2.10 The random thermal motions of electrons in the absence (*left*) and presence (*right*) of external field

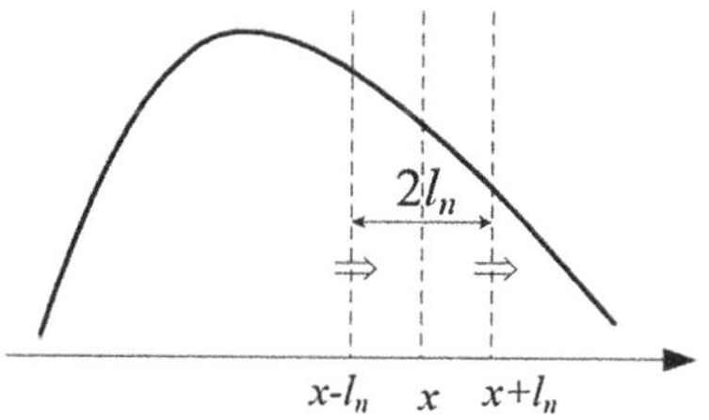

Fig. 2.11 An inhomogeneous electron concentration versus x and a differential volume element consisting of two planes at $x \pm l_n$ centered around x with l_n denoting the electron diffusion length

$$v_{dp} \approx \frac{q\mathrm{E}}{m_p}\tau_p \equiv \mu_p\mathrm{E}, \mu_p \equiv \frac{q\tau_p}{m_p} \tag{2.26}$$

Note that the mean collision time, τ_n or τ_p in turn depends upon the total velocity and the medium in which the transport occurs. The constant mobility independent of E, as defined in (2.25) and (2.26), is valid in the small field regime.

Diffusion Coefficient, D: The diffusion coefficient connects the output flux of particles to the input concentration gradient. Given a non-uniform electron density as sketched in Fig. 2.11, the electron flux at x can be conveniently analyzed by introducing the mean free path of electrons l_n on both sides of x. The mean free path is the average distance a particle travels before encountering the collision. Thus, electrons can be treated as free particles in the volume elements from $x - l_n$ to x and from x to $x + l_n$. The net number of electrons crossing the x plane per unit area from left to right is then given by

$$\dot{N} = N_{LR} - N_{RL} = \frac{1}{2}n(x - l_n)l_n - \frac{1}{2}n(x + l_n)l_n \tag{2.27}$$

where 1/2 factor accounts for the random motion of electrons in equilibrium, that is, only one half of the electrons are moving from left to right or vice versa. Since the mean free path l_n can be taken small, compared with the typical spatial range of variation in concentration, one can Taylor expand the electron concentrations at $x \pm l_n$ up to the first order as

$$n(x \pm l_n) \approx n(x) \pm \frac{\partial n}{\partial x}l_n \tag{2.28}$$

By inserting (2.28) into (2.27), one can write

$$N = -l_n^2 \frac{\partial n(x)}{\partial x} \tag{2.29}$$

Now the flux of electrons from left to right at x is, by definition, given by dividing N in (2.29) by the mean collision time, τ_n:

$$F_n \equiv \frac{N}{\tau_n} = -\frac{l_n^2}{\tau_n}\frac{\partial n(x)}{\partial x} \equiv -D_n\frac{\partial n(x)}{\partial x}, \ D_n \equiv \frac{l_n^2}{\tau_n} \tag{2.30}$$

In this way, the output flux is related to the input concentration gradient via the diffusion coefficient, D_n. The coefficient in turn is specified by the mean free path l_n and the collision time τ_n. These two physical quantities, l_n and τ_n, are related by

$$l_n = v_T\tau_n \tag{2.31}$$

where v_T is the thermal speed of electrons. Although the total electron velocity consists of thermal and drift velocities, the former is much greater than the latter, unless the electric field is extremely high. Thus, the latter term has been neglected in (2.31). Moreover, as discussed in Chap. 1, v_T is to be given in terms of the thermal energy via the equipartition theorem:

$$m_n v_T^2/2 = k_B T/2 \tag{2.32}$$

Einstein Relation: Now that both mobility and diffusion coefficients of the electron are specified by (2.25) and (2.30), respectively, the relationship between the two can be found by considering the ratio:

$$\frac{D_n}{\mu_n} = \frac{l_n^2/\tau_n}{q\tau_n/m_n} = \frac{k_B T}{q} \tag{2.33}$$

Here, (2.31) has been used for relating the mean free path to the thermal velocity. Similarly, one can write

$$\frac{D_p}{\mu_p} = \frac{l_p^2/\tau_p}{q\tau_p/m_p} = \frac{k_B T}{q}, \ l_p = v_T\tau_p \tag{2.34}$$

Equations (2.33) and (2.34) are known as the Einstein relation, relating the two transport coefficients.

High-Field Mobility: As pointed out, in the high-field regime, the mean collision time, hence the drift velocity depends on E, and therefore the mobility is not a constant but becomes a function of E. This can be understood in light of the response function. A response function, connecting the input to the output generally depends on the input as its magnitude increases. For the case of mobility, the input is the electric field; hence, μ naturally depends on E. This means that as the channel length of the device is shortened the drift velocity should be increasingly dependent E. In silicon, for example, the drift velocity can be expressed empirically by

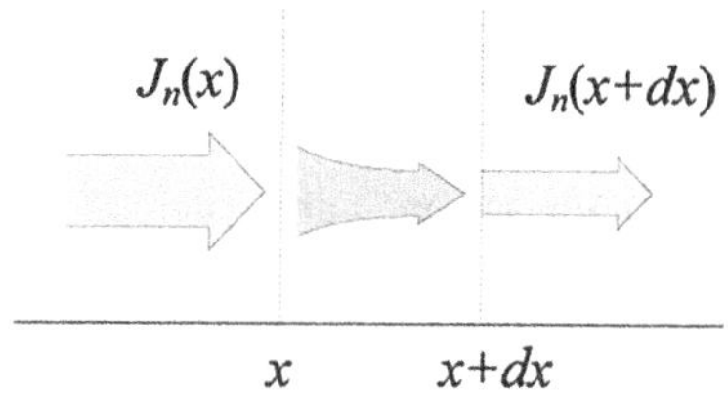

Fig. 2.12 A sketch for the charge conservation in 1D differential volume element

$$v_d = \frac{\mu \mathrm{E}}{1 + \mu \mathrm{E}/v_s} \tag{2.35}$$

where μ is the low field mobility and v_s the saturation velocity in silicon. As clear from (2.35), v_d increases linearly with E at a slope given by μ in low field regime, but in the limit of large field, v_d saturates at a level given by v_s. In between the two limits, v_d increases sub-linearly with E.

Drift and Diffusion Current: Now that the mobility and diffusion coefficient have been discussed, the current densities associated are briefly considered. Naturally, the current density of electron consists of the drift and diffusion components. The drift current is contributed by the drift velocity v_{dn}, and the electrons acquire in between collisions driven by the external electric field E, i.e.,

$$J_{n\,\mathrm{drift}} \equiv (-q)nv_{dn}, \; v_{dn} = \mu_n \mathrm{E} \tag{2.36}$$

The diffusion current is driven by the spatial gradient of the electron concentration, i.e.,

$$J_{n\,\mathrm{diff}} \propto -\frac{dn}{dx} \equiv (-q)D_n\left(-\frac{dn}{dx}\right) \tag{2.37}$$

where the electron diffusion coefficient, D_n, connects the flux with the concentration gradient. The total electron current density of the electron is thus given by

$$J_n = qn\mu_n E + qD_n \nabla n \tag{2.38}$$

Likewise, the hole current density is given by

$$J_p = qp\mu_p E - qD_p \nabla p \tag{2.39}$$

Continuity Equation: Given a volume element dx at x, the time rate of the change of the charge density therein is due to the net current density flowing into the element, as shown in Fig. 2.12. Thus, one can write

$$\frac{\partial}{\partial t}(-qn\mathrm{d}x) = J_n(x) - J_n(x + dx) = \frac{\partial J_n}{\partial x} \tag{2.40}$$

where $J_n(x + dx)$ has been Taylor expanded at x.

Thus, generalizing (2.40) to 3D, the continuity equation for the electron charge density reads as

$$\frac{\partial n}{\partial t} = -\frac{1}{q}\nabla \cdot J_n, J_n = q\mu_n n\mathrm{E} + qD_n\nabla n \tag{2.41}$$

Likewise, one can write

$$\frac{\partial p}{\partial t} = -\frac{1}{q}\nabla \cdot J_p, J_p = q\mu_p p\mathrm{E} - qD_p\nabla p \tag{2.42}$$

2.5 Tunneling

The tunneling phenomenon is one of the unique features of quantum mechanics and is rooted in the wave nature of particles. The tunneling consists essentially of a particle transmitting through a potential barrier higher than its kinetic energy. The probability of a particle with kinetic energy E to tunnel through a potential barrier of height V and thickness d is approximately given by

$$T \approx \exp -2\left[\frac{2m}{\hbar^2}(V - E)\right]^{1/2} d \tag{2.43}$$

Thus, given a potential barrier $V(x)$, one can decompose it into a juxtaposition of barrier elements with height Vj and width dx and find the total net probability by multiplying the individual probabilities, obtaining

$$T = \exp -\frac{2\sqrt{2m}}{h}\int_{x_1}^{x_2} dx[(V(x) - E)]^{1/2} \tag{2.44}$$

where x_1, x_2 are determined by the condition, $V = \mathrm{E}$.

Direct Tunneling Probability: Given a potential barrier of height V at the interface between the semiconductor and dielectric material, it is transformed by the applied electric field E into

$$V(x) = V - q\mathrm{E}x \tag{2.45}$$

where q the magnitude of the electron charge. Upon inserting (2.45) into (2.44) and performing the integration, one finds

$$T = \exp -\left\{\frac{4(2m)^{1/2}}{3q\mathrm{E}h}\left[(V - E)^{3/2} - (V - E - q|\mathrm{E}|d)^{3/2}\right]\right\} \tag{2.46}$$

where d is the barrier thickness. Clearly, T is mainly determined by the applied electric field, E, d, and the incident energy E of the incident electron. The tunneling through the trapezoidal barrier is known as the direct tunneling. The direct tunneling of electrons through the dielectric layer imposes fundamental limitations on down scaling the semiconductor devices into nanoregimes.

Fowler–Nordheim Tunneling. When the potential barrier is of a triangular shape instead, (2.46) reduces to

$$T = \exp - \frac{4(2m)^{1/2}}{3q\mathrm{E}h}(V - E)^{3/2} \tag{2.47}$$

and is known as the Fowlwer–Nordheim (F–N) tunneling. Understandably, T is shown to depend primarily on E and incident energy E. The F–N tunneling fundamentally limits the device operation and downscaling because it enhances the gate leakage in logic devices. However, it also provides the driving force for nonvolatile memory cells.

2.6 The Applications of Tunneling

The tunneling is extensively utilized in various semiconductor and optoelectronic devices. For example, the tunneling underpins the operation of tunnel diodes, Schottky-Ohmic contacts, resonant tunneling devices, single electron transistors, nonvolatile memory cells. In addition, the tunneling offers a convenient means for the scanning tunneling microscopy. More importantly, the resonant tunneling provides the physical basis whereby electrons or holes move as free particles in solids. Some of these applications are briefly discussed.

Direct and F–N Tunneling in MOSFET: As pointed out, the tunneling through trapezoidal or triangular potential barriers play critical roles in the operation of MOSFETs. The barrier in this case is provided by the silicon dioxide layer deposited between the gate electrode and the silicon substrate. Without the bias between the gate electrode and the substrate, the silicon dioxide can effectively block the tunneling of electrons or holes from the gate to the semiconductor or vice versa. When the gate voltage is applied, however, the barrier is transformed into a trapezoidal or triangular shape, and consequently, tunneling probability is greatly enhanced. As a result, the leakage current is enhanced greatly, degrading the performance of MOSFETs. However, the F–N tunneling constitutes the driving force for the operation of nonvolatile memory cells, e.g., flash EEPROM. In these memory cells, electrons are tunneled into the floating gate from the channel for programming. Also, the electrons stored in the floating gate are tunneled out for erasing.

Scanning Tunneling Microscopy: The tunneling provides an efficient means of probing the surface morphology with atomic-scale resolution. The working principle of the microscope simply consists of keeping the tunnel current between the probe tip and the surface atoms fixed at given bias. Or the distance of the tip is kept fixed with respect to the sample surface. As discussed, the F–N or direct tunneling probability is dictated by (1) the distances involved and (2) the voltages or electric fields applied. Hence, for the case of the former scheme, to keep the tunnel current constant at given voltage, the distance of the probe tip with respect to the surface atom has to be maintained at fixed distance. This requires an adjustment of the height of the tip as it scans the surface. The changing height of the tip versus the scanning reveals the surface morphology with about 0.1 nm in accuracy. For the latter case, the varying tunneling current at fixed voltage is translated into the surface morphology.

Field Emission Display: The image information to be displayed is transferred in the form of applied voltages from the driver circuitry to the array of the metallic tips, forming the pixels. The voltages transferred induce the field crowding at the metallic tip, enhancing the tunneling probability. The electrons thus tunneled out from the metallic tips carry the image information to the screen for display.

Ohmic Contact: The tunneling provides the means of implementing the Ohmic contact, an essential component of semiconductor devices, connecting them to the outside world with negligible resistance. Such a contact can be fabricated using the interface between the metal and the semiconductor. At the interface, a potential barrier is formed resulting from the difference between two work functions involved. However, the width of the barrier can be made very thin by heavily doping the semiconductor, rendering the potential barrier nearly transparent to tunneling electrons or holes. Thus, under bias, electrons and holes are readily injected from metal into semiconductor or vice versa with negligible voltage drop at the interface.

2.7 Problems

2.1 The Boltzmann distribution function for a system of free particles is given by

$$f_0(v) = N \exp - \left[(v_x^2 + v_y^2 + v_z^2)/2k_B T \right]$$

(a) Carry out the integration,

$$\begin{aligned} I &= \int_{-\infty}^{\infty} dv f_0(v) \\ &= N \int_{-\infty}^{\infty} dv_x e^{-(v_x^2/k_B T)} \int_{-\infty}^{\infty} dv_y e^{-(v_y^2/k_B T)} \int_{-\infty}^{\infty} dv_z e^{-(v_z^2/k_B T)} \end{aligned}$$

by using the formula,

$$\int_{-\infty}^{\infty} dxe^{-ax^2+bx} = \sqrt{\frac{\pi}{a}}e^{b^2/4a}$$

and find the normalization constant N by putting $I = 1$.

(b) Show that

$$\langle v_x^2 \rangle = \int_{-\infty}^{\infty} dv v_x^2 f_0(v)$$

$$= N \int_{-\infty}^{\infty} dv_x e^{-(v_x^2/k_BT)} v_x^2 \int_{-\infty}^{\infty} dv_y e^{-(v_y^2/k_BT)} \int_{-\infty}^{\infty} dv_z e^{-(v_z^2/k_BT)}$$

$$= k_BT/m$$

2.2 Starting from the general representation of the electron concentration,

$$n = \int_{E_C}^{E_C+\Delta E_c} dE g_{n3D}(E) f_n(E)$$

(a) Compact the expression of n in terms of the Fermi 1/2 integral given in (2.9a) by using the new variable of integration given in (2.9b).
(b) Derive (2.11a) by using a new variable of integration, $\eta = \xi^2$.
(c) Derive (2.13a) and (2.15a) by repeating a similar analysis.

2.3 Find Fermi potentials φ_{Fn} in silicon for the doping levels $N_D = 10^{15}$, 10^{17}, 10^{18} cm^{-3} by using $n_i = 1.45 \times 10^{10}$ cm^{-3}.

2.4 Consider an intrinsic silicon in between two silicon dioxide layers W distance apart.

(a) Find the ground and first excited state of electrons for $W = 10{,}100$ nm.
(b) Find the electron concentrations in the conduction band for $W = 100{,}10$ nm at room temperature.
(c) Compare the results obtained in (b) with n_i in bulk silicon. For simplicity of analysis, you may take the quantum well by an infinite square well potential.

Take $m_n/m_0 = 1.1$.

2.5 Consider a quantum wire made up of intrinsic silicon of cross-sectional area, W^2. The quantum wire is surrounded by silicon dioxide.

(a) Find the ground and first excited sub-bands.
(b) Find the electron concentrations in the conduction band for $W = 100, 10\,\text{nm}$ for $T = 300\,\text{K}$. For simplicity of analysis, you may approximate the quantum well by an infinite square well potential. Take $m_n/m_0 = 1.1$.

Suggested Reading

1. Kim, D. M. (2010). *Introductory quantum mechanics for semiconductor nanotechnology*. New Jersey: John Wiley-VCH.
2. Sze, S. M., & Ng, K. K. (2006). *Physics of semiconductor devices* (3rd ed.). New Jersey: Wiley-Interscience.
3. Streetman, B. G., & Banerjee, S. (2005). *Solid state electronic devices* (6th ed.). New Jersey: Prentice Hall.
4. Blakemore, J. S. (2002). *Semiconductor statistics*. New York: Dover Pubns.
5. Pierret, R. F. (2002). *Advanced semiconductor fundamentals, modular series on solid state devices* (2nd ed., Vol. VI). New Jersey: Prentice Hall.
6. Muller, R. S., Kamins, T. I., & Chan, M. (2002). *Device electronics for integrated circuits* (Third Sub edition ed.). New Jersey: John Wiley & Sons.
7. Pierret, R. F. (1988). *Semiconductor fundamentals, modular series on solid state devices* (Second edition ed., Vol. I). New Jersey: Prentice Hall.
8. Yariv, A. (1982). *An introduction to theory and applications of quantum mechanics*. New Jersey: John Wiley & Sons.
9. McKelvey, J. P. (1982). *Solid state and semiconductor physics*. Huntington: Krieger Pub Co.
10. Grove, A.S. (1967). *Physics and technology of semiconductor deices*. New Jersey: John Wiley & Sons.

Chapter 3
P–N Junction Diode: I–V Behavior and Applications

Dae Mann Kim, Bong Koo Kang and Yoon-Ha Jeong

Abstract The p–n junction diode is a simple two-terminal solid-state switch, but the theories underlying its operation encompass the central core of the semiconductor device physics. Thus, the I–V modeling of the junction diode should provide a convenient basis for modeling other kinds of semiconductor devices, including the silicon nanowire field effect transistor (SNWFET). Additionally, the p–n junction is used extensively as photodiodes, solar cells, light-emitting and laser diodes, etc. and constitutes a key element of MOSFET. This chapter is addressed to the I–V modeling and applications of the p–n junction diode and should thus provide a general background for discussing SNWFETs in the chapters to follow.

Abbreviation

SNWFET	Silicon nanowire field effect transistor
LED	Light-emitting diode
LD	Laser diode

D. M. Kim (✉)
Korea Institute for Advanced Study, Seoul, Korea
e-mail: dmkim@kias.re.kr

B. K. Kang · Y.-H. Jeong
POSTECH, Pohang, Korea
e-mail: bkkang@postech.ac.kr

Y.-H. Jeong
e-mail: yhjeong@postech.ac.kr

D. M. Kim and Y.-H. Jeong (eds.), *Nanowire Field Effect Transistors: Principles and Applications*, DOI: 10.1007/978-1-4614-8124-9_3,

3.1 General Background

The p–n junction consists of n and p type semiconductors in equilibrium contact. There are two kinds of junctions, homo and hetero junctions. In the former, the two bandgaps in contact are the same, while in the latter the bandgaps are different. In this chapter the homojunction is singled out for discussion, but the results obtained can also be applied to the heterojunctions. The general I–V characteristics of the diode are shown in Fig. 3.1. Under a forward bias, i.e., a positive voltage applied to the p side, a large current is drawn from p to n. Under a reverse bias, on the other hand, a minimal current flows from n to p. Hence, the diode works as a switch.

The operation of the diode is based on the interplay between the equilibrium and non-equilibrium. Specifically, the input bias is used to push the junction away from the equilibrium to the non-equilibrium. Then the junction reacts to push itself back to the equilibrium, thereby inducing the output current. Thus, it is necessary to first consider the physics of the equilibrium and the non-equilibrium.

Equilibrium in Single Semiconductor: As discussed in Chap. 2, in thermal equilibrium, the carrier concentrations, n and p are quantified by a single Fermi level, E_F. Also, the current density is commensurate with the slope of E_F, which is shown as follows. The 1D current density of electrons is given from (2.38) by

$$J_n = q\mu_n nE + qD_n \frac{dn}{\partial x} \tag{3.1}$$

Here the electric field, E, is to be expressed in terms of the electrostatic potential, φ, i.e., $E = -\partial\varphi/\partial x$. Also, φ in turn represents the electron potential energy when multiplied by the charge of the electron, $-q$, so that $-q\varphi$ is identical to and parallels with E_C, E_V or the midgap energy E_i. Thus, one can write

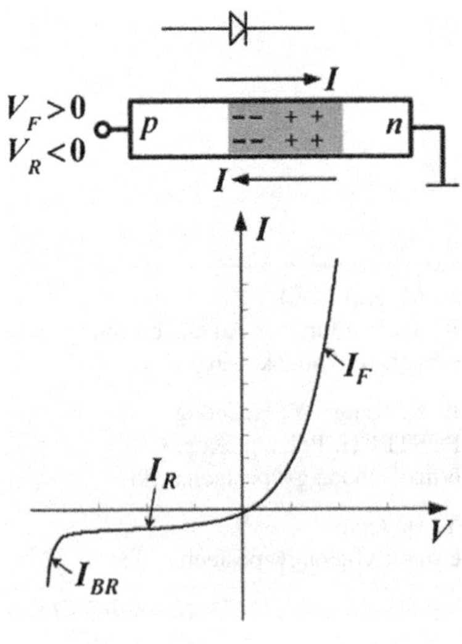

Fig. 3.1 The p–n junction diode and I–V *curve*, consisting of the forward, reverse and breakdown currents

$$E = -\frac{\partial \varphi}{\partial x} \equiv \frac{1}{q}\frac{\partial E_i}{\partial x} \tag{3.2}$$

Also, for non-degenerate case, n is analytically expressed in terms of E_i and E_F [see (2.11)]. Hence with the use of (2.11), (3.2) and the Einstein relation, (2.33), J_n in (3.1) can be compacted as

$$J_n = \mu_n n \frac{\mathrm{d}E_F}{\mathrm{d}x} \tag{3.3}$$

(problem 3.1). One can likewise express the hole current density as

$$J_p = \mu_p p \frac{\mathrm{d}E_F}{\mathrm{d}x} \tag{3.4}$$

Since no current flows in equilibrium, the Fermi level should therefore be flat.

Equilibrium in a Composite Semiconductor System: A composite semiconductor system consists of two different semiconductors in equilibrium contact, as shown in Fig. 3.2.

The electron flux from left to right, F_{RL}, is then balanced by the reverse flux from right to left, F_{RL}. Now, F_{LR} is specified from the Pauli exclusion principle by (1) the number of electrons occupying the quantum states on the left at a given energy, E, and (2) the empty quantum states on the right with the same E for the electrons to move in, i.e.,

$$F_{LR} \propto n_L(E) v_R(E) \tag{3.5a}$$

Here the electron density on the left at E

$$n_L = g_L(E)\mathrm{d}E f_L(E) \tag{3.5b}$$

is given by the density of states, $g_L(E)$ and the Fermi occupation factor, while the density of vacant states on the right,

$$v_R = g_R(E)\mathrm{d}E[1 - f_R(E)] \tag{3.5c}$$

is given in terms of the density of states and the probability that the state is not occupied by the electron.

Hence, one can explicitly specify (3.5a) by

$$F_{LR} \propto g_L(E)\mathrm{d}E f_L(E) g_R(E)\mathrm{d}E[1 - f_R(E)] \tag{3.6}$$

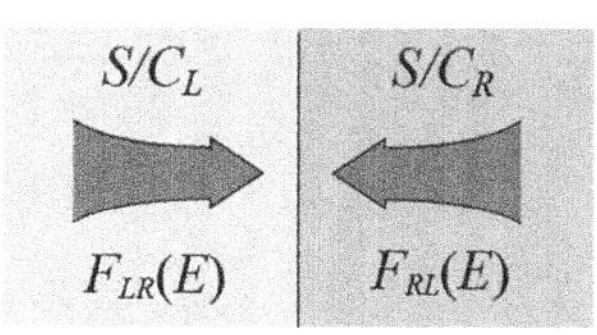

Fig. 3.2 Two semiconductors in equilibrium contact. The electron flux from *left* to *right* is balanced by its inverse flux from *right* to *left* in equilibrium

Similarly, F_{RL} is given by

$$F_{RL} \propto g_R(E)\mathrm{d}Ef_R(E)g_L(E)\mathrm{d}E[1 - f_L(E)] \tag{3.7}$$

Since $F_{LF} = F_{RL}$ in equilibrium, it follows from (3.6) and (3.7) that $f_L(E) = f_R(E)$,i.e.,

$$\frac{1}{1 + e^{(E-E_{FL})/k_BT}} = \frac{1}{1 + e^{(E-E_{FR})/k_BT}} \tag{3.8}$$

The only way to satisfy (3.8) is to have identical Fermi levels on both sides, that is,

$$E_{FL} = E_{FR} \tag{3.9}$$

Therefore, E_F should line up and be flat in equilibrium.

3.2 p–n Junction in Equilibrium

Junction Band Bending: The behaviors of E_F discussed in the preceding section are next applied to considering the junction band bending. When *p*- and *n*-type semiconductors are brought together into an equilibrium contact, the band bending should ensue for the reasons as follows. Before the contact, $E_F$on the n side lies above E_i, while $E_F$on the p side lies below E_i, as shown in Fig. 3.3. Thus, the difference between the two Fermi levels is given by the sum of two Fermi potentials.

Upon the equilibrium contact, however, E_F should line up and be flat. Otherwise, the current would flow in violation of the equilibrium condition [see (3.3), (3.4)]. The only way to satisfy the condition is for the band to bend by an amount,

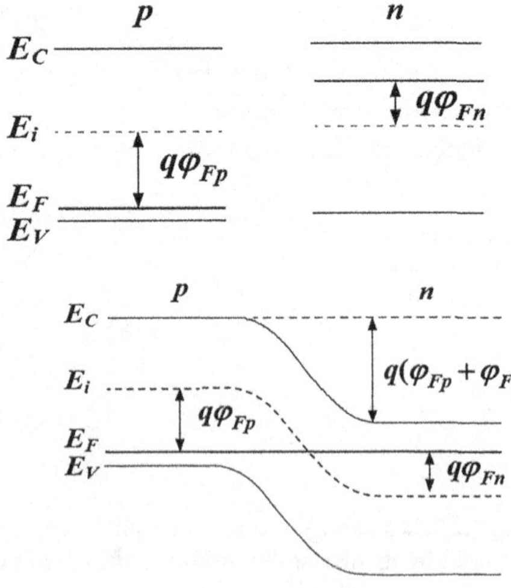

Fig. 3.3 The energy band of the p–n junction before (*top*) and after (*bottom*) the contact

$$q\varphi_{bi} = q\varphi_{Fn} + q\phi_{Fp} \tag{3.10}$$

where $\varphi_{bi}$is the built-in potential given by the sum of Fermi potentials, φ_{Fn}, φ_{FP}.

Electrostatics for Band Bending: The required band bending is supported by the electrostatics entailed in the equilibrium contact. Upon contact, the electrons should diffuse from the high concentration *n* region to the low concentration *p* region. By the same token, holes diffuse from *p* to *n* region. These diffusions leave behind the uncompensated donor and acceptor ions, qN_D^+, qN_A^-in *n* and *p* regions, respectively, near the junction interface (see Fig. 3.4). Once the dipolar space charge is formed, it induces the space potential energy $-q\varphi$ that connects the misaligned $E_C$and $E_V$in the junction region, accounting for the band bending. The E field thus induced acts in turn as the source for the opposite flow of carriers via drift, the electrons drifting from *p* to *n*, and the holes drifting from *n* to *p* regions, respectively. These drift fluxes should balance the diffusion fluxes, so that there is no net flux of electrons or holes in equilibrium.

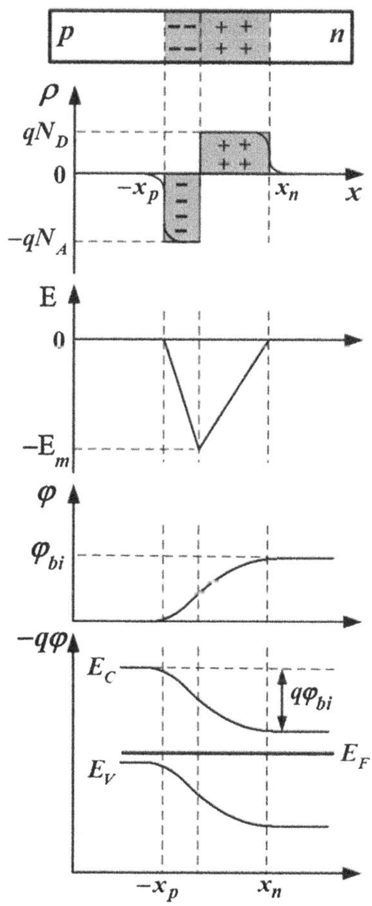

Fig. 3.4 The dipolar space charge, ρ, field, E, and potential, φ

The Coulomb's law for the dipolar space-charge layer is given by

$$\frac{\partial}{\partial x}\mathrm{E} = \frac{\rho}{\varepsilon_S}, \quad \rho = \begin{cases} qN_D, & 0 \le x \le x_n \\ -qN_A, & -x_p \le x \le 0 \end{cases} \tag{3.11}$$

where ε_S is the permittivity of the semiconductor. Here the space charge, ρ, is taken as a step function in completely depleted approximation, as depicted in Fig. 3.4. This simplification accurately accounts for the real space-charge profile for non-degenerate carrier concentrations (problem 3.2). In this concentration regime, E_F ranges in the energy gap below E_C and above E_V by a few thermal energies, k_BT, as discussed in Chap. 2. In this case, practically all of N_D and N_A atoms are ionized, and x_n and x_p accurately demarcate the junction region from n and p bulk regions, respectively. Given this representation of ρ, one can readily find the space-charge field as a linear function of x as

$$\mathrm{E}(x) = \begin{cases} (qN_D/\varepsilon_S)(x - x_n), & 0 \le x \le x_n \\ -(qN_A/\varepsilon_S)(x + x_p), & -x_p \le x \le 0 \end{cases} \tag{3.12}$$

Here, the boundary conditions of $E(x)$ are given by $\mathrm{E}(x_n) = \mathrm{E}(-x_p) = 0$, since E should not penetrate into n and p bulk regions. Also, $E(x)$ should be continuous everywhere, and this condition applied at the interface, $x = 0$ yields

$$qN_Dx_n = qN_Ax_p \tag{3.13}$$

Thus, the total number of electrons and holes spilled over from n to p and from p to n regions, respectively, are the same. Also the magnitude of the maximum space-charge field at $x = 0$is given by

$$\mathrm{E}_{\max} = qN_Dx_n/\varepsilon_S = qN_Ax_p/\varepsilon_S \tag{3.14}$$

Built-in Potential and Depletion Depth: Now that $\mathrm{E}(x)$ has been found, the electrostatic potential can be obtained as a quadratic function of x. But the quantity of interest is the difference in potential between $-x_p$and x_n,which should amount to the built-in potential, φ_{bi}. Evidently finding $\varphi_{bi}$is equivalent to finding the area of the triangle formed by $\mathrm{E}(x)$curve in Fig. 3.4, and one can write

$$\varphi_{bi} = \frac{1}{2}\mathrm{E}_{\max}\mathrm{W}, \quad \mathrm{W} \equiv \mathrm{x_n} + \mathrm{x_p} \tag{3.15}$$

where W is the junction depletion depth. The depth, W, in turn can be expressed in terms of x_n or x_p alone by using (3.13):

$$W = x_n(1 + N_D/N_A) = x_p(1 + N_A/N_D) \tag{3.16}$$

Hence, with the use of (3.16) for W, the maximum field in (3.14) can be found in terms of W:

$$\mathrm{E}_{\max} = \frac{qN_AN_D}{\varepsilon_S(N_A + N_D)}W \tag{3.17}$$

and therefore the built-in potential is given from (3.15) by

$$\varphi_{bi} \equiv \frac{1}{2}\mathrm{E}_{\max} W = \frac{q}{2\varepsilon_S}\frac{N_A N_D}{N_A + N_D} W^2 \tag{3.18}$$

Also, by using (2.23) and (2.24), φ_{bi}can be specified in terms of doping levels N_D and N_A in the regime of the non-degenerate carrier concentration as

$$\varphi_{bi} \equiv \varphi_{Fn} + \varphi_{Fp} = \frac{k_B T}{q} \ln\left(\frac{N_A N_D}{n_i^2}\right) \tag{3.19}$$

In this manner, the junction parameters are all specified in the completely depleted approximation.

Equilibrium Carrier Concentrations: The electron concentration, n, in the depletion depth, W, ranges from the majority carrier concentration, n_{nO}, in the n bulk region to the minority carrier concentration, n_{p0}, in the p bulk region:

$$n_{n0} = n_i e^{q\varphi_{Fn}/k_B T}, \quad n_{p0} = n_i e^{-q\varphi_{Fp}/k_B T} \tag{3.20}$$

so that $n_{p0}/n_{n0} = \exp -[q\varphi_{bi}/k_B T]$. Similarly, the hole concentration, p, ranges from p_{p0} in the p bulk region to p_{n0} in the n bulk region:

$$p_{p0} = n_i e^{q\varphi_{Fp}/k_B T}, \quad p_{n0} = n_i e^{-q\varphi_{Fn}/k_B T} \tag{3.21}$$

Hence, the ratio between p_{nO} and p_{pO} is same as that between n_{pO} and n_{nO}, i.e., $p_{n0}/p_{p0} = \exp -[q\varphi_{bi}/k_B T]$.

3.3 The Junction Under Bias

Forward and Reverse Bias: When the junction is forward biased with a positive voltage, V,applied to the p side, the band in the p bulk is, by definition, lowered by definition with respect to the n bulk by an amount qV. Hence, the band bending reduces from the equilibrium value, $q\varphi_{bi}$to $q(\varphi_{bi} - V)$,as sketched in Fig. 3.5. However, the location of E_F with respect to E_C, E_V should remain the same in both n and p regions, since the equilibrium carrier concentrations are preserved therein. These two requirements cannot be met with the single E_F as in equilibrium. Instead, the two quasi-Fermi levels are required, one for electron, E_{Fn}, and the other for hole, E_{Fp}. Also, the reduced band bending by an amount qVshould be accompanied by reduced W and $E_{\max}$ (Fig. 3.5).

In the quasi-equilibrium approximation, E_{Fn} is taken flat in both n and junction regions and to merge with E_{Fp} in the p bulk region, where a single Fermi level should suffice to quantify n and p. By the same token, E_{Fp} is flat in p and junction regions and merges with E_{Fn} in the n bulk. As a consequence, a loop is formed in W.

When the junction is reverse biased by applying negative voltage, $-V$, on the p side, the band therein is raised by qV and the junction band bending should

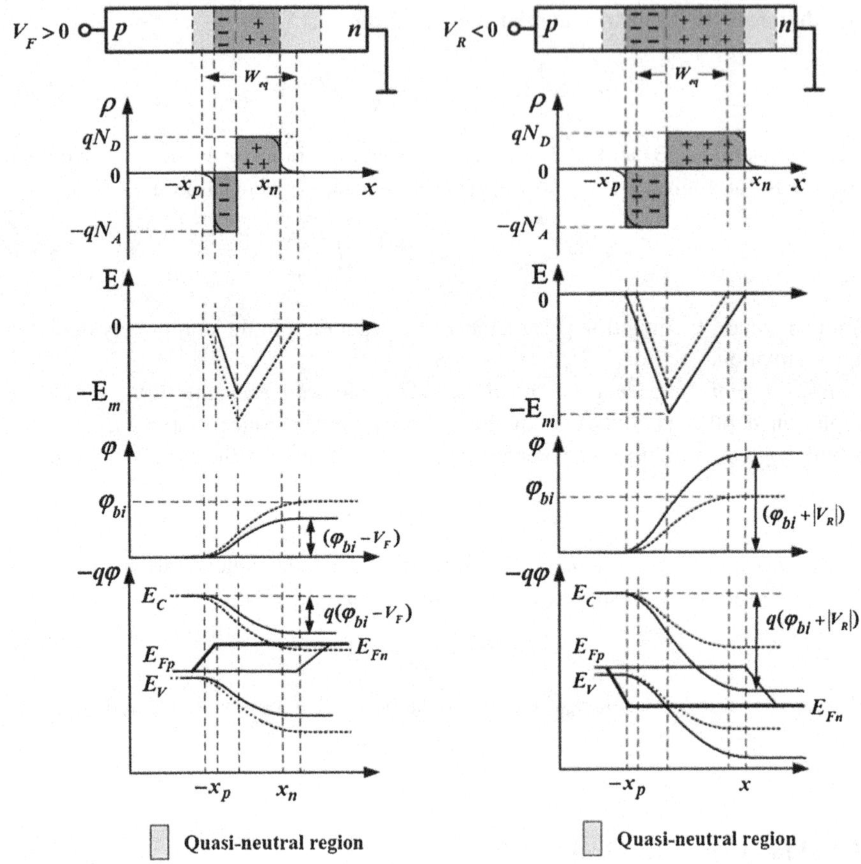

Fig. 3.5 The p–n junction under forward (*left*) and reverse (*right*) biases. The band bending, *W*, E, and φ are reduced and increased under forward and reverse biases, respectively, and the electron and hole quasi-Fermi levels split in the junction by qV

therefore increase from $q\varphi_{bi}$to $q[(\varphi_{bi} - (-V))]$. Concomitantly, *W* and $E_{\max}$ should also increase (Fig. 3.5). The increase or decrease of *W* and $E_{\max}$ should thus depend on the bias and is given from (3.17), (3.18) by

$$E_{\max}(V) = \frac{qN_D N_A}{\varepsilon_S(N_A + N_D)} W(V) \tag{3.22}$$

with

$$W(V) = \left[\frac{2\varepsilon_S(N_A + N_D)}{qN_A N_D}(\varphi_{bi} - V)\right]^{1/2} \tag{3.23}$$

where *V* is positive or negative depending on the forward or negative biases, respectively.

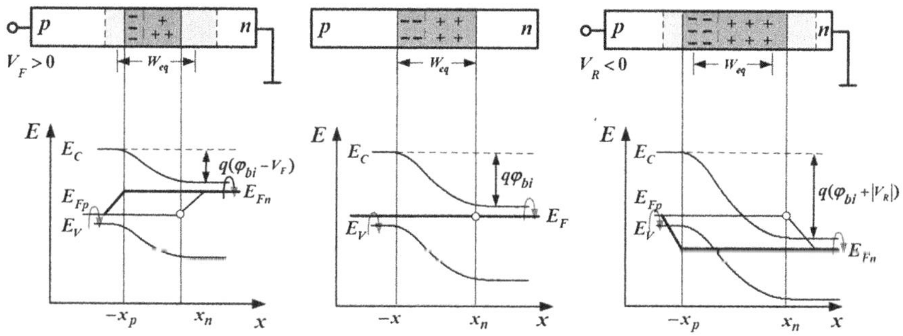

Fig. 3.6 The junction band bending in equilibrium (*middle*) and under forward (*left*) and reverse (*right*) biases. Two quasi-Fermi levels split in the depletion depth and merge in the quasi-neutral regions

Charge Injection and Extraction. As pointed out, the control of n and p is a key element in the operation of the junction diode or for that matter any active devices. The overall picture of n and p in equilibrium and under bias is sketched in Fig. 3.6. Obviously, the two quasi-Fermi levels in the junction depletion region should split by

$$E_{Fn} - E_{Fp} = qV \tag{3.24}$$

However, E_{Fn} and E_{Fp} merge gradually just outside of x_n and x_p in the regions called the quasi-neutral regions.

Since E_C and E_V vary versus x in the depletion region, n and p should also vary therein. For the non-degenerate case, one should replace E_F by E_{Fn} and E_{Fp} and specify the $n\,p$ product as

$$n(x)p(x) = n_i e^{[E_{Fn}-E_i(x)]/k_BT} n_i e^{[E_i(x)-E_{Fp}]/k_BT} = n_i^2 e^{qV/k_BT} \tag{3.25}$$

It is therefore clear that the charge is injected above the equilibrium value under forward bias, while it is extracted under reverse bias, as shown in Fig. 3.7. The injection and extraction of minority carriers under bias are the driving forces of the diode operation, as will be discussed in a few sections to follow.

3.4 The Ideal Diode I–V Model

Overview: As noted, there is no net flux of electrons or holes in equilibrium. For the case of electrons, for example, the diffusion flux from n to p region is balanced by the drift flux from p to n region, driven by the space charge field in the junction region. However, when the junction is driven away from equilibrium, a reactive process ensues to drive the system back to equilibrium. Specifically, the charge injection is accompanied by the recombination of electrons and holes, so that the

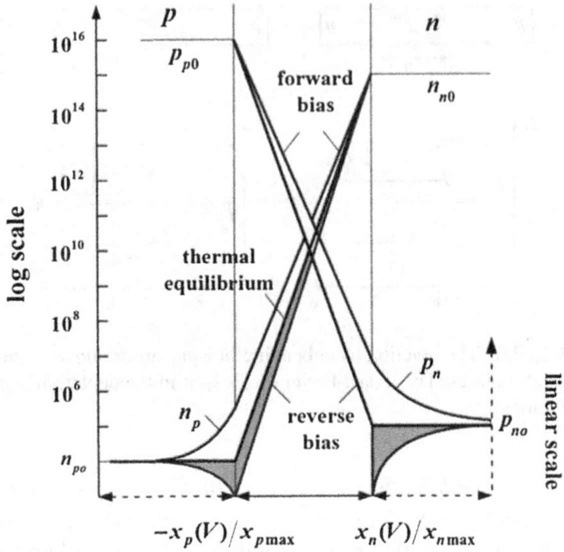

Fig. 3.7 The injection and extraction of electrons and holes in the junction region including the quasi-neutral regions, under forward and reverse biases

excess charge carriers are reduced. By the same token, the charge extraction is accompanied by the generation of electron–hole pairs. These two reactive processes give rise to the forward and reverse currents. In this manner, the diode operation is based on the interplay between the equilibrium and the non-equilibrium.

In spilling over from n to p regions, the electrons have to overcome the potential barrier of height, $q\varphi_{bi}$,in equilibrium while they can freely roll down the barrier from p to n regions, irrespective of the barrier height (see Figs. 3.3 and 3.5). This detailed balancing between diffusion and drift is broken under bias. Under forward bias, the potential barrier is lowered, as a consequence of which there should be more diffusion flux than the drift flux.

With this general overview in mind, the diode I–V behavior is quantified by using the theory by Shockley. In his model, Shockley introduced a few simplifications. First, the abrupt deletion approximation has been used, in which the dipole charge layer, ρ, is taken constant at the level, given by N_D and N_A and ending abruptly at x_n and $-x_p$ (Fig. 3.4). Also, the charge injection is confined to a low level, so that the minority carrier concentrations, n_p and p_n, are much smaller than the majority carrier concentrations, p_{p0} and n_{n0}, respectively. Finally, the doping levels, N_D and N_A, are confined to the non-degenerate level. For simplicity, the junction interface region is taken ideal, and the surface state enhanced recombination or generation processes occurring therein are neglected.

In the I–V modeling, the p–n junction is divided into three regions, as shown in Fig. 3.6: (a) the depletion region, W, in which the dipolar charge, ρ, supporting the band bending is formed, (b) the two quasi-neutral regions where E_{Fn} and E_{Fp} gradually merge and (c) the n and p bulk regions.

Forward and Reverse Currents: Under a forward bias, the diffusion flux is greater than the drift flux, as discussed, providing excess electrons and holes in the junction interface, as shown in Fig. 3.7. For the case of holes, the excess concentration, p_n,in the quasi-neutral region on the n side evolves in time, dictated by the generalized continuity equation,

$$\frac{\mathrm{d}p_n}{\mathrm{d}t} = -\frac{\mathrm{d}}{\mathrm{d}x}\left(p_n\mu_p\mathrm{E} - D_p\frac{\mathrm{d}p_n}{\mathrm{d}x}\right) - \frac{p_n - p_{n0}}{\tau_p} \tag{3.26}$$

Here, the recombination for the excess hole concentration has been empirically introduced in the continuity Eq. (2.42).

In the steady state, $\partial p_n/\partial t = 0$ and the space charge field, E, in the quasi-neutral region is small, so that (3.26) simplifies as

$$\frac{\mathrm{d}^2p_n}{\mathrm{d}x^2} - \frac{p_n - p_{n0}}{L_p^2} = 0, \;\; L_p \equiv (D_p\tau_p)^{1/2} \; for \quad x \geq x_n \tag{3.27}$$

Here, L_p thus introduced denotes the hole diffusion length.

The boundary conditions for $p_n(x)$ to satisfy are

$$p_n(x_n) = p_{n0}e^{qV/k_BT}, \quad p_n(x \to \infty) = p_{n0} = n_i^2/N_D$$

The first condition accounts for the charge injection at the edge of the junction x_n, while the second condition states that p_n in the n bulk should be equal to its equilibrium value p_{n0}. The solution of (3.27) satisfying the boundary conditions is thus given by

$$p_n(x - x_n) = p_{n0}(e^{qV/k_BT} - 1)e^{-(x-x_n)/L_p} + p_{n0}, \;\; x \geq x_n \tag{3.28}$$

where L_p is the hole diffusion length.

Having found $p_n(x)$, the resulting diffusion current density is obtained as

$$J_p(x) \equiv qD_p\left(-\frac{\mathrm{d}p_n}{\mathrm{d}x}\right) = \frac{qD_pp_{n0}}{L_p}\left[e^{qV/k_BT} - 1\right]e^{-(x-x_n)/L_p}; \quad x \geq x_n \tag{3.29}$$

Similarly, the electron current density in the quasi-neutral region on the p side is obtained as

$$J_n(x) \equiv -qD_n\left(-\frac{\mathrm{d}n_p}{\mathrm{d}x}\right) = \frac{qD_nn_{p0}}{L_n}\left[e^{qV/k_BT} - 1\right]e^{(x+x_p)/L_p}; \quad x \leq -x_p \tag{3.30}$$

The total forward current is then contributed by the two current densities of electrons and holes, and one may take the total current as the sum of J_n and J_p, evaluated at x_n, and $-x_p$, respectively:

$$I_{\mathrm{ideal}} = I_n(-x_p) + I_p(x_n) = I_S(e^{qV/k_BT} - 1) \tag{3.31a}$$

where

$$I_S = \left(\frac{qD_n n_{p0}}{L_n} + \frac{qD_p p_{n0}}{L_p}\right)A_J = qn_i^2\left(\frac{D_n}{L_n N_A} + \frac{D_p}{L_p N_D}\right)A_J \tag{3.31b}$$

is the saturation current. Here A_J denotes the area of the diode cross section and minority carrier concentrations, n_{p0} and p_{n0} have been specified by the well known relationship:

$$n_{p0} = \frac{n_i^2}{p_{p0}} = \frac{n_i^2}{N_A}, \quad p_{n0} = \frac{n_i^2}{n_{n0}} = \frac{n_i^2}{N_D}$$

The Eq. (3.31a, 3.31b) represents the ideal diode I–V model.

Clearly, the diode current is contributed by both electrons and holes; hence, the p–n junction diode is a bipolar device. Also, under the forward bias, the current exponentially increases, which is driven by exponentially injected minority carriers. Moreover, electrons and holes diffuse in opposite directions, but the two currents add up because of the opposite polarity of the charges carried by electrons and holes. The forward current flows from p to n.

The two diffusion currents depend sensitively on x as clear from (3.29) and (3.30). However, the majority carrier drift currents also vary in each quasi-neutral region, in such a manner that the total current is maintained at a constant level. The voltage required to induce the required drift currents takes up a minute fraction of the total applied voltage because of the large majority carrier concentration taking part in the drift. Naturally, the constant current level throughout the junction region is an indispensible requirement for the steady state operation.

Under a reverse bias, in which $V < 0$,both p_n and n_p are depleted in respective quasi-neutral regions, and electrons and holes diffuse in the reverse direction. Equation (3.31a, 3.31b) also accounts for the reverse current flowing from n to p regions, provided V is replaced by $-V$.

3.5 The Non-ideal I–V Model

In the ideal I–V model, the injection and extraction of electrons or holes are considered. In practice, however, the recombination (r) or generation (g) of electrons and holes constantly takes place in the junction region, thereby giving rise to additional current components. The r, g processes are based on the band-to-band or trap-assisted excitation or recombination of electrons and holes, as sketched in Fig. 3.8. The resulting recombination current I_R is to be specified by the theories developed by Shockley and Hall or Reed and is given by

$$I_R \approx qU_R W A_J, \quad U_R = \frac{1}{2\tau} n_i e^{qV/2k_B T} \tag{3.32}$$

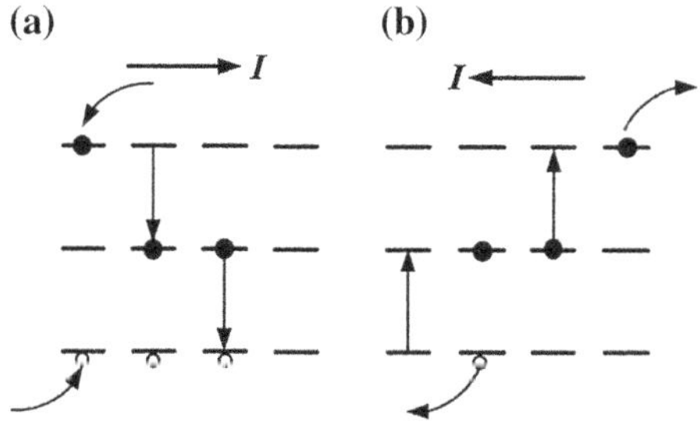

Fig. 3.8 The cyclic trap-assisted recombination (**a**) and generation (**b**) of electron–hole pairs

Here, $WA_J$is the volume of the depletion region where the recombination occurs, and $U_R$is the recombination rate per unit volume with τ denoting the recombination lifetime. Understandably, $I_R$is commensurate with the amount of excess charge carriers injected under the forward bias. Similarly, the generation current, I_G,is given by

$$I_G \approx qU_G WA_J, \quad U_G = \frac{n_i}{2\tau} \tag{3.33}$$

with U_G denoting the generation rate per unit volume.

Thus, the general non-ideal I–V behavior including I_R and I_G is given by

$$I = \begin{cases} I_{\text{ideal}} + I_R; & V > 0 \\ I_{\text{ideal}} + I_G; & V < 0 \end{cases} \tag{3.34}$$

where terms appearing in (3.34) have all been specified in preceding sections.

Note that I_R and I_G are contributed via the band-to-band or trap-assisted generation or recombination of electron hole pairs. In practice, I_R and I_G are primarily due to the latter process, the mechanisms of which are sketched in Fig. 3.9. Under the forward bias, the excess electrons and holes are supplied from n and p bulk to the junction regions and are recombined in two steps in succession. An incoming excess electron is captured by a trap, and the trapped electron in turn captures a hole via recombination. This two-step process is repeated for some of the electron–hole pairs injected, completing thereby the current loop for I_F.

Similarly, the generation current, I_G,can be specified in terms of the alternating emissions of holes and electrons in succession, as shown in the same figure. An electron in the valence band gets trapped in a trap site, emitting a hole. The trapped electron is in turn promoted into the conduction band. As a result, an e–h pair is generated, and both carriers are swept across the depletion region, holes to the p region and electrons to the n region, driven by strong junction electric field again to complete the loop for I_R.

The currents, I_R and I_G can be empirically put into the ideal I–V expression as

$$I = I_S\left[\exp(\frac{qV}{mk_BT}) - 1\right], \quad I_S \approx I_{S\text{ideal}} + A_J\frac{qn_i}{2\tau}W \tag{3.35}$$

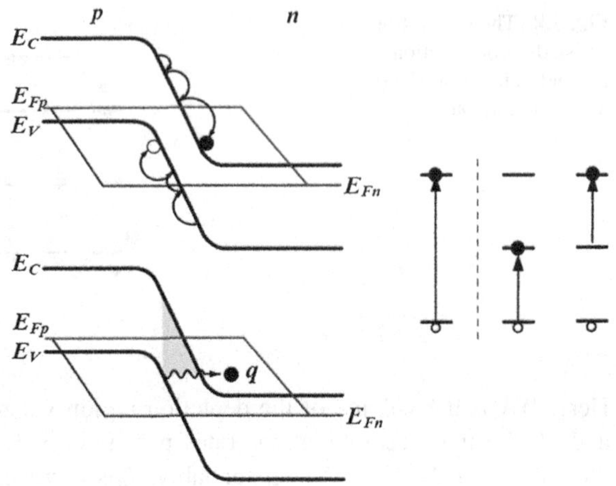

Fig. 3.9 The diode breakdown: the band-to-band or trap-assisted generation of e–h pairs (*right*) followed by the impact ionization occurring in cascade (*top left*) [reprinted from suggested reading 1]; the Fowler–Nordheim tunneling of valence band electrons to conduction band (*bottom left*) [reprinted from the same]

In this representation, the saturation current, I_S, includes I_G empirically, while $I_R$is incorporated into the ideality factor, m, which varies from 1 to 2 to fit the observed forward current. In addition, the explosive increase of $I_R$beyond a critical voltage, called the breakdown voltage, V_{BR} is due to a few basic physical processes, as discussed below. The mechanisms of breakdown are graphically illustrated in Fig. 3.9.

Avalanche breakdown: With increased band bending and strong space charge field induced by V_R, electrons gain kinetic energies as they roll down the junction potential hill sufficient to ionize the host atoms. As a result, e–h pairs are generated. Such impact ionization processes occurring in cascade give rise to the explosive growth of the e–h pairs, hence the breakdown current. The process is known as the avalanche breakdown.

Tunneling and Zener breakdown: The tunneling is a process unique in quantum mechanics as discussed in the chapter 2. The electrons are prohibited to reside in the energy gap, E_G, so that the gap acts as a potential barrier to electrons in the valence band. But the valence electrons can tunnel through the barrier to the conduction band and roll down the band to contribute to the reverse current as sketched in Fig. 3.9. Under a reverse bias, the potential barrier takes up a triangular shape of height E_G and width narrowed by the strong electric field, E, in the junction region (Fig. 3.9). Hence, the F–N tunneling prevails, and the probability is given from (2.47) by

$$T \propto \exp - \frac{4(2m_n)^{1/2}}{3qE\hbar} E_G^{3/2}$$

and increases exponentially with increasing E or the reverse voltage, V_R. The tunneling induced breakdown is known as Zener breakdown.

3.6 Applications of Diode

The p–n junction diode is utilized extensively, and a few applications are discussed in this section. The list of applications chosen for discussion includes the photodiode, solar cell, light-emitting diode, and laser diode. All of the applications chosen for discussion are based on the interaction of light with the diode. Thus, the optical absorption or emission in semiconductors is first discussed.

3.6.1 Optical Absorption and Emission

Figure 3.10 shows the conduction and valence bands in direct and indirect bandgap semiconductors. Also shown in the figure are the dispersion curves of electrons in the conduction band and those of holes in the valence band. Since electrons and holes move freely in the conduction and valence bands, the energy of these charge carriers, E, consists of the kinetic energy

$$E = \hbar^2 k^2/2m_j, \quad j = n, p \tag{3.36}$$

Here, m_j denotes the effective mass of the electron or hole, $\hbar = h/2\pi$ with h denoting the universal Planck constant, and the linear momentum, p, is given in terms of the wave vector, k. In a direct bandgap semiconductor, the minimum and maximum points in two dispersion curves coincide, while they do not in indirect semiconductor.

As discussed in Chap. 2, the quantum states of electrons above $E_F$are mostly empty, while those below $E_F$are practically all filled up in the non-degenerate concentration regime. Also, an electron when promoted to the conduction band, moves freely with an effective mass, m_n,near the bottom of conduction band.

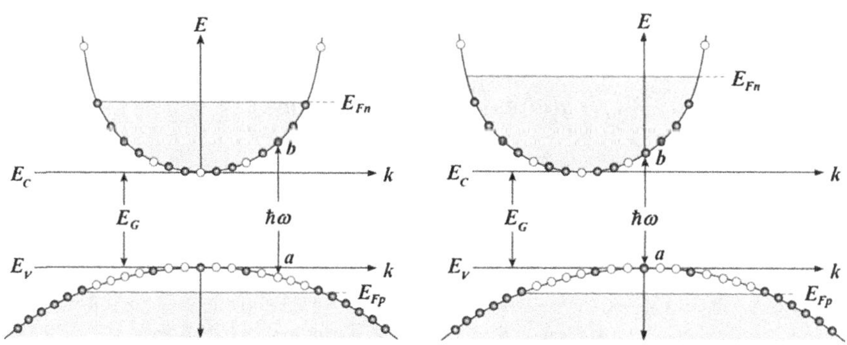

Fig. 3.10 The electron and hole dispersion curves in conduction and valence bands in a direct-bandgap (*left*) and indirect-bandgap (*right*) semiconductors. The generation of e–h pair via the absorption of a photon is also shown

Likewise, the hole in the valence band moves with the effective mass, m_p, near the top of the valence band. With increasing kinetic energy, the electron in the conduction band moves up the dispersion curve, while the hole in the valence band moves down the dispersion curve in a symmetric manner. The motions of the electrons and holes are the mirror images of each other.

The quantum mechanical description accounting for the light absorption or emission is a rather involved. Thus, the main highlights of the results are summarized in this section, relegating the detailed discussion to the books on quantum mechanics listed. Thus, consider the light incident on the semiconductor with frequency ν. The light is generally taken as a wave oscillating with ν, but it is also represented, due to its corpuscular nature, by the ensemble of photons of energy $h\nu$ streaming with the velocity of light. The absorption then consists of an electron promoted from the valence to conduction band by absorbing the energy of a photon, provided the photon energy is greater than the bandgap, $h\nu > E_G$ (Fig. 3.10). By the same token, the emission consists of the reverse process, i.e., the recombination of an electron with a hole, emitting a photon to conserve the energy.

Now, the transitions made by electrons in the absorption or emission of radiation occur vertically at a fixed wave vector, k, as shown in Fig. 3.10. This is due to the fact that the wavelength of light in visible frequency range under consideration is much larger than the lattice spacing, as discussed in detail in the reference books listed. This in turn points to the fact that in the indirect bandgap semiconductor, a certain amount of thermal energy is required to enable the vertical transition of an electron, emitting or absorbing photons. This additional requirement renders the emission probability smaller than that in the direct-bandgap semiconductor, in which there is no such requirement.

Linear Attenuation Coefficient: The light incident on and traversing through the absorbing medium is attenuated as

$$I(z) = I_0 e^{-\alpha z} \tag{3.37a}$$

where the linear attenuation coefficient, α, is given by the power absorbed per unit volume per incident power flux of the light, i.e., the Poynting vector:

$$\alpha(\omega) \equiv \frac{\hbar\omega N/V}{c\varepsilon \mathrm{E}_0^2}, \quad \hbar\omega = h\nu \tag{3.37b}$$

Here, N is the number of photons with the energy $h\nu$ absorbed in the volume V, and the Poynting vector is expressed in terms of the light intensity, E_0^2,velocity of light, c, and the permittivity of the medium, ε.

The absorption coefficient of light in semiconductor is likewise specified by

$$\alpha(\omega) = A^*(\hbar\omega - E_G)^{1/2}, \quad A^* = \omega\mu^2 m_r^{3/2}/\sqrt{2}\pi\hbar^3 c\varepsilon \tag{3.38a}$$

where μ is the dipole moment of the atom constituting the medium, E_G the bandgap and m_r reduced mass of the effective masses m_n and m_p:

$$1/m_r \equiv 1/m_n + 1/m_p \tag{3.38b}$$

as is discussed in detail again in the books referred to. It is therefore apparent from (3.38a, 3.38b) that the photon energy should be greater than the bandgap of the semiconductor for the absorption to occur. Also, the absorption coefficient increases with increasing photon energy as more electron states above E_C participate in the absorption process. Now that the absorption and emission of light has been summarized, the application of the junction diode is considered next.

3.6.2 Photodiodes

The photodiode is the junction diode used, however, for detecting the optical signal. The photodiode operates in the reverse bias mode, so that the detection speed is enhanced while the background noise is suppressed. The background noise arises from the reverse current of the diode, which is small. Thus, consider a reverse biased p–n junction with photons incident on it with the energy greater than the bandgap as sketched in Fig 3.11. The photons generate e–h pairs when absorbed via the band-to-band excitation. The e–h pairs thus generated in the junction region are separated automatically because of the junction band bending. Specifically, the electrons roll down the potential hill, while holes roll up the hill, driven by the strong built-in space charge field. The built-in junction field is further reinforced by the reverse bias. As a result, an output photocurrent flows in the reverse direction from n to p regions.

Generation of e–h pairs: To discuss the photocurrent, it is necessary to consider the generation of e–h pairs. The generation rate at the sample depth, z, is given by

$$g(z) = g_0 e^{-\alpha z}; \quad g_0 = \alpha[I_0(1 - R)/h\nu] \tag{3.39}$$

where α is the absorption coefficient (3.38a, 3.38b), I_0 the light intensity, and R the reflection coefficient. The absorption rate is thus specified by the product of α and the flux of photons reaching the depth, z, i.e., $(I_0/h\nu)(1 - R)\exp -\alpha z$. The photocurrent

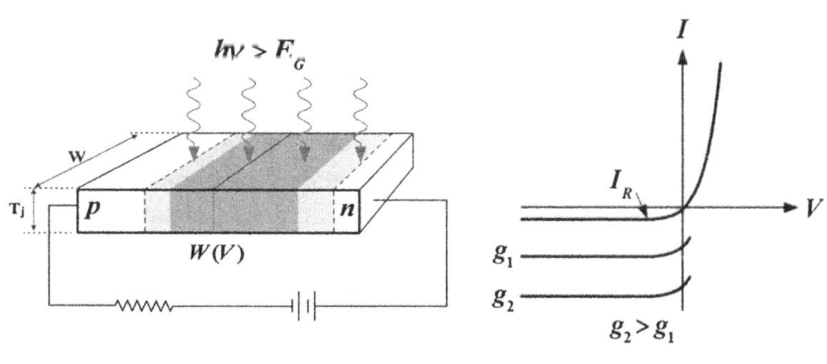

Fig. 3.11 The diode under illumination (*left*) and the resulting photocurrent (*right*)

is then contributed by the e–h pairs, optically generated and swept out of the junction region to complete the current loop, that is, the electrons are swept to the n region and holes to p regions. The resulting photocurrent therefore flows in the reverse direction and is given by integrating $g(z)$ over the volume of the junction:

$$I_{dr} = -qWw \int_0^{T_j} dz g(z) \tag{3.40}$$

where $T_j$and w are the diode thickness and width, respectively (Fig. 3.11). Upon inserting (3.39) into (3.40) and performing the integration, one finds

$$I_{dr} = -qA\tilde{g}_0 W, \quad \tilde{g}_0 = g_0\left[(1 - e^{-\alpha T_j})/\alpha T_j\right], \quad A = T_j \times w \tag{3.41}$$

with A denoting the diode cross section.

The light is also absorbed in the two quasi-neutral regions, and e–h pairs generated therein also contribute to the photocurrent. For instance, consider the e–h pair generated in the quasi-neutral region on the n side (see Figs. 3.5, 3.6 and 3.7). The electron then drifts out to the n region to join the majority carriers therein, while the holes diffuse toward the edge of the junction depletion region, x_n, where the holes are depleted under the reverse bias. Upon reaching x_n, the holes join those generated in depletion depth W and are swept out to the p region, adding to the output current, given by

$$I_{p,\text{diff}} = -qA\tilde{g}_0 L_p, \quad D_p \tau_p = L_p^2 \tag{3.42}$$

Here, L_p is the hole diffusion length given in terms of D_p and the lifetime, τ_p. The diffusion length represents the average distance a carrier diffuses before recombination (problem 3.7). Therefore, those holes generated within L_p from x_n all reach the junction edge, x_n, to be swept across the depletion depth. One can similarly obtain the photocurrent contributed by the electrons generated in the quasi-neutral region on the p side as

$$I_{n,\text{diff}} = -qA\tilde{g}_0 L_n, \quad L_n^2 = D_n \tau_n \tag{3.43}$$

with L_n denoting the electron diffusion length. The total photocurrent is given by the sum of (3.41), (3.42) and (3.43):

$$I_{ph} = -I_l, \quad I_l = I_{dr} + I_{n,\text{diff}} + I_{p,\text{diff}} \tag{3.44}$$

The I–V behavior of the photocurrent is shown in Fig. 3.11. Here, the reverse current of the diode constitutes the background noise, and the photocurrent increases in proportion to the input light intensity, as it should. Note that I_l is nearly flat with respect to the reverse voltage, V_R. This is because once generated, electrons and holes are swept out of the junction depletion region by the built-in electric field, regardless of V_R.

3.6.3 Photovoltaic Effect and Solar Cell

Photovoltaic Effect: The solar cell is a most important application of the p–n junction diode and is based on the photovoltaic effect. The effect refers to the physical processes in which an incident light generates a voltage across a certain portion of the illuminated region. The p–n junction provides such effect. The device physics underlying the junction solar cell also applies to other types of the solar cell, and therefore it is important to have a clear understanding of its operation in the p–n junction. The operation principle is substantially same as that of the photodiode, aside from the bias regime used.

The photovoltaic effect is again triggered by incident photons, generating e–h pairs via the band-to-band excitation near the junction interface, as shown in Fig. 3.12. The e–h pairs thus generated subsequently undergo the drift in opposite directions. Specifically, the electrons roll down the potential hill toward the *n* region, while holes roll up the hill toward the *p* region, driven by the junction field. In this manner, the photo-generated e–h pairs are separated from each other. As a consequence, excess holes and electrons pile up on p and n sides, respectively (Fig. 3.12). The resulting space charge induces a forward voltage, V, which in turn gives rise to the forward current, I_F, from *p* to *n* regions. However, the photocurrent as contributed by the e–h pairs separated and recombined through the external circuit flows in the opposite direction from *n* to *p* regions. Hence, the total current consists of two current components given by

$$I = I_F - I_l, \quad I_F = I_S(e^{qV/k_BT} - 1) \tag{3.45}$$

where the first term is the usual diode forward current, taken to be ideal for simplicity, and the second term is the photocurrent flowing in the opposite direction.

Now, one can find the open-circuit voltage, V_{oc}, by setting $I = 0$ in (3.45), obtaining

$$V_{oc} = \frac{k_BT}{q}\ln\left(\frac{I_l}{I_S} + 1\right) \approx \frac{k_BT}{q}\ln\frac{I_l}{I_S} \tag{3.46}$$

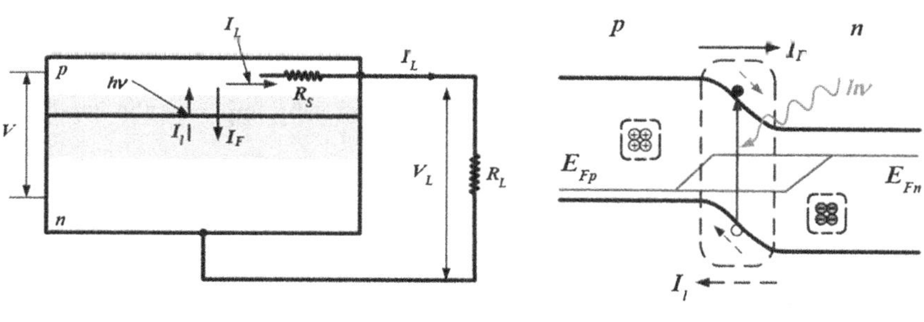

Fig. 3.12 Solar cell and equivalent circuit (*left*) and photo-generation and separation of e–h pairs (*right*) [reprinted from the suggested reading 1]

Also, one can find the short-circuit current, I_{sc}, by putting $V = 0$in (4.9), obtaining

$$I_{sc} = -I_l \tag{3.47}$$

The I–V curve given in (3.45) is in the fourth quadrant on the diode I–V plane, as it should be, and intersects with the voltage and current axes at V_{oc} and I_l, respectively. The region inside such I–V curve represents the maximum power rectangle.

When a load resistance is connected to the diode (Fig. 3.12), a forward voltage, V, is set up across the junction, but the net current, I_l–I_F flows against the forward voltage, V,in the reverse direction. Therefore, $IV < 0$and the power is extracted. Equivalently, the solar radiation sets up the load voltage, V_L, and at the same time drives the current, I_l, across the load. In this manner, the solar energy is converted to the electrical energy.

The efficiency of the converting the solar energy into the electrical power is a key parameter and is mainly determined by two factors: (1) the efficiency of light absorption and (2) the magnitude of the voltage induced at the load, V_L. Ideally, the cell should absorb the entire spectrum of the solar radiation, but in practice, the photon energy, $h\nu$, should be greater than the energy gap of the cell, as clear from (3.38a, 3.38b). This means that greater fraction of the solar spectrum is absorbed in smaller bandgap material. However, a large bandgap induces larger V_{oc}, hence larger V_L (problem 3.8).

The energy conversion can be analyzed explicitly with the use of a simple circuit, shown in Fig 3.12. Here, R_L and R_S are the load and series resistances of the diode, respectively, and $I_L$is the load current. With the absorption of light, the photocurrent, I_l, the forward current, I_F, and voltage, V, are all generated at the same time. Because the series resistance is small, the load voltage can be taken equal to V, i.e.,

$$V_L = V - I_L R_S \approx V \tag{3.48}$$

Now the load current is given by the difference between the photocurrent, I_l, and forward current, I_F:

$$I_L \approx I_l - I_s(e^{qV_L/k_BT} - 1) \tag{3.49}$$

The output power is thus given by

$$P = I_L V_L \approx V_L\left[I_l - I_s(e^{qV_L/k_BT} - 1)\right] \tag{3.50}$$

One can then estimate the maximum output power by imposing the condition, $\partial P/\partial V_L = 0$,obtaining

$$V_{Lm} = (k_BT/q)\ln[(1 + I_l/I_S)/(1 + qV_{Lm}/k_BT)] \tag{3.51}$$

Here, V_{Lm} is the load voltage at which the maximum power is extracted. One can thus find V_{Lm} from (3.51) and by using (3.46):

$$V_{Lm} = V_{oc} - \frac{k_B T}{q} \ln[1 + V_{Lm}/(k_B T/q)] \approx V_{oc}, \ V_{oc} \propto I_S^{-1} \tag{3.52}$$

Clearly, the optimal load voltage, V_{Lm}, is determined primarily by V_{oc}. Once V_{Lm} is found, the corresponding load current is obtained from (3.49) and (3.52) as

$$I_{Lm} = I_l - I_S(e^{qV_{Lm}/k_B T} - 1) \approx I_l(1 - \frac{k_B T/q}{V_{Lm}}) \tag{3.53}$$

where use has been made of $I_l/I_S \gg V_{Lm}/(k_B T/q) \gg 1$. Clearly, I_{Lm} is contributed primarily by the photocurrent, I_l, as it should.

It is therefore clear that the absorption of the solar radiation is an essential factor for the cell efficiency. Also, attaining a large load voltage, V_{Lm}, is also essential, for which large bandgap is required to reduce the saturated current level, I_S, and to enhance V_{oc}, as evident from (3.46), (3.52). Therefore, an optimal combination of small- and large-bandgap materials is crucial for attaining the high-efficiency solar cell.

3.6.4 Light-Emitting Diode

The solar cell and the photodiode operate based on (1) the absorption of light and concomitant generation of e–h pairs and (2) the separation and eventual recombination of e–h pairs in the external circuits provided. The reverse process of injecting electrons and holes in the p–n junction, followed by the radiative recombination, has also been extensively utilized in such devices as light-emitting diodes and laser diode. These two light sources constitute the key element in the fiber optical communication. Moreover, since LEDs and LDs are based on the electrical pumping of excess electrons and holes, these two light sources can be readily incorporated into the optoelectronic network. In addition, LED has become one of the promising light sources, with the long lifetime and low power usage. Thus, these two photonic devices are briefly discussed qualitatively as the final example of the diode application.

The two light sources are again based on the p–n junction, in particular, the p–n junction in a direct-bandgap material such as GaAs, doped heavily with donor and acceptor atoms. In this case, the Fermi level, E_F, is raised above the conduction band in the n region, while it is lowered below valence band. Consequently, the conduction and valence bands overlap in the junction region.

Therefore, under a forward bias, a large number of excess electrons and holes are injected into the junction region and they recombine, emitting the radiation. LEDs operate based on this process of the electrical pumping, followed by the radiative recombination. If the pumping rate is sufficient, the junction is turned into an active layer supporting the lasing action. The laser diode operates based on the electrical pumping sufficient to offset the loss.

A factor crucial for efficient operation of these two light sources is the luminescence efficiency, that is, the efficiency of the radiative recombination,

$$\eta = \frac{1/\tau_r}{1/\tau_r + 1/\tau_{nr}}$$

Here, $1/\tau_r$ and $1/\tau_{nr}$ represent the radiative and non-radiative recombination rates with τ_r and τ_{nr} denoting the radiative and non-radiative lifetimes, respectively. The high luminescence efficiency is generally attained in direct bandgap material, since the optical transitions therein are the first order process. In indirect bandgap semiconductors, on the other hand, the optical transitions are the second order process and the efficiency is generally low.

Naturally, a factor crucial for the operation of the laser diode is the injection current level by which to attain the sufficient gain to offset the loss. Once this threshold condition is met, a steady state operation of LD ensues when the saturated gain is balanced by the loss in the cavity:

$$\frac{g(\omega)}{1 + I/I_S} = \alpha_T, \quad g(\omega) \propto I_F$$

Here α_T is the net loss factor, resulting from the scattering and imperfect cavity mirror reflectivity and $g(\omega)$ the gain factor arising from the radiative recombination. Also, I represents the laser intensity and I_S the saturated intensity, a parameter inherent in the lasing medium. Obviously, the gain should be commensurate with the forward pumping current, I_F. The threshold condition is specified by $g(\omega) = \alpha_T$. When the condition is met, the laser intensity, I, starts to grow and in the presence of I, the saturated gain should balance α_T for the steady state operation of the laser diode. The corresponding intensity is then given by

$$I = I_S\left(\frac{g(\omega)}{\alpha_T} - 1\right)$$

Finally, the low power laser diodes can be fabricated in superlattice heterostructures with built-in quantum wells. When the quantum wells for both electrons and holes are introduced in the junction region, the electrons and holes are injected into the respective sublevels in the quantum well. In this case, the excess electrons and holes can have long interaction time for radiative recombination rather than being swept out of the junction region before the recombination.

3.7 Problems

3.1 (a) Starting from the electron current density expression in (3.1) derive (3.3) by using (3.2) and the Einstein relation, $qD_n = \mu_n k_B T$and $n = n_i \exp(E_F - E_{Fi})/k_B T$.

(b) Derive (3.4) by repeating similar steps for holes.

3.2 The profile of the dipole space charge, ρ was taken as a step function in the completely depleted approximation. Estimate the validity of this approximation by finding the widths in the region near $x_n$and $-x_p$,in which the carrier concentrations are not negligible and by comparing the widths with the typical values of x_n and x_p. The width can be approximated by finding the range of n varying from n_{n0} to 1 % of n_{n0}. The width for the hole concentration can be treated in the similar way.

3.3 A p$^+$n step junction in silicon is doped with $N_A = 2 \times 10^{18}\,\mathrm{cm}^{-3}$, $N_D = 1 \times 10^{17}\,\mathrm{cm}^{-3}$.

(a) Find x_n, x_p, $E_{\max}$, W and φ_{bi}.

(b) At which reverse bias will the junction undergo breakdown if the maximum field for breakdown is 3×10^5 V/cm?

3.4 Find the Zener breakdown voltages in silicon and germanium, if the same breakdown field of 3×10^5 V/cm is assumed.

3.5 (a) Is it possible to achieve the junction band bending greater than the energy gap of the semiconductor in contact?

(b) Estimate the donor and acceptor doping levels which will make $q\phi_{bi} \approx E_G$in Si.

3.6 The diffusion equation of the holes injected into the quasi-neutral region on the n side under illumination is given by

$$\frac{d^2 p_n}{dx^2} - \frac{p_n - p_{n0}}{L_p^2} + \frac{\tilde{g}_0}{D_p} = 0, \quad L_p^2 = D_p \tau_p$$

where $\tilde{g}_0$is the generation rate of holes as given in (3.41). The first two terms in the equation are the usual diffusion equation as considered in Chap. 2 in the absence of the incident light.

(a) Show that the solution of the equation subject to the boundary conditions, $p_n(x = x_n) = 0$and $p_n(x \to \infty) = p_{n0} + \tilde{g}_0 \tau_p$is given by $p_n(x) = (p_{n0} + \tilde{g}_0 \tau_p) \quad \{1 - \exp - [(x - x_n)/L_p]\}$

(b) Find the hole diffusion current given in (3.29) by using the solution in (a) and by performing the operation

$$J_{p,\mathrm{diff}} \equiv -qD_p \frac{\partial p_n(x = x_n)}{\partial x}$$

(c) Derive (3.10) for the electron diffusion current by using a similar analysis.

3.7 The excess hole concentration profiles in the quasi-neutral region on the n side is given by

$$p_n(x - x_n) \propto \exp - [(x - x_n)/L_p]$$

(a) Find the average value,

$$<x - x_n> \; = \frac{\int_{x_n}^{\infty} dx e^{-(x-x_n)/L_p}(x - x_n)}{\int_{x_n}^{\infty} dx e^{-(x-x_n)/L_p}}$$

(b) Interpret your result.

3.8 By using the theory of the p–n junction, show that a large bandgap is required to attain the large open-circuit voltage, V_{oc}.

3.9 What is the maximum possible band bending of the p–n junction in equilibrium?

(a) Is it possible to achieve the junction band bending by an amount greater than the energy gap of the semiconductor in contact, i.e., $q\varphi_{bi} > E_G$?

(b) Estimate the donor and acceptor doping levels which will make $q\phi_{bi} \approx E_G$ in silicon

3.10 The laser diode is degenerately doped in both p and n regions.

(a) Estimate the donor and acceptor doping levels, for which the conduction and valence bands are overlapped by an amount, 0.2 eVin silicon and gallium arsenide.

(b) Estimate electron and hole fluxes under forward bias.

Suggested Reading

1. Kim, D. M. (2010). *Introductory quantum mechanics for semiconductor nanotechnology*. John Wiley-VCH.
2. Sze, S. M., & Ng, K. K. (2006). *Physics of semiconductor devices*. 3rd edn, Wiley-Interscience.
3. Streetman, B. G., & Banerjee, S. (2005). *Solid state electronic devices* (6th ed.). Englewood Cliffs; Prentice Hall.
4. Muller, R. S., Kamins, T. I., & Chan, M. (2002) *Device electronics for integrated circuits*. Third Sub edn, Wiley.
5. Bhattacharya, P. (1996). *Semiconductor optoelectronic devices* (2nd ed.). Englewood Cliffs: Prentice Hall.
6. Neudeck, G. W. (1989). *The bipolar junction transistor, modular series on solid state devices* (2nd ed., Vol. III). Englewood Cliffs: Prentice Hall.
7. Yariv, A. (1982). *An Introduction to theory and applications of quantum mechanics*. Wiley.
8. Grove, A. S. (1967). *Physics and technology of semiconductor devices*. Wiley.

Chapter 4
Silicon Nanowire Field-Effect Transistor

Dae Mann Kim, Bomsoo Kim and Rock-Hyun Baek

Abstract The field effect transistor was conceived in 1930s and was demonstrated in 1960s. Since then, MOSFET emerged as the mainstream driver for the digital information technology. Because of the simplicity of structure and low cost of fabrication, it lends to a large scale integration for the multifunctional system-on-chip (SOC) applications. Moreover, the device has been relentlessly downsized for higher performance and integration. The physical barriers involved in downscaling the device have prompted the development of process technologies. There has also been the development of device structures from 3D bulk to the gate-all-around nanowire. This chapter is addressed to the discussion of the silicon nanowire field effect transistor (SNWFET). The discussion is carried out in comparison and correlation with the well known theory of MOSFET. The similarities and differences between the two FETs are highlighted, thereby bringing out features unique to SNWFET. Also, an emphasis is placed upon the underlying device physics rather than the device modeling per se. The goal of this chapter is to provide a background by which to comprehend the theories being developed rapidly for SNWFETs.

Abbreviation

SOC	System-on-chip
SNWFET	Silicon nanowire field-effect transistor

D. M. Kim (✉) · B. Kim
Korea Institute for Advanced Study (KIAS), Seoul, Republic of Korea
e-mail: dmkim@kias.re.kr

B. Kim
e-mail: bomsoo@kias.re.kr

R.-H. Baek
SEMATECH, Albany, NY, USA
e-mail: rock-hyun.baek@sematech.org

D. M. Kim and Y.-H. Jeong (eds.), *Nanowire Field Effect Transistors: Principles and Applications*, DOI: 10.1007/978-1-4614-8124-9_4,

4.1 MOSFET

Overview: In this section the theory of MOSFET is compactly summarized to be used as the reference for discussing silicon nanowire field-effect transistor (SNWFET). MOSFET is a three-terminal, unipolar and normally off device and has been successfully scaled down to about 10 nm channel length. Also, the device has provided convenient platforms for a number of applications e.g., memory cells, sensors and solar cells, etc. Of the two types of MOSFETs, NMOS is singled out for discussion. The results obtained can readily be applied to PMOS by replacing the roles of electrons with those of holes.

NMOS I–V Behavior: Fig. 4.1 shows the cross section, consisting of the n^+ source and drain and the n^+ polysilicon gate, which is electrically insulated by SiO_2. The source and drain are separated by the *p*-type substrate, so that n^+-p and p-n^+ junctions are formed back to back. With the gate voltage off ($V_{GS} = 0$) and the drain voltage on ($V_{DS} > 0$), the p-n^+ junction at the drain end gets reverse biased, cutting off the current (off state), as detailed in Chap. 3.

However, with V_{GS} on at a value greater than the threshold voltage, i.e., $V_{GS} > V_{Tn}$, the channel is inverted. In which case, the n^+-p junction barrier at the source end is lowered and electrons are injected from the source into the channel and contribute to the drain current, I_D. Figure 4.2 shows the transistor I–V curves and transfer characteristics. Each $I_D - V_D$ curve divides into the triode and

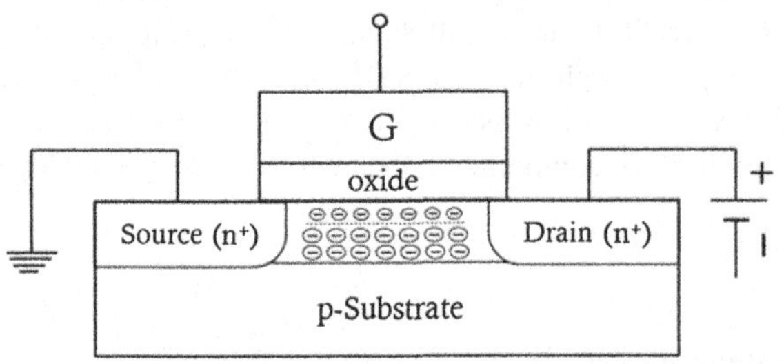

Fig. 4.1 The cross-sectional view of NMOS, consisting of the *p*-substrate, n^+ source, drain and gate electrodes

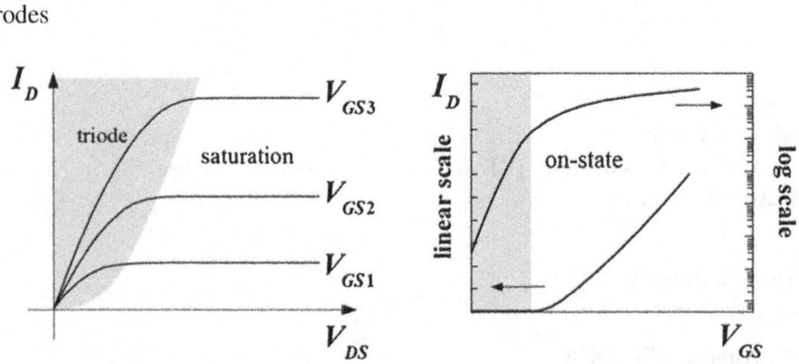

Fig. 4.2 The transistor I–V curves (*left*) and transfer characteristics (*right*) of the *n*-channel MOSFET

saturation regions. The ON-to-OFF current ratio, typically 10^6 or greater, is a parameter gauging the device as a switch.

The standard long channel I–V behavior is described by the SPICE model, level 1:

$$I_{DS} = \frac{W}{L} C_{OX} \mu_n (V_{GS} - V_{Tn} - \frac{1}{2} V_{DS}) V_{DS}, \quad 0 \leq V_{DS} \leq V_{DSAT} \equiv V_{GS} - V_T \tag{4.1}$$

Here, μ_n is the electron mobility, V_{Tn} the threshold voltage, V_{GS} the gate-to-source voltage and the ratio between width and length of the channel, *W/L* is called the aspect ratio. The oxide capacitance per unit area is given in terms of the oxide permittivity, ε_{OX}, and thickness, t_{OX}, as $C_{OX} = \varepsilon_{OX}/t_{OX}$. The triode and saturation regions of I_{DS} are demarcated by $V_{DSAT}(= V_{GS} - V_{Tn})$, at which the channel is pinched off.

Equivalently I_{DS} can be compacted into a simpler form as

$$I_{DS} = Q_{nL} v_D, \; Q_{nL} = W C_{OX}(V_{GS} - V_{Tn} - V_{DS}/2), \; v_D = \mu_n (V_D/L) \tag{4.2}$$

Here, $v_D = \mu_n(V_D/L)$ is the drift velocity of the electron, and Q_{nL} is the average line charge induced under the gate. In this representation, I_{DS} is shown to be contributed by Q_{nL} sweeping across the channel with v_D. In device saturation, where $V_{DSAT} = V_{GS} - V_{Tn}$ (4.2) is reduced to

$$I_{DSAT} - Q_{nSAT} v_D, \; Q_{nSAT} \equiv W C_{OX}(V_{GS} - V_{Tn})/2 \tag{4.3}$$

with Q_{nSAT} denoting the average line charge in saturation.

4.1.1 Channel Inversion in NMOS

Consider the NMOS system as shown in Fig. 4.3 together with respective work functions. The work function of the semiconductor is the sum of the affinity factor, $q\chi$ and $E_C - E_F$ with $q\chi$ denoting the energy required to excite an electron from E_C to the vacuum level. When the three components are brought together in equilibrium contact, the Fermi level should line up and be flat, which necessitates the band bending, as discussed in Chap. 3.

Equilibrium Band Bending: The band bending occurs via the exchange of electrons between the n^+ gate and the p substrate. Specifically, the Fermi level, E_F, of the n^+ gate electrode is higher than that of the substrate; hence, electrons spill over from the gate to substrate. This leaves behind the positive charge sheet at the surface of the gate electrode, which in turn pushes holes in the p substrate away from the interface, thereby exposing acceptor ions uncompensated. Consequently, a dipolar space charge is formed, and the band bends downward as shown in Fig. 4.4. The total band bending is given by the work function difference between the gate electrode and the p substrate, and the bending occurs in both substrate and oxide.

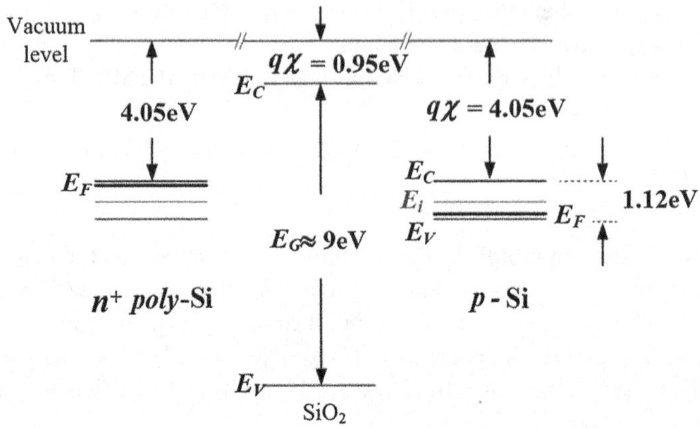

Fig. 4.3 NMOS system: n^+ poly-Si, SiO_2 and silicon *p*-substrate. Also shown are the affinity factor, $q\chi$ and Fermi levels

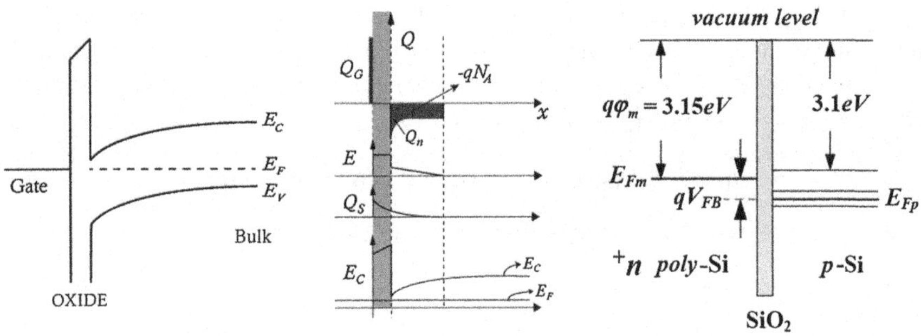

Fig. 4.4 NMOS system in equilibrium contact: The band bending (*left*) and space charge, field and potential underlying the bending (*middle*) are shown. Also sketched is the flat band voltage at which the band flattens out

Flat Band Voltage: The band bending is flattened out with the application of the flat band gate voltage, given by the difference between the work functions of the gate electrode and the p substrate:

$$\begin{aligned} qV_{\mathrm{FBn}} &\equiv q\chi + E_C - E_{\mathrm{Fn}} - (q\chi + E_C - E_{\mathrm{Fp}}) \\ &= E_{\mathrm{Fp}} - E_{\mathrm{Fn}} \end{aligned} \tag{4.4}$$

Since $E_{\mathrm{Fp}} < E_{\mathrm{Fn}}$ in this case $V_{\mathrm{FBn}} < 0$ and with V_{FBn} applied, the positive charge sheet in the gate electrode is annulled, and the space charge disappears together with the band bending.

Surface Charge: By using V_{FBn}, one can introduce the charging voltage, V'_G, which is dropped in both the oxide and the substrate,

$$V'_G \equiv V_G - V_{\mathrm{FBn}} = V_{\mathrm{OX}} + \varphi_S \tag{4.5}$$

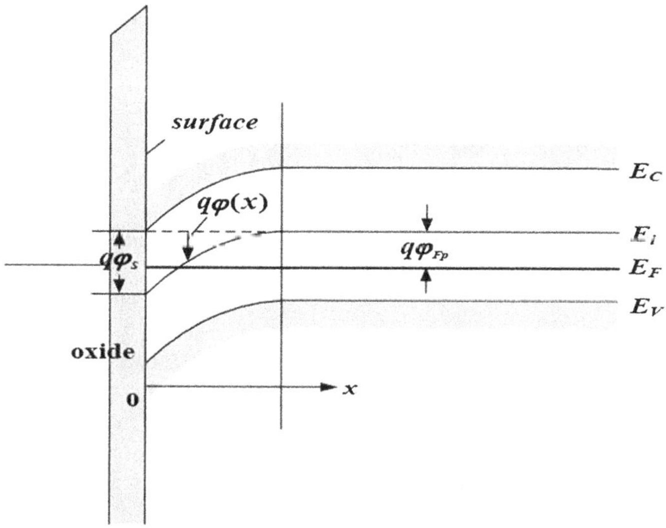

Fig. 4.5 The band bending, $q\varphi(x)$, in the n-channel MOSFET; $q\varphi_S$ and $q\varphi_{FP}$ denote the surface and Fermi potentials, respectively

Here, φ_S is the surface potential, i.e., $\varphi(x = 0)$. With $V'_G(>0)$ on, the band bends down, as indicated in Fig. 4.5. The resulting potential, φ, is found by solving the Poisson equation:

$$\frac{d^2\varphi(x)}{dx^2} = -\frac{\rho(x)}{\varepsilon_S}, \ \rho(x) = q[(p_p(x) - N_A^- - n_p(x)] \tag{4.6a}$$

where the space charge, ρ, in the substrate is made up of the hole, acceptor ion and electron charges. In the p bulk the charge neutrality holds true, i.e., $p_{p0} = N_A^- + n_{p0}$, but near the surface n, p become x-dependent and is given by

$$p_p(x) = p_{p0}e^{-\beta\varphi(x)}, \ n_p(x) = n_{p0}e^{\beta\varphi(x)}, \ \beta \equiv q/k_BT \tag{4.6b}$$

(see Fig. 4.5). Hence, upon inserting (4.6b) into (4.6a) together with $N_A^- = p_{p0} - n_{po}$ there results,

$$\frac{d^2\varphi(x)}{dx^2} = -\frac{\rho(x)}{\varepsilon_S}, \ \rho(x) = q\left[p_{p0}(e^{-\beta\varphi} - 1) - n_{p0}(e^{\beta\varphi} - 1)\right] \tag{4.7}$$

Note here that without the band bending, that is, for $\varphi = 0$ $\rho(x) = 0$, as it should. Also, Eq. (4.7) is strongly nonlinear and is difficult to solve. However, one can perform the first integration by recasting (4.7) by multiplying both sides by $d\varphi$ as

$$\int_0^{-E} EdE = -\frac{1}{\varepsilon_S}\int_0^{\varphi} \rho(\varphi)d\varphi, \quad E \equiv -\frac{d\varphi}{dx} \tag{4.8}$$

(problem 4.1). Here the integrations start from edge of the bulk substrate. Hence, upon inserting the expression for ρ in (4.7) into (4.8) and carrying out the integration, one finds

$$\mathrm{E}(x) \equiv -\frac{\mathrm{d}\varphi}{\mathrm{d}x} = \sqrt{2}\frac{k_B T}{q}\frac{1}{L_D}F(\beta\varphi, n_{po}/p_{p0}), \; L_D \equiv \left(\frac{k_B T \varepsilon_S}{q^2 p_{po}}\right)^{1/2} \tag{4.9a}$$

where L_D thus defined is the Debye length and the F function is given by

$$F(\beta\varphi) \equiv \left[(e^{-\beta\varphi} + \beta\varphi - 1) + e^{-2\beta\varphi_{\mathrm{Fp}}}(e^{\beta\varphi} - \beta\varphi - 1)\right]^{1/2}, \; e^{-2\beta\varphi_{\mathrm{Fp}}} = n_{p0}/p_{p0} \tag{4.9b}$$

with φ_{Fp} denoting the hole Fermi potential in the p substrate.

The surface field, E_S at $x = 0$, can therefore be specified in terms of the surface potential, φ_S. Once E_S is found, the surface charge, Q_S is obtained from the well known boundary condition at the oxide interface, i.e.,

$$Q_S \equiv -\varepsilon_S \mathrm{E}_S(\varphi_s) = -\varepsilon_S\sqrt{2}\frac{k_B T}{q}\frac{1}{L_D}F(\beta\varphi_s) \tag{4.10}$$

Weak and Strong Inversion: Figure 4.6 shows Q_S as a function of φ_S and the space charge associated. For $V_G = V_{\mathrm{FB}}$, there is no band bending, that is, $\varphi_S = 0$, hence $Q_S = 0$. For $\varphi_S < 0$, the band bends up and the first term in (4.9b) becomes dominant, and holes as the majority carrier are accumulated at the surface. In the depletion region, in which $0 < \varphi_S < \varphi_{\mathrm{Fp}}$, the band bends down, supported mainly by the acceptor ions and Q_S consists of the uncompensated ions. In the weak inversion region, $\varphi_{\mathrm{Fp}} < \varphi_S < 2\varphi_{\mathrm{Fp}}$, electrons begin to populate the interface region. For $\varphi_S \approx 2\varphi_{\mathrm{Fp}}$ $n_s \approx p_{p0}$, and from this point on, the increase of Q_S with increasing φ_S is primarily due to the electron charge induced, and the channel is thus inverted. In this regime, the electrons are concentrated practically at the surface and do not contribute significantly to the band bending, pinning φ_S approximately at a constant level. Therefore, the condition

$$\varphi_S = 2\varphi_{\mathrm{Fp}}$$

represents the onset of the strong inversion.

MOS capacitor: Naturally, the channel is inverted by the gate voltage, V_G, and is due essentially to the capacitive coupling. To analyze the channel inversion in terms of the capacitive charging, consider the charging voltage divided in the gate oxide and the p substrate:

$$V_G' \equiv V_G - V_{\mathrm{FB}} = V_{\mathrm{OX}} + \varphi_S, \quad V_{\mathrm{OX}} \equiv \frac{|Q_S|}{C_{\mathrm{OX}}} \tag{4.11}$$

Since Q_S is a function of φ_S (see 4.10) φ_S is specified in terms of V_G or vice versa. The capacitor, C, connecting the gate electrode and the p substrate consists of the oxide (C_{OX}) and surface (C_S) capacitors connected in series, as shown in Fig. 4.7:

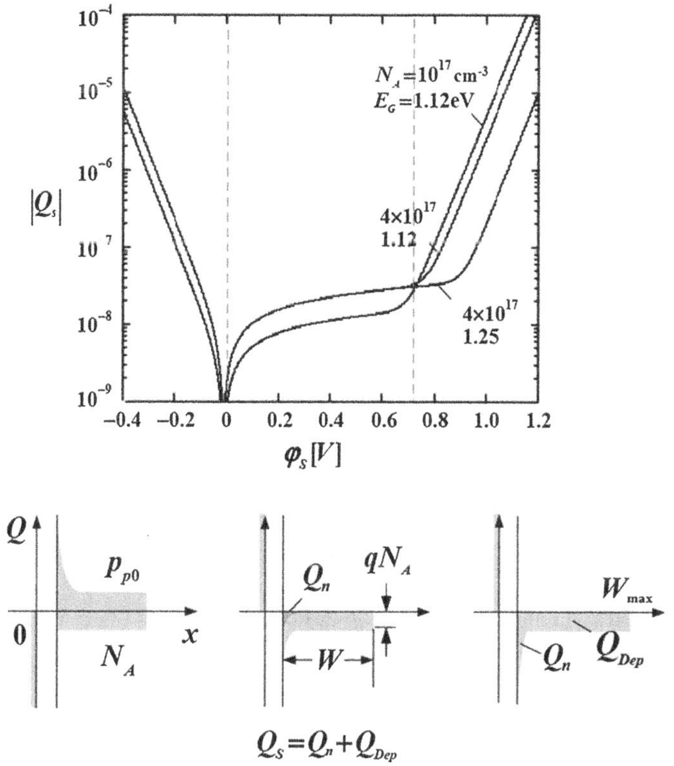

Fig. 4.6 The surface charge, Q_S, is shown versus the surface potential, φ_S, in NMOS with doping level and band gap as parameters. Also shown are the total surface charge, consisting of uncompensated acceptor ions and electrons induced under the gate electrode in accumulation, depletion and inversion regimes

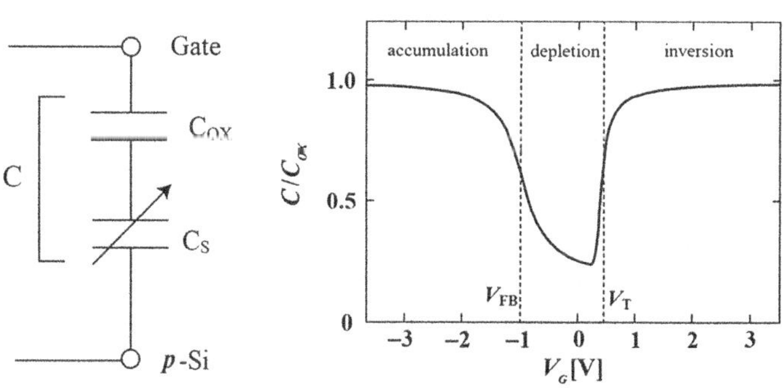

Fig. 4.7 The total capacitor of NMOS, consisting of oxide (C_{OX}) and the surface (C_S) capacitors connected in series (*left*). Also shown is the total capacitance in accumulation, depletion and inversion regions (*right*)

$$\frac{1}{C} = \frac{1}{C_{OX}} + \frac{1}{C_S}, \quad C_S \equiv \frac{\partial |Q_S|}{\partial \varphi_S} \tag{4.12}$$

Here, C_S accounts for the change in the surface charge, Q_S with φ_S. Since Q_S is a function of φ_S and since φ_S in turn depends on V_G (see 4.11), C_S is a variable capacitance, as shown in Fig. 4.7. In accumulation and inversion regions, $C_S \gg C_{OX}$, (see 4.9, 4.10), so that $C \approx C_{OX}$. In depletion and weak inversion regions, on the other hand, $C_S \ll C_{OX}$ and C is mainly determined by C_S. Also indicated in the figure is the location of the threshold voltage, V_T. Ideally, only the mobile electron charge should be induced by V_G, but the channel inversion requires the surface band bending, which is supported by the ionic charge. Hence, the ionic charge is inseparably coupled to the channel inversion.

4.1.2 I–V Modeling

ON Current: The lumped view of the drain current was discussed in Sect. 4.1 and is now analyzed, based on the channel inversion. The total surface charge consists of the electron (Q_n) and fixed ionic (Q_{DEP}) charges, i.e.,

$$Q_S \equiv -C_{OX}V_{OX} = Q_n + Q_{DEP} \tag{4.13}$$

and terminates the field lines emanating from the positive charge sheet on the gate electrode (see Fig. 4.4). Thus, the key to modeling I–V is to untangle Q_n from Q_{DEP}.

Now, given the substrate doping level, N_A, the depletion charge is given by $Q_{DEP} = -qN_A W_D$ with W_D denoting the depletion depth (see Fig. 4.4). Also, the surface potential, φ_S, is supported by Q_{DEP} and is given by $\varphi_S = qN_A W_D^2/2\varepsilon_S$, in the completely depleted approximation. Therefore, Q_{DEP} can be expressed in terms of φ_S as

$$Q_{DEP} = -qN_A W_D = -(2\varepsilon_S q N_A \varphi_S)^{1/2} \tag{4.14}$$

Hence, upon inserting (4.11) for V_{OX} and (4.14) for Q_{DEP} into (4.13), there results

$$Q_n = -C_{OX}(V_G - V_{FB} - \varphi_S - \gamma\varphi_S^{1/2}); \; \gamma_n \equiv (2\varepsilon_S q N_A)^{1/2}/C_{OX} \tag{4.15}$$

Here, the constant, γ_n, is known as the body effect coefficient. Hence, Q_n at the onset of the strong inversion is obtained by replacing φ_S by $2\varphi_{Fp}$ as discussed:

$$Q_n = -C_{OX}(V_{GS} - V_{Tn}), \; V_{Tn} = V_{FB} + 2\varphi_{Fp} + \gamma_n(2\varphi_{Fp})^{1/2} \tag{4.16}$$

In this manner, the capacitive charging of Q_n is quantified.

When the drain voltage, V_{DS}, is turned on, it is distributed in the channel. Thus, at a channel position at y with the channel voltage, $V(y)$ $(0 \le V(y) \le V_D)$ the condition of the channel inversion, $\varphi_S = 2\varphi_{FP}$ has to be generalized, incorporating the splitting of E_{Fn} and E_{Fp} caused by $V(y)$. However, for the simplicity of discussion, the role of $V(y)$ is taken simply to reduce the effective gate voltage at y and represent $Q_n(y)$ as

$$Q_n(y) = -C_{OX}(V_{GS} - V - V_{Tn}) \tag{4.17}$$

Transistor I–V: With the use of $Q_n(y)$ thus specified, the I–V behavior is derived next. In the channel element from y to $y + dy$, the channel voltage, V, drops by dV, which is given by

$$dV \equiv I_D dR, \ dR \equiv \frac{dy}{W\mu_n|Q_n|} \tag{4.18}$$

where the resistivity has been expressed in terms of Q_n. Therefore, one can recast (4.18) into two integrations, one involving y and the other V, i.e.,

$$\int_0^L I_D dy = \int_0^{V_D} dV \mu_n W|Q_n| \tag{4.19}$$

One can then perform the integrations in (4.19) by using (4.17) for Q_n, and the fact that I_D is constant throughout the channel, obtaining

$$I_D = \frac{W}{L}\mu_n C_{OX}(V_{GS} - V_{Tn} - \frac{1}{2}V_{DS})V_{DS}, \ V_{DS} \le V_{GS} - V_{Tn} \tag{4.20a}$$

When the channel is pinched off at the drain end, $Q_n(L) \approx 0$ and the I–V relation given in (4.20a) ceases to be valid. The pinch-off voltage, V_{DSAT} is found by putting Q_n to zero in (4.17), i.e., $V_{DSAT} = V_{GS} - V_{Tn}$ and the saturation current at V_{DSAT} is therefore pinned at

$$I_{DSAT} = \frac{W}{2L}\mu_n C_{OX}(V_{GS} - V_{Tn})^2, \ V_{DSAT} \equiv V_G - V_{Tn} \tag{4.20b}$$

Equation (4.20) agrees with (1.1), and the parameters such as μ, V_{Tn} can further be refined to fit the measured data.

Subthreshold Current: The subthreshold current, I_{SUB}, bridges I_{OFF} and I_{ON} in the V_G range, $0 < V_G < V_T$, or in φ_S range, $0 < \varphi_S < 2\varphi_{Fp}$. In this regime, the second term of Q_S in (4.9b) is dominant and one can Taylor expand Q_S in (4.10) centered around the second term, obtaining

$$Q_S \equiv Q_{DEP} + Q_n \approx -(2qN_A\varepsilon_S\varphi_S)^{1/2}\left(1 + \frac{1}{2}\frac{e^{\beta(\varphi_S - 2\varphi_{Fp})}}{\beta\varphi_S}\right), \ \beta = q/k_BT \tag{4.21}$$

where the Debye length, L_D, in (4.9a) has been spelled out (problem 4.2). Evidently, the first term in (4.21) represents Q_{DEP}, while the second term denotes the surface charge of electron, Q_n.

As clear from (4.21), Q_n is exponentially enhanced with increasing φ_S in this regime of the weak inversion. Also, in the presence of the substrate bias, V_B and channel voltages at the source and drain, V_S and V_{DS} and the expression of Q_n should incorporate the splitting of the quasi-Fermi levels. Thus, one can write

$$Q_n(0) = -qN_A L_D \left(\frac{1}{2\beta\varphi_{\mathrm{SS}}}\right)^{1/2} e^{\beta(\varphi_{\mathrm{SS}}-2\varphi_{\mathrm{Fp}})}, \ \varphi_{\mathrm{SS}} \equiv \varphi_S - (V_S - V_B) \tag{4.22a}$$

at the source and

$$Q_n(L) = -qN_A L_D \left(\frac{1}{2\beta\varphi_{\mathrm{SD}}}\right)^{1/2} e^{\beta(\varphi_{\mathrm{SD}}-2\varphi_{\mathrm{Fp}})}, \ \varphi_{\mathrm{SD}} \equiv \varphi_{\mathrm{SS}} - V_{\mathrm{DS}} \tag{4.22b}$$

at the drain. It is therefore clear from (4.22) that Q_n decreases exponentially from the source to the drain. Hence, I_{SUB} should be driven by diffusion, and one can write

$$\begin{aligned} |I_{\mathrm{SUB}}| &\approx WD_n \frac{Q_n(0) - Q_n(L)}{L} \\ &\approx \frac{W}{L} D_n q N_A L_D \left(\frac{1}{2\beta\varphi_{\mathrm{SS}}}\right)^{1/2} e^{\beta(\varphi_{\mathrm{SS}}-2\varphi_{\mathrm{Fp}})}\left(1 - e^{-\beta V_{\mathrm{DS}}}\right) \end{aligned} \tag{4.23}$$

The subthreshold current thus derived bridges I_{ON} and I_{OFF}.

PMOS I–V Behavior: Naturally, the PMOS I–V modeling can be similarly carried out by interchanging the roles of electrons and holes. One can thus derive the hole surface charge, obtaining

$$Q_p = C_{\mathrm{OX}}(|V_{\mathrm{GS}}| - V_{\mathrm{Tp}} - |V|) \tag{4.24a}$$

where the hole threshold voltage is given by

$$V_{\mathrm{Tp}} \equiv V_{\mathrm{FB}} + \varphi_{\mathrm{Fp}} + \gamma_p \varphi_{\mathrm{Fp}}^{1/2}, \ \gamma_p \equiv (2\varepsilon_S q N_D)^{1/2}/C_{\mathrm{OX}} \tag{4.24b}$$

(problem 4.3). Here, V_{FB} (>0) is the work function difference between the p^+ poly gate and n substrate, and the Fermi potential, φ_{Fn}, of the n substrate and the body coefficient, γ_p, are specified in terms of the donor concentration, N_D. Once Q_p is obtained, the I–V relation can be derived in a manner similar to NMOS.

Quantum Modifications: The channel inversion necessitates the band bending, downward in NMOS and upward in PMOS. As a result, the quantum wells are formed for electrons and holes, respectively, as shown in Fig. 4.8. This means that the inverted electrons and holes are 2D particles spatially confined in the direction normal to the surface, but moving freely on the interface plane. The resulting quantum mechanical modifications have to be taken into account.

First, the electrons and holes reside in the quantized sublevels or subbands, as sketched in Fig. 4.8. These sublevels are characterized by 2D density of states, as detailed in Chap. 1. Therefore, the statistics of the channel inversion should differ from what has been discussed for the 3D bulk MOSFET. Additionally, the

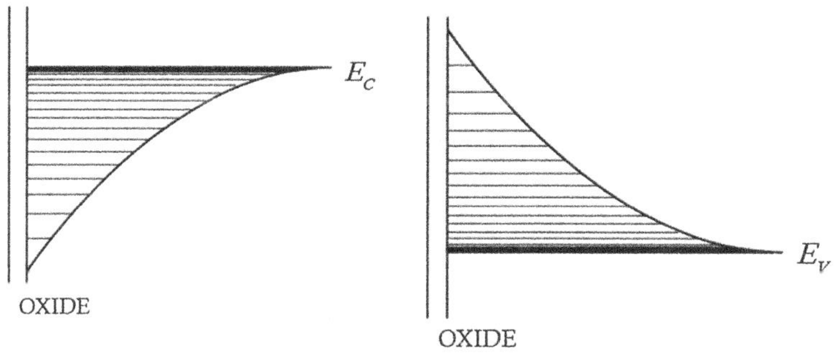

Fig. 4.8 The quantum well and the subbands of electrons (*left*) and holes (*right*), induced at the oxide interface by the respective gate voltages

probability density of the carrier wave function in each sublevel is peaked away from the oxide interface. This suggests that electrons or holes are inverted away from the interface, thereby increasing the effective oxide thickness and reducing the capacitance of the gate dielectric. Also, due to the discrete sublevels formed in the quantum well, the substrate band gap is in effect broadened. Consequently, the quantum modification reduces in essence the efficiency of channel inversion.

4.2 Silicon Nanowire Field Effect Transistor

Overview: The SNWFET holds up a promising potential as a driver of the nanoelectronics and is discussed in this section. In so doing, the well known theory of MOSFET is utilized both as the reference and the general background. In particular, the similarities and differences existing between SNWFET and MOSFET are highlighted, thereby bringing out the features unique to SNWFET. In addition, the ballistic transport operative in short channel SNWFETs is considered. For brevity, the discussion is confined to the n-channel SNWFETs. However, the results obtained can be readily applied to the p-channel FET by interchanging the roles of electrons and holes.

Also, the I–V modeling is focused on intrinsic SNWFET, for simplicity. An interesting feature of such FETs is that both n- and p-channel FETs can be fabricated with the use of the same nanowire by doping the source, drain and gate with donors and acceptors, respectively, as shown in Fig. 4.9. As a consequence, n^+-i and i-n^+ junctions and p^+-i and i-p^+ junctions are built in back to back, respectively. Hence, both types of SNWFETs are unipolar and normally off devices. Figure 4.10 shows a typical I–V and transfer characteristics from SNWFET with the channel length approximately 100 nm long. Clearly, the I–V behavior in such a long channel SNWFET is generally similar to that of MOSFET, which indicates that the physical principles underlying the operation are substantially same in both kinds of long channel FETs.

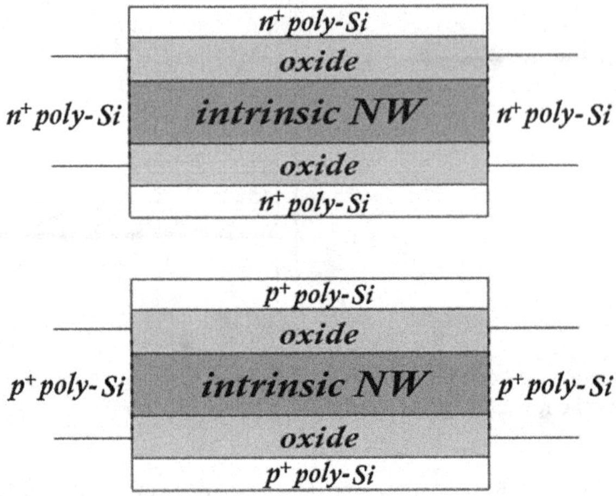

Fig. 4.9 The cross-sectional view of the *n*-type (*top*) and *p*-type (*bottom*) SNWFET

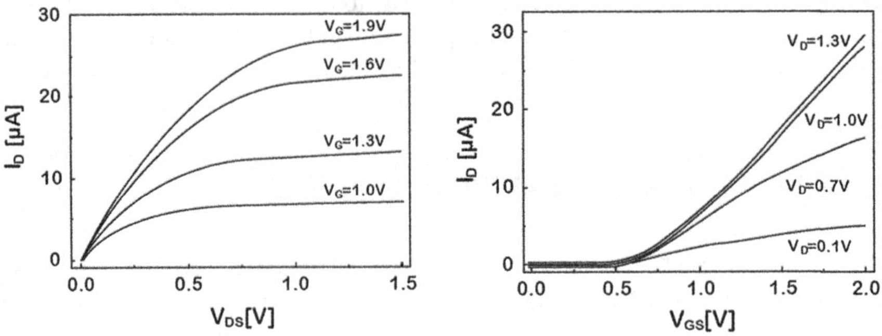

Fig. 4.10 The transistor I–V and transfer characteristics of silicon nanowire field-effect transistor with 100 nm channel length

4.2.1 The n-channel SNWFET

Equilibrium Contact: The equilibrium contact of the n^+ ploy-Si gate—SiO_2—intrinsic nanowire system is essentially the same as in NMOS, aside from the fact that the p substrate in NMOS is replaced by the intrinsic nanowire, as shown in Fig. 4.11. Nevertheless, the electrons are transferred from the gate electrode to the nanowire, again due to the difference in the Fermi levels. Inside the nanowire, the electrons reside in the sublevels therein and are not necessarily concentrated near the oxide interface as in the case of NMOS. This is because of the wave nature of electrons. Specifically, the probability density of electrons in each sublevel can be taken approximately uniform across the cross section of the nanowire, as was discussed in Chap. 1.

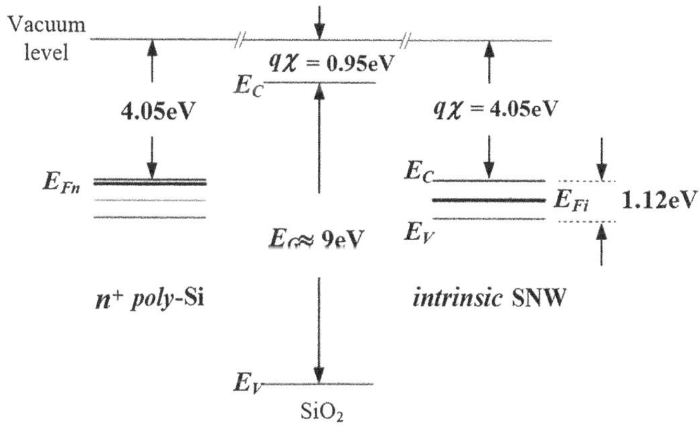

Fig. 4.11 n^+ poly-Si–SiO_2—intrisic Si nanowire system. Also shown are the affinity factors and the Fermi levels

Subband Spectra in Nanowire: The discussion of the channel inversion requires the quantum treatment of the electron states by solving the Schrödinger equation, coupled self-consistently with the Poisson equation in the effective mass approximation. Such an analysis has been carried out in the literature, and the electrons in the nanowire can be taken to move freely along the direction z of the wire, while confined in the x, y plane. The energy eigenequation for bound states is thus given by

$$\left[-\frac{\hbar^2}{2m_n}\left(\frac{\partial^2}{\partial x^2}+\frac{\partial^2}{\partial y^2}\right)+V\right]u(x,y)=Eu(x,y) \tag{4.25}$$

where V represents the two-dimensional quantum well with a finite barrier height. The eigenequations and eigenvalues in the quantum wire with the rectangular cross section have been analyzed in detail in Chap. 1. The general features of the subbands are well represented by the simple analytical expression obtained for the infinite barrier height:

$$E_n=\sum\nolimits_j E_j n_j^2,\quad E_j=\hbar^2\pi^2/2m_n l_j^2;\quad j=x,\,y;\quad n_j=1,2,\ldots \tag{4.26}$$

Here, n_j is the quantum number, l_j the width of the rectangle in the j direction and E_j the jth sublevel energy

One can likewise analyze the subbands in a cylindrical nanowire with radius R by recasting (4.25) into the cylindrical coordinate frame, obtaining

$$\left[-\frac{\hbar^2}{2m_n}\left(\frac{1}{r}\frac{\partial}{\partial r}r\frac{\partial}{\partial r}+\frac{1}{r^2}\frac{\partial^2}{\partial\varphi^2}\right)+V(r)\right]u(r,\varphi)=Eu(r,\varphi) \tag{4.27a}$$

where

$$V(r) = \begin{cases} 0 \text{ for } r \le R \\ V \text{ for } r > R \end{cases} \tag{4.27b}$$

As usual, one can look for the solution in the form

$$u(r, \varphi) \propto e^{\mathrm{i}n\varphi} g(r) \tag{4.28}$$

in which case, (4.27a) reduces to the Bessel differential equation for $g(r)$ inside the well:

$$r^2 \frac{\mathrm{d}^2 g(r)}{\mathrm{d}r^2} + r \frac{\mathrm{d}g(r)}{\mathrm{d}r} + (k^2 r^2 - n^2) g(r) = 0,\ k^2 \equiv \frac{2mE}{\hbar^2} \tag{4.29a}$$

The solutions are therefore given in terms of the Bessel functions of the first (J_n) and second (Y_n) kinds. However, Y_n diverges for $r \to 0$ and should be discarded. Thus, the solution is given by

$$g(r) = C_1 J_n(kr),\ \ r \le R \tag{4.29b}$$

Outside the quantum well, there ensues the modified Bessel differential equation

$$r^2 \frac{\mathrm{d}^2 g(r)}{\mathrm{d}r^2} + r \frac{\mathrm{d}g(r)}{\mathrm{d}r} - (\kappa^2 r^2 + n^2) g(r) = 0,\ \kappa^2 \equiv \frac{2m}{\hbar^2}(V - E) \tag{4.30a}$$

The solutions are therefore given by the modified Bessel functions of the first (I_n) and second (K_n) kinds. However, I_n diverges for $r \to \infty$ and should be discarded and one can write

$$g(r) = C_2 K_n(\kappa r),\ \ r > R \tag{4.30b}$$

Once the energy eigenfunctions are found, the energy eigenvalues are obtained as usual by imposing the boundary conditions, namely that the eigenfunctions and its first derivatives be continuous at $r = R$. Here, again the general features of the sublevels are well represented by the analytic expressions valid for the infinite barrier height. In this limit, the eigenenergies are given by

$$E_n = E_0(3/2 + n + 2s)^2,\ \ E_0 = \hbar^2 \pi^2 / 8 m_n R^2\,,\ \ n, s = 0, 1, 2, \cdots \tag{4.31}$$

Here, n, s denotes the quantum numbers and R the radius of the cylindrical nanowire (problem 4.4).

Figure 4.12 shows the typical sublevel spectra which have been found numerically for the oxide barrier of 3.1 eV in intrinsic silicon nanowires with cross sections different in area and shape. Clearly, a few general trends of the sublevel spectra emerge from the figure. First, the spectra are different between the rectangular and square cross sections, particularly in lower lying sublevels, although the cross-sectional area is the same. This is due to the subband energy levels depending sensitively on the width in each direction of the rectangle see (4.26).

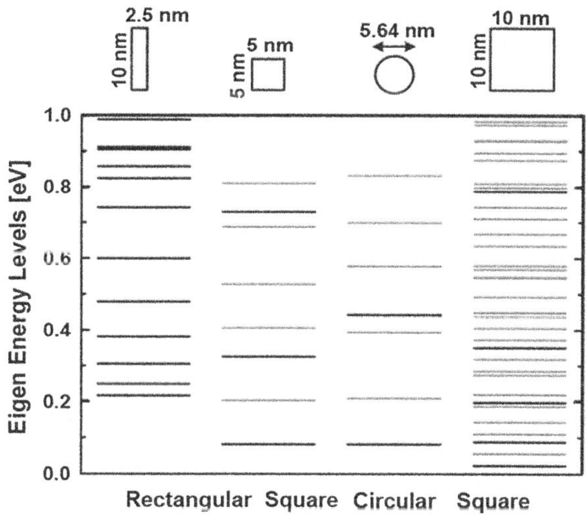

Fig. 4.12 The subband spectra in intrinsic silicon nanowire surrounded by the gate oxide for rectangular, square and circular cross sections

Also, the spectra in the square and circular cross sections with the same area are about the same, in general agreement with the results obtained in (4.26) and (4.31). More important, there is a significant difference in the sublevel spectra between the small- and large-area cross sections, as expected.

Surface Charge and 1D Density: Now that the sublevel spectra have been found, the surface charge of electrons induced is considered next. The first step for obtaining Q_n consists of finding the 1D density of electrons, n_{1D}, which is specified by

$$n_{1D}(\varphi) = \sum_{n=1}^{N} \int_{E_C+E_n}^{E_C+\Delta E_C} d\varepsilon\, g_{1D}(\varepsilon) F_n(\varepsilon)\,, \; g_{1D}(\varepsilon) = (\sqrt{2m_n}/\pi\hbar)/\varepsilon^{1/2} \qquad (4.32a)$$

Here, g_{1D} is the 1D density of states [see (1.48)], N the total number of subbands in the quantum well and ΔE_C the conduction band width. The Fermi occupation factor for the electrons in the nth sublevel with energy, E_n is given by

$$F_n(E) = \frac{1}{1 + \exp[(E - E_{Fi} - q\varphi)/k_B T]}\,, \; E = \varepsilon + E_C + E_n \qquad (4.32b)$$

Here, E_{Fi} is the intrinsic Fermi level of the nanowire, and $q\varphi$ is the bulk band bending. The electron energy, ε, in the nth subband ranges from $E_C + E_n$ to $E_C + \Delta E_C$, and the difference, $E - E_{Fi}$ is reduced by the band bending, $q\varphi$.

The integration of (4.32a) can be carried out either numerically or analytically by expanding the Fermi function. Once $n_{1D}(\varphi)$ is found, the 3D density, n_{3D}, is obtained by dividing $n_{1D}(\varphi)$ with the cross-sectional area, A of the nanowire. The surface field can then be obtained by using the well known relation used in (4.10):

$$\mathrm{E}(\varphi) = \sqrt{2}(q/\varepsilon_S)^{1/2}[N(\varphi)]^{1/2} \qquad (4.33a)$$

where the integration over the space charge density in (4.8) for obtaining E in the analysis of MOSFET is equivalent to finding the excess electron density induced by the bulk band bending:

$$N(\varphi) = \int_0^{\varphi} [n_{3\text{D}}(\varphi) - n_{3\text{D}}(0)]\text{d}\varphi, \quad n_{3\text{D}}(\varphi) = n_{1\text{D}}(\varphi)/A \tag{4.33b}$$

The surface charge, Q_n, can therefore be found in terms $\text{E}(\varphi)$ by again using the well-known boundary condition, (4.10), i.e.,

$$Q_n(\varphi) \equiv -\varepsilon_S \text{E}(\varphi) \tag{4.34}$$

In this manner, $Q_n(\varphi)$ is found in terms of the bulk band bending $q\varphi$ and the properties of the nanowire such as the shape and size of the cross section. Note that Q_n attained in the p substrate of NMOS is specified by the surface potential, $\varphi(x = 0)$. Whereas, Q_n in the nanowire is given in terms of the bulk band bending φ for the reasons discussed.

4.2.1.1 Channel Inversion

Capacitive Coupling: The channel inversion is next addressed to by using Q_n thus found. Consider the charging voltage divided into the gate oxide and the nanowire

$$V_G' \equiv V_G - V_{\text{FB}} = V_{\text{OX}} + \varphi, \; V_{\text{OX}} \equiv \frac{|Q_n(\varphi)|}{C_{\text{OX}}} \tag{4.35}$$

where V_{FB} is the flat band voltage associated with the n^+ gate and intrinsic nanowire, as depicted in Fig. 4.11. In intrinsic nanowire, there is no fixed ionic charge and therefore the surface charge, Q_S, consists solely of Q_n. Hence, the surface capacitance, C_S, should differ appreciably from that of NMOS. Nevertheless, the channel inversion via the capacitive coupling can be analyzed by using (4.35) with Q_n given in (4.34).

As in the case of NMOS, Q_n can be found explicitly as a function of V_G from (4.35). Figure 4.13 shows Q_n thus found plotted versus V_G for various kinds of the nanowire cross sections. Also shown in the figure is the $Q_n - V_G$ curve of an NMOS, for comparison. As expected, $Q_n - V_G$ curves in intrinsic nanowire do not exhibit the transition region demarcating the strong and weak inversion. Instead, Q_n in the nanowire increases exponentially in the small V_G regime. In this regime, the electron concentration is low, so that it requires relatively large bulk band bending, $q\varphi$, for the induced electrons to terminate the gate field lines. However, when V_G exceeds a certain value, the electron concentration attains such a level that any further increase in V_G and concomitant increase in the gate field lines can be compensated by electrons with relatively small changes in $q\varphi$. That is to say, $q\varphi$ is approximately pinned while supplying electrons sufficient to terminate the gate field lines. Hence, in this V_G range, Q_n should increase approximately in

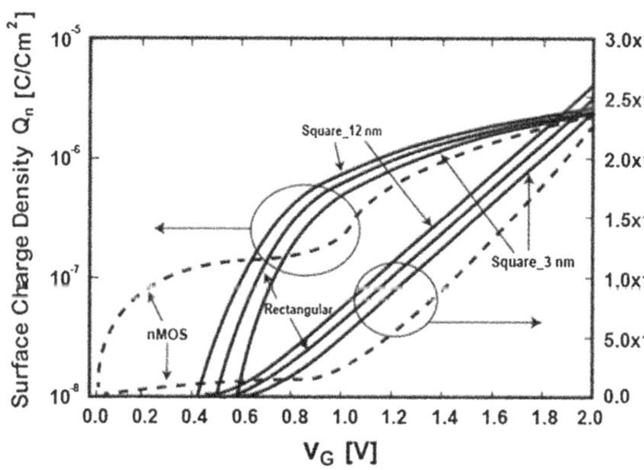

Fig. 4.13 The electron surface charge density is shown versus the gate voltage in nanowire with rectangular (3 × 12 nm) and square (3 × 3 nm, 12 × 12 nm) cross sections. Also plotted for comparison is the electron surface charge in NMOS with the substrate doping of $N_A = 10^{17}$ cm^{-3}

linear fashion with V_G just as in the case of NMOS after the onset of the strong inversion. Consequently, Q_n versus V_G curves naturally divide into the subthreshold and linear regimes as in the case of NMOS.

Moreover, it is clear from Fig. 4.13 that more surface charge, Q_n, is induced at given V_G in the larger cross section, and the threshold voltage therein is reduced. Here, V_T is defined as the value of V_G inducing a specified level of the drain current at given V_D, a procedure often used in the I–V characterization. According to this definition, V_T is simply the value of V_G which inverts a specified level of Q_n. The larger Q_n and smaller V_T with increasing cross-sectional area are consistent with the sublevel spectra shown in Fig. 4.12. As pointed out, the sublevels in larger cross section are more densely distributed at the energy level lower than those in smaller cross sections. Hence, more electrons should be induced for given V_G.

4.2.1.2 I–V Behavior in Long Channel *n*-type SNWFET

Overview: The modeling of I–V behavior in SNWFET has been extensively investigated from various standpoints. Understandably, the emphasis has been placed on the carrier scattering and transport. The carriers in the long channel are transported mainly by the drift-diffusion, while the ballistic transport is prevalent in the short channel. Thus, the carrier transport in FETs consists in general of the mixture of the drift–diffusion and ballistic transports. The I–V behavior in SNWFET is discussed from various standpoints in a few sections to follow, fusing together the two modes of the transport. In addition, the Landauer formulation of the ballistic transistor as applied to SNWFET is briefly touched upon. The discussion starts out with the long channel FET.

Naturally, a key quantity involved in the I–V modeling is the surface charge, Q_n, which is induced by V_G. As detailed in the preceding section, Q_n can be taken to increase linearly with the gate overdrive, and one can thus write

$$Q_n = C_{\text{eff}}[V_G - V_{\text{Tn}} - V] \tag{4.36}$$

where V is the channel voltage at y and C_{eff} the total capacitance of the nanowire. The capacitance, C_{eff}, is somewhat smaller than C_{OX} above V_T for the reasons as follows. As well known, C_{eff} consists of the oxide (C_{OX}) and surface (C_S) capacitances, connected in series, i.e., $1/C_{\text{eff}} = 1/C_{\text{OX}} + 1/C_S$. In NMOS, the electrons are induced practically at the oxide interface. Consequently, C_S is much larger than C_{OX}, and therefore C_T is practically identical to C_{OX} above V_T. In nanowire FET, however, the distribution of the induced electrons is nearly uniform across the cross section of the nanowire. This is due to the wave nature of electrons as pointed out earlier. Hence, the ratio of C_S with respect to C_{OX} is not as large as in NMOS; hence, C_{eff} is somewhat smaller than C_{OX}. However, the ratio C_S/C_{OX} is still large enough to keep C_{eff} practically constant at a level slightly below C_{OX}.

In addition, V_T in nanowire depends primarily on the geometry of the cross section, while in NMOS, V_T is determined mainly by the doping level of the substrate and the bandgap of the substrate. Obviously, this difference arises from the fact that in nanowire Q_n sensitively depends on the shape and size of the cross section, while in NMOS, Q_n is dictated by the doping level of the donor atoms and the band gap of the substrate. Moreover, in long channel nanowire FETs, the carriers are generally transported via the drift and diffusion. Hence, for modeling the I–V behavior in SNWFET, one can follow the same steps as used in NMOS, obtaining (4.20). In this case, the tools developed for SPICE model for fitting the I–V data can also be used.

4.2.1.3 I–V Behavior in Short Channel *n*-type SNWFET

Overview: An attractive feature of FETs is its scalability down to nano regimes. In such short channel FETs, the ballistic transport is prevalent. In ultrascaled MOSFETs, for example, the carrier transport has been taken up by the ballistic component by as much as 50 %. The ballistic efficiency in SNWFETs is believed to be comparable to or even higher than that of MOSFET. It is therefore important to consider the ballistic nanowire FETs.

Figure 4.14 shows the typical band diagram along the channel under bias. The maximum point of the band is located near the source end, the height of which is

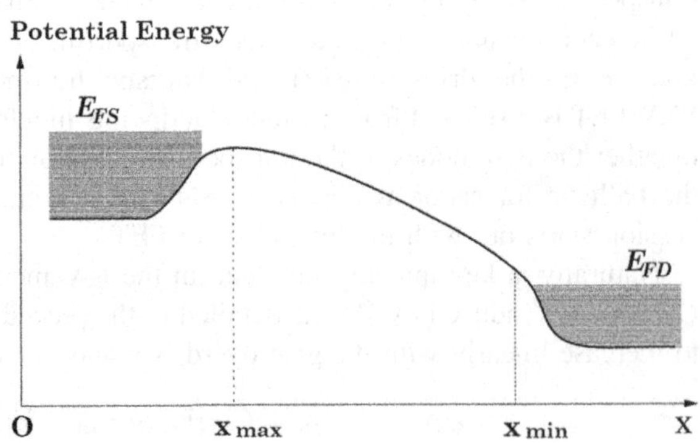

Fig. 4.14 The schematic view of the band bending in FETs is shown under the drain bias

determined by the n^+-i band bending at the source junction and also the voltages applied at the gate and the drain. The band diagram points specifically to the fact that the drain current is contributed by electrons injected from the n^+ source electrode and transported down the channel, subject, however, to the backscattering. Naturally, the height of the band maximum is reduced with increasing V_{GS}, so that more electrons are injected into the channel. In the following, the I–V behavior in short channel FETs is discussed from various standpoints, using the theories developed recently.

1. *One-Flux Scattering Theory*:

To facilitate the discussion, the one-flux scattering theory by Lundstrom is introduced first, in which the saturated drain current is given by

$$I_{\mathrm{DSAT}} = Q_{\mathrm{nLS}} v_{\mathrm{eff}} \tag{4.37a}$$

Here, Q_{nLS} is the line charge induced at the source end and is given in terms of the total capacitance C_{eff} per unit area and the channel width, W_{eff}, of the nanowire as

$$Q_{\mathrm{nLS}} = C_{\mathrm{eff}} W_{\mathrm{eff}} (V_G - V_{\mathrm{Tn}}) \tag{4.37b}$$

And v_{eff} is the effective velocity with which the electrons are transported down the channel subject to the backscattering and is given by

$$v_{\mathrm{eff}} = v_{\mathrm{inj}} \eta, \; \eta \equiv \left(\frac{1 - r_c}{1 + r_c} \right) \tag{4.37c}$$

where r_c is the backscattering coefficient and v_{inj} is the injection velocity of electrons.

Evidently, the two representations of I_{DSAT} given in (4.37) for SNWFET and (4.3), for MOSFET, respectively, are of the same format and characterize I_{DSAT} in similar contexts. However, there are a few differences existing between the two. Note that Q_{nLS} in (4.3) is the average line charge under the gate electrode, given by the average gate overdrive, $(V_{\mathrm{GS}} - V_{\mathrm{Tn}})/2$. Whereas Q_{nLS} in (4.37b) is the line charge induced specifically at the source end of the channel. In ballistic FETs, the line charge at the source end is a key parameter dictating the drain current. Additionally, the velocity, v_D, appearing in (4.3) is the drift velocity as characterized by the small field mobility, μ_n, while v_{eff} in (4.37c) is dictated primarily by v_{inj}.

It is interesting that η appearing in (4.37c) also represents the ballistic efficiency. In the limit of small r_c at which $\eta \cong 1$, nearly all of the injected electrons traverse the channel via the ballistic transport with v_{inj}. By the same token, in the opposite limit of $\eta \cong 0$, the transport is dominated by the drift–diffusion instead. For the general case of r_c ranging from 0 to 1, η is essentially an indicator showing the degree of mixing between the two modes of transport.

Obviously r_c depends on the channel length, L, and therefore η is also dependent on L. The dependency is taken into account in the one-flux scattering theory as follows. In thermal equilibrium or in the limit of small longitudinal channel field the backscattering coefficient as a function of L is given by

$$r_{c0} = \frac{L}{L+\lambda} \tag{4.38}$$

where λ is the mean free path. With λ thus introduced, it follows from (4.37) and (4.38) that η gauges the mode of carrier transport at a given channel length, L. For $\lambda \ll L$, for example, $\eta \cong 0$; hence, carriers are driven by the drift and diffusion, as it should. For $\lambda \gg L$, on the other hand, $\eta \cong 1$ and the ballistic transport becomes prevalent, as expected.

Under the bias, the backscattering coefficient, r_c, can be treated in such a way that the two competing modes of transport are naturally fused to represent the effective carrier transport. Now, the coefficient, r_c, is given by

$$r_c = \frac{l}{l+\lambda} \tag{4.39}$$

In this representation, L has been replaced by the critical length, l, over which the electron gains the kinetic energy equal to the thermal energy, k_BT.

With the use of l introduced in this manner, the expression of I_{DSAT} given in (4.37a) is naturally generalized to incorporate both the drift–diffusion and ballistic transport and that as a function of L. This can be shown as follows. When V_{DS} is applied, the band bends down from the source to the drain, as sketched in Fig. 4.14, and the injected electrons roll down the potential hill in the channel. Once the electron surpasses the distance l, Price observation assures that the electron can be taken to proceed to the drain, irrespective of the backscattering. This is equivalent to the electron traversing the channel essentially via the ballistic transport.

Next, to be specific, the critical distance, l, is to be expressed in terms of the thermal energy, k_BT, and the longitudinal channel field, $\mathrm{E}(0^+)$, at the band maximum:

$$q\mathrm{E}(0^+)l \equiv k_BT \tag{4.40a}$$

Also the mean free path, λ, is specified approximately via the thermal velocity, v_T, and the mean collision time, τ_n, or equivalently the low field mobility, μ_n:

$$\lambda = v_T\tau_n = v_T(m_n\mu_n/q) \tag{4.40b}$$

where (2.25) has been used to replace τ_n by μ_n.

By inserting (4.40) and (4.39) into (4.37) one can easily obtain (problem 4.5),

$$I_{\mathrm{DSAT}} = Q_{\mathrm{nLS}}\frac{1}{\frac{1}{v_T}+\frac{1}{\mu_n\mathrm{E}(0^+)}} \tag{4.41}$$

In this representation of $I_{\rm DSAT}$, both the drift–diffusion and ballistic transports are naturally fused in. Additionally, the relative importance of two competing modes of transport versus the channel length, L, is also accounted for. Thus, $I_{\rm DSAT}$ derived in (4.41) is applicable to a variety of short-channel FETs. It should be noted that since the longitudinal channel field scales with L, *i.e.*, $\mathrm{E}(0^+) \propto 1/L$, $\mu_n\mathrm{E}(0^+) \gg v_T$ in short channel FETs, and therefore the FET operates in ballistic mode with v_T serving as the saturation velocity. By the same token, in a long channel FET, $\mu_n\mathrm{E}(0^+) \ll v_T$, in which case the electrons traverse the channel via the usual drift velocity, v_D. In this latter limit, (4.41) converges to the SPICE I–V model given in (4.3).

However, the two line charges in (4.37b) and (4.3) are different in contents, as pointed out. In Eq. (4.3), $Q_{\rm nL}$ is the line charge averaged over the channel. The averaging procedure is required in the drift–diffusion formulation, since the transit time of electrons across the channel is long, so that the quasi-equilibrium settles in under the gate electrode. In this case, Q_n at the channel position, y, should be governed by the local gate overdrive. In the scattering theory formulation, on the other hand, I_{DS} is primarily determined by $Q_{\rm nLS}$ at the source end, sweeping across channel with a high ballistic efficiency. The two differing approaches should be duly taken into account for characterizing $I_{\rm DS}$ in SNWFETs with varying channel lengths.

2. *Apparent Mobility Model*:

The discussion of the apparent mobility model by Shur is in order at this point. According to the model, the total mobility consists of the usual low field mobility, μ_n, and the ballistic mobility, $\mu_{\rm ball}$, connected in series

$$\frac{1}{\mu} = \frac{1}{\mu_n} + \frac{1}{\mu_{\rm ball}}, \; \mu_{\rm ball} = \kappa L \tag{4.42}$$

where $\mu_{\rm ball}$ is taken commensurate with the channel length, L. The total mobility, μ, thus introduced is called the apparent mobility. One can then express $I_{\rm DSAT}$ with the use of μ as,

$$I_{\rm DSAT} = Q_{\rm nLS}\mu\mathrm{E}(0^+), \; \mu = \frac{\mu_n\mu_{\rm ball}}{\mu_n + \mu_{\rm ball}} \tag{4.43}$$

In the limit of short channel length, $\mu_{ball} \ll \mu_n$, and μ is reduced to $\mu_{\rm ball}$. In this case, the ballistic transport prevails with $\mu_{\rm ball}E(0^+)$ providing the saturation velocity. In the other limit, μ is reduced to μ_n, and $I_{\rm DSAT}$ is therefore driven by the drift–diffusion. It is therefore clear that the one-flux scattering theory and the apparent mobility model lead essentially to the same representation of $I_{\rm DSAT}$.

3. *Landauer Formulation of Ballistic FET*:

The continued scaling down of the semiconductor devices has pushed the device dimensions into the mesoscopic regime in between the atomic and the microscopic regimes. In such short channel devices, the mean free path, typically 50 nm long, cannot be taken much shorter than the channel length. By the same token, the carrier relaxation and the coherence lengths that are closely linked to the mean free path cannot be regarded much smaller than the channel length as well. Therefore, the wave nature of electrons and the mesoscopic scattering must be taken into consideration in modeling the I–V behavior. It thus behooves to consider the Landauer formulation for treating the ballistic FETs.

In the Landauer formulation, the drain current is taken due primarily to the net flux of electrons from the source to the drain via the ballistic transport. The drain current is thus given by

$$I_{\mathrm{DS}} = \frac{2q}{h}\sum_i \int_{E_i}^{E_u} dE[F(E,E_{\mathrm{FS}}) - F(E,E_{\mathrm{FD}})]T_i(E) \tag{4.44a}$$

Here the first term represents the electron flux from the source to the drain, while the second term denotes the flux in the reverse direction. Also, i is the summation index of the subbands with the energy level, E_i, in the well, and E_u is the upper limit of the integration given by the conduction band width, i.e., $E_u = E_C + \Delta E_C$. Naturally, the Fermi functions

$$F(E,E_{Fj}) = \frac{1}{1+\exp\left(\frac{E-E_{Fj}}{k_B T}\right)}, \; j = S, D \tag{4.44b}$$

are characterized by the Fermi levels at the source and drain ends, respectively. Under a drain bias, the two quasi-Fermi levels should split by the amount, qV_{DS}, i.e., $E_{\mathrm{FD}} = E_{\mathrm{FS}} - qV_{\mathrm{DS}}$, as detailed in Chap. 3. The factor T_i denotes the transport coefficient of electrons in the ith subband, and I_{DS} is thus contributed separately by electrons in each subband. In short channel FETs, one can put $T_i(E) \approx 1$ since the band bending in the channel is usually gradual, so that the backscattering therein is to be neglected.

The gist of (4.44) can be seen by considering the case of small drain voltage. One can then Taylor expand the Fermi function at the drain, retaining only the first expansion term and write

$$F(E,E_{\mathrm{FS}}) - F(E,E_{\mathrm{FS}} - qV_{\mathrm{DS}}) \approx -\frac{\partial F(E,E_{\mathrm{FS}})}{\partial E} qV_{\mathrm{DS}} \tag{4.45a}$$

Thus, the difference between the two fluxes is shown to be commensurate to the first derivative of F, which is well approximated by the delta function,

$$\frac{\partial F(E,E_{\mathrm{FS}})}{\partial E} \approx \delta(E - E_{\mathrm{FS}}) \tag{4.45b}$$

Upon inserting (4.45) into (4.44) there results

$$I_{\mathrm{DS}} = G\sum_i g_i V_{\mathrm{DS}}, \; G \equiv \frac{2q^2}{h} \tag{4.46}$$

In this manner, I_{DS} is naturally shown to be specified by the quantum conductance, G, and the sum of the separate contributions from all subbands, including the degeneracy, g_i, therein.

For an arbitrary V_{DS}, I_{DS} can also be specified as follows. For this purpose, one may first introduce a new variable of integration

$$\eta = E/k_BT$$

and compact the expression of I_{DS} in (4.44a) as

$$I_{DS} = G\left(\frac{k_BT}{q}\right)\tilde{M} \tag{4.47a}$$

where the form factor, $\tilde{M}$, reads as

$$\tilde{M} = \sum_i \int_{\eta_i}^{\eta_{max}} d\eta\left[\frac{1}{1+e^{(\eta-\eta_{FS})}} - \frac{1}{1+e^{(\eta-\eta_{FS}+qV_{DS}/k_BT)}}\right],\ \eta_{FS} = E_{FS}/k_BT \tag{4.47b}$$

where, $\eta_i = E_i/k_BT$ and $\eta_{max} = (E_C + \Delta E_C)/k_BT$. Obviously, to evaluate $\tilde{M}$, it is essential to find the relative location of E_{FS} with respect to say E_C under a given V_{GS}.

The difference, $E_C - E_{FS}$, varying as a function of V_{GS} is clearly illustrated in Fig. 4.15. In the figure are shown the band diagrams of the n^+ gate, gate oxide and intrinsic nanowire both in equilibrium and under a bias. In equilibrium, the band

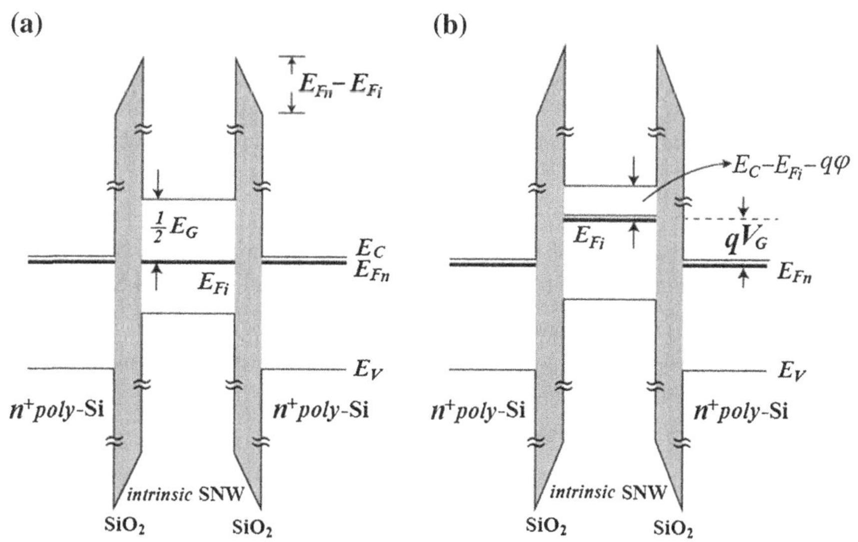

Fig. 4.15 The energy band diagram of n^+ poly gate, silicon dioxide and intrinsic silicon nanowire, in equilibrium (*left*) and under a positive gate bias (*right*). Here, $q\varphi$ denotes the bulk band bending induced by the gate voltage

bending should occur mainly in the gate oxide by an amount $E_{\mathrm{Fn}} - E_{\mathrm{Fi}}$ so that the Fermi level lines up and be flat (Fig. 4.15a). Under a gate bias, V_{GS}, the band in the n^{+} gate electrode is further lowered and the two Fermi levels split (Fig. 4.15b). Then, V_{GS} is dropped in the gate oxide and also in the nanowire, inducing the bulk band bending, $q\varphi$, therein. As a result, $E_{\mathrm{Fn}} - E_{\mathrm{Fi}}$ is reduced from its equilibrium value by the amount $q\varphi$, and one can therefore write

$$E_C - E_{\mathrm{FS}} = E_C - E_{\mathrm{Fi}} - q\varphi \tag{4.48}$$

Now, $q\varphi$ is in turn specified in terms of V_{GS} by means of (4.35), as discussed in detail already. Thus, the difference, $E_C - E_{\mathrm{FS}}$ can be found explicitly as a function of V_{GS}, and one can therefore evaluate $\tilde{M}$ and find I_{DS} as a function of V_{GS} and V_{DS}. It has therefore become clear that the I–V modeling in Landauer formulation consists of solving the coupled equations, (4.47) and (4.35).

The resulting features of I–V curves in ballistic FETs can also be surmised on a general ground with the use of (4.35) and Fig. 4.13. For small V_{GS}, hence small φ, Q_n is at the low level, and most of the charging voltage is taken up by φ, the reasons of which have already been detailed in connection with Fig. 4.13. As a consequence, $E_C - E_{\mathrm{FS}}$ shrinks rapidly, enhancing I_{DS} exponentially. This small V_{GS} regime should therefore correspond to the subthreshold regime. Once φ exceeds a certain value, the growth of φ slows down, but I_{DS} still increases with V_{DS} for given V_{GS}, reproducing thereby the triode regime. Beyond a certain value of V_{DS}, the flux of electrons from the drain to the source becomes negligible, rendering I_{DS} insensitive to further increase of V_{DS} and the device enters the saturation mode.

In this chapter, the I–V behavior of SNWFET has been discussed in comparison with theory of MOSFET, thereby bringing out the features unique to SNWFET. The I–V behavior has also been discussed in long- and short-channel SNWFETs. The I–V modeling in short-channel FETs have been discussed from various standpoints by using the one-flux scattering theory, the apparent mobility model and the Landauer formulation.

Problems

4.1 (a) Multiply both side of the Poissson equation given in (4.7) and prove that the left hand side of (4.7) reduces to the left hand side of (4.8).

(b) Perform the integrations in (4.8) and derive (4.9).
(c) Use the result obtained for discussing the threshold voltages as a function of the doping level and the band gap.
(d) Repeat the corresponding analysis for PMOS.

4.2 Derive (4.21) by Taylor expanding Q_S given in (4.10).

4.3 The I–V modeling in p-type FETs can be done by following the similar steps as used in n-type FETs and also by replacing the roles of electrons with those of holes.

(a) Derive the surface charge and the threshold voltage of the PMOS given in (4.24).
(b) Derive I–V relation and compare the results with that of NMOS.

4.4 Starting from the energy eigenequation in the cylindrical nanowire given in (4.27) derive (4.29a) and (4.29b). Also derive the expression of the energy eigenvalue given in (4.31), which is valid for the infinite barrier potential.

4.5 (a) Derive the saturated drain current given in (4.41) by inserting the expressions of the critical distance, l and the mean free path, λ given in (4.40).

(b) Discuss the mode of carrier transport as a function of the channel length, L, in the silicon nanowire FET with the cross-sectional areas of $5 \times 5\,\text{nm}$ and $10 \times 10\,\text{nm}$ and for L ranging from 10 to 100 nm. You can use the results obtained for the case of the infinite barrier height for the simplicity of discussion.

Suggested Reading

1. Kim, D. M. (2010). *Introductory quantum mechanics for semiconductor nanotechnology*. USA: John Wiley-VCH.
2. Lundstrom, M. S. (2006). *Fundamentals of carrier transport* (2nd ed.,). Cambridge University Press.
3. Datta, S. (1995). *Electronic transport in mesoscopic structures*. Cambridge University Press.
4. Lundstrom, M. S., & Ren, Z. (2002). Essential physics of carrier transport in nanoscale MOSFETs. IEEE Transactions on Electron Devices, ED *49*, 133–141.
5. Lundstrom, M. S. (1997). Elementary scattering theory of the Si MOSFET. *IEEE Electron Device Letters, 18*, 361–363.
6. Natori, K. (2008). Compact modeling of ballistic nanowire MOSFETs. *IEEE Transactions Electron Devices*, ED *55*, 2877–2885.

Chapter 5
Fabrication of Nanowires and Their Applications

Yang-Kyu Choi, Dong-Il Moon, Ji-Min Choi and Jae-Hyuk Ahn

Abstract The processing and application of nanowires and the field-effect transistors fabricated therein are discussed in this chapter. The pattern definition of nanowires as a key element of the top-down processing technique is described in conjunction with the optical, electron beam, and spacer lithography. Also, the process details of various kinds of FETs, such as FinFETs, Mug-FET, and gate-all-around nanowire FETs, are presented. An emphasis is placed upon fabricating the suspended nanowires in silicon and other compound semiconductors, and the Bosch process and the stiction problem of nanowires are also addressed to. In addition, the nanowire-based applications are elaborated such as biristor and FinFACT. Also, the bottom-up approaches for processing CNTs and various kinds of nanowires in silicon as well as compound semiconductors are elaborated. Finally, the transfer printing techniques are highlighted, enabling the fabrication of electronic circuits on flexible and diverse substrates by utilization of the releasable nano- and microstructures from donor substrates.

Abbreviation

CMOS	Complementary metal-oxide-semiconductor
RIE	Reactive ion etching
ICP	Inductively coupled plasma
DUV	Deep ultraviolet
VLSI	Very-large-scale integration
EUV	Extreme ultraviolet
LDD	Lightly doped drain
CVD	Chemical vapor deposition
ALD	Atomic layer deposition
UTB	Ultrathin-body

Y.-K. Choi (✉) · D.-I. Moon · J.-M. Choi · J.-H. Ahn
Department of Electrical Engineering, Korea Advanced Institute of Science and Technology (KAIST), Daejeon 305-701, Korea
e-mail: ykchoi@ee.kaist.ac.kr

D. M. Kim and Y.-H. Jeong (eds.), *Nanowire Field Effect Transistors: Principles and Applications*, DOI: 10.1007/978-1-4614-8124-9_5,

SOI	Silicon-on-insulator
FinFET	Fin field-effect transistor
SCE	Short-channel effects
3-D	Three-dimensional
2-D	Two-dimensional
LP	Low-pressure
RTA	Rapid thermal annealing
Mug-FET	Multiple-gate FET
GAA	Gate-all-around
SiNW	Silicon nanowire
BOX	Buried oxide
Biristor	Bistable resistor
MOSFET	Metal-oxide-semiconductor field-effect transistor
BJT	Bipolar junction transistor
FinFACT	Fin flip-flop actuated channel transistor
NEMS	Nano-electro-mechanical system
SWCNT	Single-walled carbon nanotube
LB	Langmuir-Blodgett
AC	Alternating current
VLS	Vapor-liquid-solid
MBE	Molecular beam epitaxy
TP	Transfer printing
TFT	Thin-film transistors
TMAH	Tetra-methyl ammonium-hydroxide
PDMS	Polydimethylsiloxane
DOF	Depth of focus

5.1 Nanowire Fabrication by Top-Down Approach

5.1.1 Introduction

Semiconductor nanowires are of great interest due to their unique properties, arising from their small dimensions. In addition, the nanowires hold up the promising potential to be the driving force in the field of electronic, optical, chemical, biological, energy, and magnetic device applications. Semiconductor nanowires can be fabricated by using a wide range of methods, which can be categorized into two approaches: the top-down and bottom-up approaches. The former has been utilized predominantly in the semiconductor industry because it can be used to fabricate small patterns down to a few nanometers in size. Moreover, the top-down methods enable the excellent control over the feature size and placement for creating patterns on wafers with large diameters at a low cost and high throughput. The top-down method can be defined as a process that can be used to form carved structures from a larger piece of material via a subtractive method. The subtractive method consists of film deposition, photolithography, and etching as in the complementary metal-oxide-semiconductor (CMOS) technology. In this approach, mask patterns are defined in the resist material by photolithography and are transferred to the substrate by a subsequent etching process. Reactive ion etching (RIE) and inductively coupled plasma (ICP) etching are the most commonly used techniques for the pattern transfer to the underlying layers. In this section, the top-down techniques to delineate the nanowire patterns are discussed, based on the CMOS process technologies.

5.1.2 Pattern Definition Method of Nanowires

Optical lithography, also termed photolithography, is used for microfabrication to delineate designed patterns on a thin film or a bulk substrate. It uses light to transfer a geometric pattern from a photomask to a light-sensitive photoresist on a substrate. Over the last few decades, a significant development of the optical lithography technology has been achieved in conjunction with the photoresist technology and optics. The ability to project a clear image of a small feature onto a wafer is limited by the wavelength of the light. The minimum feature size that a projection system can define is specified by the Rayleigh criterion [1].

$$\mathrm{CD} = \frac{\kappa_1 \lambda}{\mathrm{NA}} \tag{5.1}$$

Here, CD is the critical dimension (target design rule) or the minimum feature size, κ_1 the coefficient that reflects process-related factors, λ the wavelength of the incident light used, and NA the numerical aperture of the lens as observed from the wafer. Clearly, the minimum feature size can be decreased by a shorter wavelength and a higher numerical aperture. In other words, the minimum feature size is limited by the wavelength and the projection system.

5.1.2.1 Limitation of Optical and Electron Beam Lithography

Current state-of-the-art optical lithography tools use deep ultraviolet (DUV) light from excimer lasers with wavelengths of 248 nm krypton fluoride laser (KrF) and 193 nm argon fluoride (ArF). Optical lithography has been extended to a feature size below 50 nm using a 193 nm ArF excimer laser aided by immersion lithography [2]. Immersion lithography is a technique that can be used to increase the NA effectively. The final projection lens of the radiation source is immersed in a medium with a higher refractive index than air, thereby enabling the use of optics with numerical apertures exceeding 1.0. However, a single optical exposure has reached its practical limit, i.e., a feature size of 40 nm. Further development of optical lithography for the next-generation very-large-scale integration (VLSI) is becoming increasingly difficult. In particular, the patterning of nanowires requires smaller feature size than current optical systems can allow. Ultimately, advanced lithography techniques will be required for ordered arrays of semiconductor nanowires with excellent control over the placement and the feature size.

Electron beam (e-beam) lithography, ion beam lithography, extreme ultraviolet (EUV) lithography, X-ray lithography are considered, among others, as possible candidates for sub-20 nm patterning. Although a short wavelength of 13.5 nm by EUV lithography and sub-1 nm by X-rays offer the potential for nanowire patterning, these techniques are under development and face challenges related to the source power, mask, and photoresist. In the case of e-beam lithography, it is widely used for the fabrication of high-resolution photomasks for current optical lithography systems. In addition, the e-beam lithography has been shown to be capable of producing sub-10 nm feature sizes. However, the throughput produced by e-beam lithography is too low for mass production due to the exposure time. The development of multiple-beam systems will improve the throughput of the e-beam lithography. But further investment is indispensable for developing a proper photoresist and for resolving the issues of pattern size uniformity stemming from electron scattering and shot noise, and other issues.

5.1.2.2 Spacer Lithography

A single optical exposure based on an ArF excimer laser with immersion lithography has reached its resolution limit of a half-pitch of 40 nm [2]. Also, EUV and the e-beam lithography have yet to be proven and require further development to compete with the conventional optical lithography and to play a role in the next-generation VLSI manufacturing processes. At this point, an alternative sub-lithographic technique must be introduced for pattern resolution beyond the lithographic limit. Thus far, sub-30 nm CMOS devices are being fabricated via double patterning as a way to extend the half-pitch while keeping the numerical aperture and wavelength constant. Non-optical lithography, including pitch splitting (double exposure and double patterning) and spacer lithography, will enable the fabrication of the high density of devices as desired by the semiconductor industry.

The process flows for pitch splitting and spacer lithography are summarized in Fig. 5.1. The double exposure method consists of two sequentially separated exposures on the same photoresist layer using two different photomasks [3]. The double exposure technique allows the manufacturing of minimally sized features as well as larger patterns composed of various shapes in a layout. Alternatively, it can consecutively define minimally sized features with twofold enlarged line-to-line spacing twice, which can enhance the lithographic resolution. As long as the double exposure can be used effectively with an allowable misalignment tolerance, which

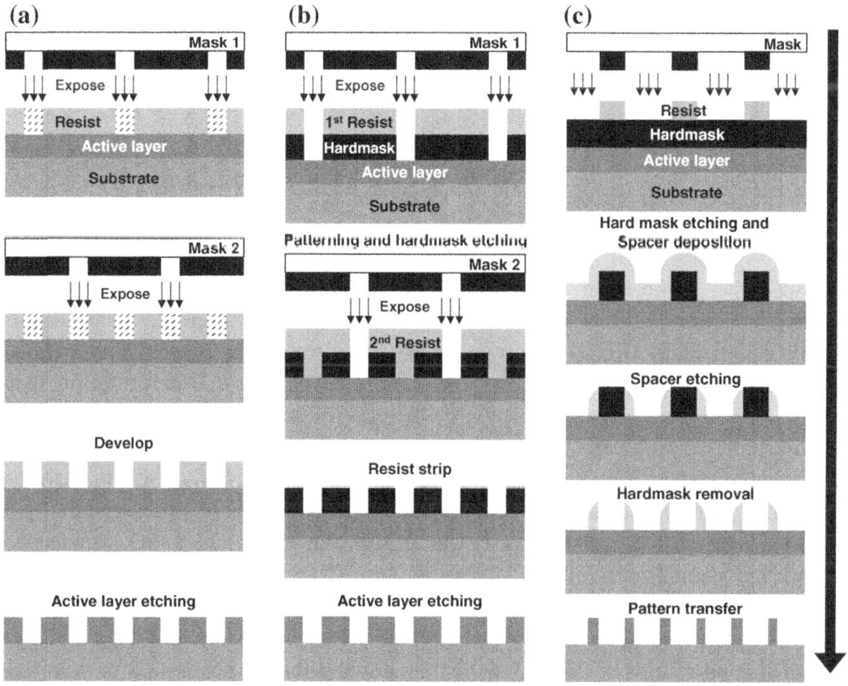

Fig. 5.1 Process flows for pitch splitting **a** double exposure, **b** double patterning, and **c** spacer lithography

should be smaller than that of the line-to-line spacing, it is a preferred patterning approach because it does not require additional subsequent process steps.

The double pattering is similar to the double exposure method. However, it uses a hardmask layer to transfer the pattern to the underlying layers [4]. Initially, the exposed photoresist patterns are transferred to an underlying hardmask layer. After the first photoresist is removed, a second layer of the photoresist is coated again. Then, a second exposure is carried out, and the exposed photoresist patterns are transferred to the underlying hardmask layer, which generates features between the features patterned on the hardmask layer. Finally, the hardmask patterns are transferred to the final layer underneath. This allows the doubling of the pattern density. Extrapolation from double patterning to multiple patterning has been considered, but the cost and yield control can still be a problem. Moreover, the double exposure and the double patterning technique cannot reduce the minimum feature size beyond the lithographic limit. That is, pitch splitting is still constrained by the resolution limitation of current optical lithography methods.

In view of these limitations, the spacer lithography has a great advantage because the minimum feature size depends not on the lithographic resolution but on the thickness of the subsequently deposited layer, used as the hardmask [5–8]. Moreover, the pattern density is inherently doubled due to the nature of the spacer.

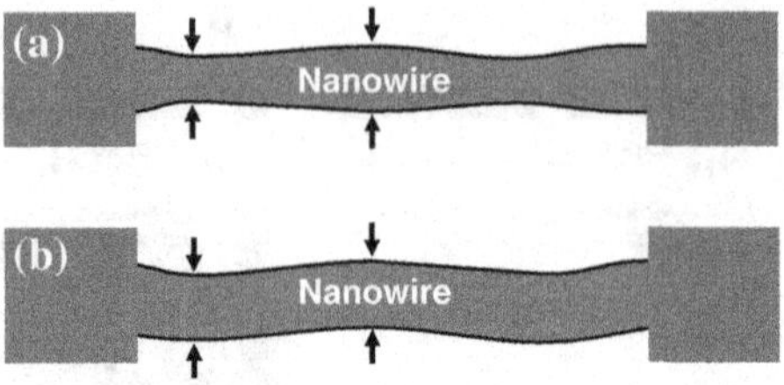

Fig. 5.2 Examples of line edge roughness in nanowires patterned by **a** conventional optical lithography and **b** spacer lithography

The spacer is a thin layer formed on the sidewalls of a pre-patterned protruding feature. For example, the spacer technique has been used to create a self-aligned lightly doped drain (LDD) junction against the preexisting gate in conventional CMOS devices. The term "spacer" was initiated by this LDD formation technology. This type of spacer is formed by the deposition of the film on a preexisting dummy pattern, followed by etching to remove all of the film material on the horizontal surfaces, leaving only the material on the sidewalls. In this case, the preexisting dummy pattern is a sacrificial layer. By removing this sacrificial pattern, only the spacers, serving as a structural or hardmask layer for a subsequent etching process, remain on both sidewalls.

It should be noted that there are two spacers for every dummy pattern; that is, the density of the hardmask is always doubled. Thus, the spacer approach is unique in that the pitch can be halved with one lithographic exposure. This favorably resolves the serious issue of the overlay between successive exposures when using double exposure and double patterning technology. The spacer materials deposited by chemical vapor deposition (CVD) or atomic layer deposition (ALD) are used as the hardmask, and its pattern fidelity is superior to the photoresist profile due to the excellent conformal step coverage. Although there is line edge roughness in the dummy pattern, the width of the spacer is uniform along the dummy pattern since this type of spacer material follows the wavy profile of the dummy pattern owing to the good conformality. In contrast, the line edge roughness of the left side and right side of the photoresist is easily mismatched in other approach [9]. In the worst case, the photoresist profile can be shaped like a bottleneck when the left edge and the right edge face each other, as shown in Fig. 5.2. Additionally, the spacer lithography can be used with other existing lithography tools i.e., mixed mode patterning to create a wide range of differently shaped patterns [5, 7].

5.1.2.3 Iterative Spacer Lithography

Additional advantage of spacer lithography is that it enables the iterative process to be applied to extremely dense patterns. Beyond the double patterning by one instance of spacer lithography, a multiple patterning methodology is essential for continued scaling, which can be achieved using the iterative spacer formation [10, 11]. For example, three instances of spacer lithography would result in an eightfold enhancement in the resolution, as illustrated in Fig. 5.3. After one instance of

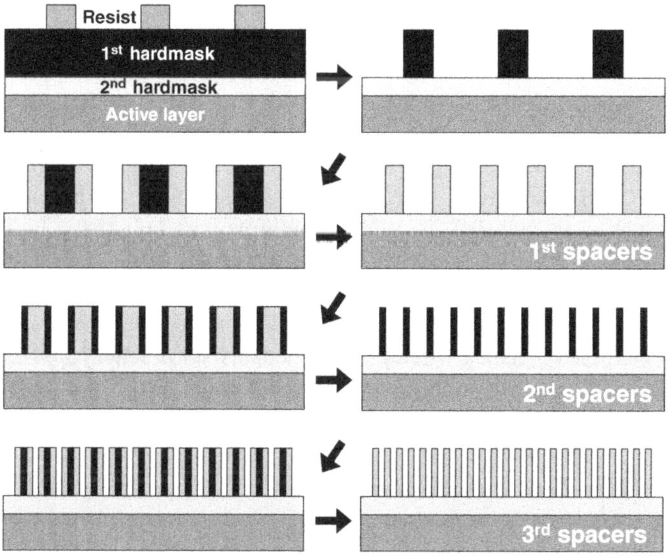

Fig. 5.3 Schematic view of iterative spacer lithography. The resolution enhancement by 8 (=2^3) can be obtained after three spacer lithography trials

spacer lithography, patterned spacers, which were initially the structural layer (hardmask), can be used in turn as the sacrificial layer for next spacer lithography. That is, the role of the spacers is reversed at each spacer lithography step from the hardmask to the sacrificial layer, and vice versa. Thus, the pattern density is doubled by each spacer lithographic process. In the spacer lithography process, it should be recalled that the feature size is determined not by the lithographic resolution but by the deposited thickness of the spacer film. Therefore, the iterative spacer lithography enables the patterning of sub-10 nm nanowires, since the deposited film thickness for the spacers can be controlled down to a few nanometers with high uniformity.

5.1.3 CMOS Technology from a Bulk to a Nanowire

Feature size scaling in CMOS technology has continued to follow Moore's law [12]. Over the past four decades, the device architecture based on a planar bulk structure has been developed by reducing the transistor gate lengths with each new generation of manufacturing technology. Junction optimization methods, such as the use of a LDD, lateral non-uniform channel doping, a shallow junction depth, and retrograded well doping with pocket and halo implants, have been utilized. Also, the strain engineering has been introduced to improve the mobility. The use of a high κ value in the gate dielectric and the metal gate electrode is now mainstream methods used for enhancing the gate controllability, lowering the gate leakage, and attaining the wide tunability of the threshold voltage. However, further aggressive device scaling based on the planar bulk structure becomes difficult and will reach the end-of-the-technology road map in the near future. As the gate length shortens, the gate control over the channel degrades due to the short-channel effects. As an alternative to the ultra-scaled CMOS device, an advanced MOSFET structure, such as the single-gate FET on the ultrathin-body (UTB), silicon-on-insulator (SOI), and the multiple-gate FET (Mug-FET), attracts a great deal of attention as future technology nodes [13–16].

It is noteworthy that the (Mug-FET) technology, which includes the FinFET and the gate-all-around FET, is based on a silicon nanowire. Therefore, it can be one of key elements in a nanowire device, especially for practical applications. For example, in 2011, Intel announced a major breakthrough and historic innovation in CMOS technology. A three-dimensional (3-D) transistor, which employs the FinFET or its derivatives relying on the silicon nanowire structure, is expected to be in mass production stage in the near future.

5.1.3.1 Fin Field-Effect Transistor

A fin field-effect transistor (FinFET) is a typical three-dimensional (3-D) FET, composed of a thin silicon channel which protrudes from a plane Si surface, like a shark's fin breaking the surface of the sea [5–7, 15–18]. The double gate straddles the Si nanowire and can therefore control the short-channel effects (SCEs) effectively while maintaining the robustness against SCEs due to the strong electrostatics over the channel potential. This device has attracted a great deal of attention because the mainstream FET structure is in a transition from a two-dimensional (2-D) planar-type structure to a 3-D structure. It has been predicted that the FinFET can drive the pace of CMOS technology advancement, fueling Moore's law for decades to come. A typical FinFET structure is shown in Fig. 5.4a. In the thin-body device of the FinFET, the body thickness (a width of the fin, W_{fin}) is a crucial parameter, governing the SCEs. A narrower W_{fin} suppresses the SCEs and off-state leakage current more effectively, especially when W_{fin} is smaller than the gate length (L_G) [19]. In contrast, the height of the fin (H_{fin}) corresponds to the channel width in a conventional 2-D planar-type FET and determines the on-state drive current. Consequently, a relatively tall FinFET with a thin body can show excellent short-channel immunity with high-performance and low-power consumption characteristics.

The schematic of the fabrication process flow used for the n-type double-gate FinFET is shown in Fig. 5.4b. A p-type SOI wafer is used as the starting material. A low-pressure (LP) CVD nitride is deposited as a hardmask. An embedded buried oxide (BOX) layer of the SOI substrate serves for the device isolation. The photolithography is applied to delineate nanowires (fins in the FinFET). Afterward, the active fin is patterned by RIE. Thermal oxidation is carried out to form the gate dielectric. For the gate electrode, in situ n^+ poly-silicon is deposited to surround

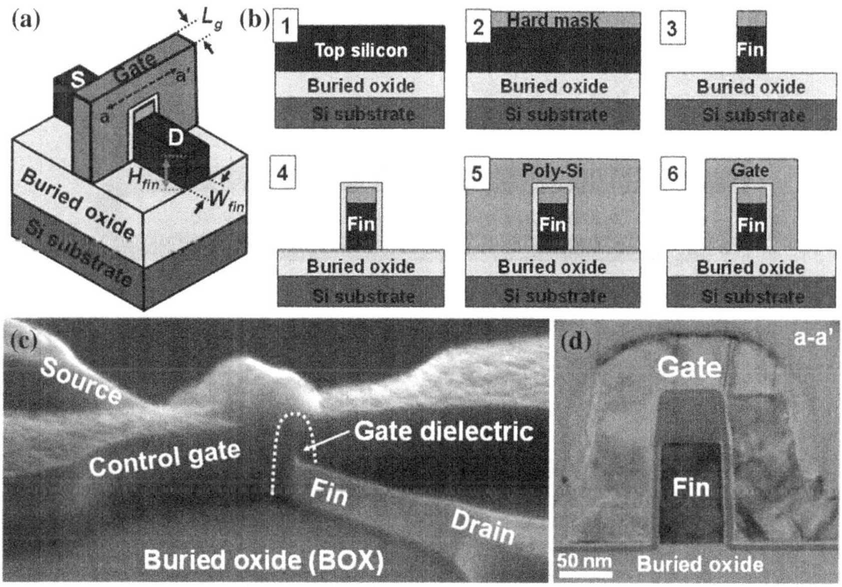

Fig. 5.4 **a** A schematic of the double-gate FinFET and **b** its process flow with a numerical order. **c** Bird's eye view and **d** cross-sectional image of the fabricated FinFET

the fin. After gate patterning by use of another photolithography and RIE process, n-type dopants are implanted to form the source/drain and activated by a rapid thermal annealing (RTA) process. Finally, the forming gas (N_2:H_2 = 9:1) annealing is followed. Figure 5.4c, d shows the bird's eye view and cross-sectional images of the fabricated device. The silicon nanowire thus formed is clearly seen in the figure.

5.1.3.2 Multiple-Gate FET

There are a couple of generic families of 3-D FETs categorized as a Mug-FET, which is nearly identical to the FinFET except for the profile of the gate surrounding the Si channel cross section. In other words, the Mug-FET generically describes any fin-based and multiple-gate transistor architecture regardless of the number of gates. For example, it can be transformed into a tri-gate, a Π-gate, a Ω-gate, or a gate-all-around (GAA) FET after structural modifications in order to provide better gate-to-channel potential controllability [20–24]. Figure 5.5 shows the evolution of the Mug-FET in the order of increasing electrostatic control of the gate. The electrostatics will improve in multiple-gate structures as the gate influences the channel potential from more than one side. Therefore, the multiple-gate structure can be scaled more aggressively than the planar 2-D structure. Among these innovative candidates, the GAA structure could become the most

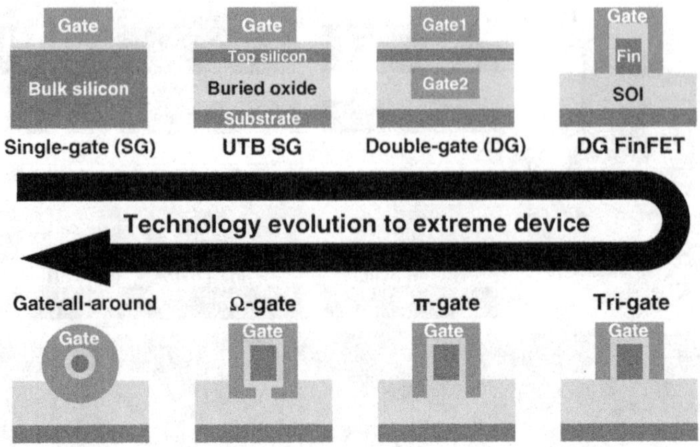

Fig. 5.5 Evolution of the device structure from a single-gate planar structure to a gate-all-around MOSFET

advanced type due to the ultimate electrostatic controllability associated with the gate completely surrounding the channel [25]. The fabrication process of the Mug-FET can be easily understood from that of the double-gate FinFET. Derivatives of the FinFETs such as a tri-gate, a Π-gate, a Ω-gate, and a GAA FET can be fabricated by the process flow of the double-gate FinFETs with minor modifications. For example, the tri-gate structure is obtained by removing the hardmask layer on top of the fin. Also, the partial or full removal of the BOX underneath the fin provides the Π-gate/Ω-gate or GAA structures.

5.1.4 Suspended Silicon Nanowire

A suspended nanowire has been used in a wide variety of devices such as sensors, actuators, energy harvesting devices, and other electronic devices. The fundamental interest in the mechanical motion and electrical functionality of a suspended nanowire has prompted the motivation to fabricate this structure. Among the many possible applications involving a nanowire-based FET, the GAA structures are most attractive and have been investigated extensively. The GAA structure has the near ultimate gate electrostatic controllability of the channel, a key element in the nanoscale CMOS technology. Also, among various GAA structures, the suspended silicon nanowire (SiNW) is most important and the process methodologies for the formation of such nanowires are discussed in this section. Various approaches for processing SiNW's are summarized in Fig. 5.6. It is also pointed out that the size of the SiNW defined by the lithographic process can be reduced further with the use of additional processes, such as hardmask trimming, sacrificial oxidation, or etching in H_2 ambient [26–30].

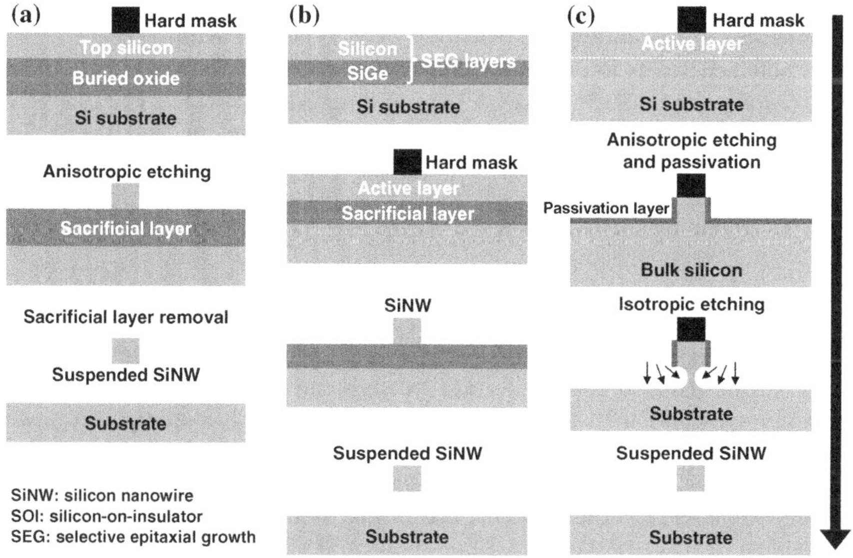

Fig. 5.6 Process flow for SiNWs: **a** SOI process, **b** SEG process, and **c** the Bosch process

5.1.4.1 Silicon-on-Insulator Substrate

The concept of the GAA structure was demonstrated by the use of a SOI wafer [24]. The top silicon of a SOI wafer is patterned as a SiNW and is then etched down to the BOX. The SiNW is then suspended from the substrate by the partial removal of the BOX. The BOX as a sacrificial layer is etched by a wet-etching process. After the formation of the suspended SiNW from the substrate, the GAA structure is formed by using the conventional MOSFET fabrication procedure.

5.1.4.2 Selective Epitaxial Process

In this approach, a SiGe layer is used as the sacrificial layer to form the suspended SiNW [26, 31–34]. Initially, SiGe and Si layers are hetero-epitaxially grown on a Si bulk wafer. Subsequently, a silicon nanowire is patterned by photolithography. The sacrificial SiGe layer is then removed by a wet- or dry-etching process. Therefore, a suspended SiNW is fabricated on a Si bulk wafer.

5.1.4.3 Bosch Process

Based on a plasma-etching method, a suspended SiNW is also formed on a Si bulk wafer [35–38]. A bulk wafer is used as the starting material. Optical lithography is employed to define the SiNW on a bulk substrate. The patterned bulk-SiNW is separated from the substrate by the Bosch process. First, in situ generated CF_4-based polymer begins to passivate the exposed Si surface during the anisotropic reactive-ion-etching (RIE) process. Thus, the partially patterned bulk-SiNW is conformally protected by the CF_4-based polymer before the subsequent SF_6 plasma etching. Then, the suspended SiNW is formed via the isotropic RIE process. The underside of the patterned bulk-SiNW is clearly etched after the Bosch process without the stiction problem arising.

5.1.4.4 Stiction Problem of Nanowire in Wet Etching

The nanowire tends to collapse toward the substrate when the restoring force is unable to overcome the capillary force, which is inevitably introduced in a wet removal step of the sacrificial layer, as depicted in Fig. 5.7. Once the nanowire collapses, the catastrophic pinning failure referred to as "stiction" occurs [39, 40]. The allowable length of a stiction-free nanowire is shortened as the nanowire is thin and narrow. Therefore, the stiction problem becomes important as the technology evolves. An elastocapillary number, N_{EC}, has been proposed to analyze the stiction of a suspended structure quantitatively [39].

$$N_{\mathrm{EC}} = \frac{128\, Eh^2 t^3}{15\gamma_1 \cos\theta_c l^4 (1 + t/w)} \left[1 + \frac{2\sigma_r l^2}{7\, Et^2} + \frac{108\, h^2}{245\, t^2} \right] \tag{5.2}$$

Here, E is the Young's modulus, σ_r the residual tensile stress, θ_c the contact angle, γ_l the liquid surface tension, h the gap distance, t the thickness of the structure, l the length of the structure, and finally, w the width of the structure. Equation (5.2) describes the ratio of the elastic restoring energy to the capillary energy. It is thus clear that the structure is free of stiction when $N_{\mathrm{EC}} > 1$, whereas it sticks to the substrate when $N_{\mathrm{EC}} < 1$. Therefore, the criteria of the stiction can be obtained by choosing the proper dimensions that satisfy $N_{\mathrm{EC}} = 1$. According to the model, a large gap distance h can increase N_{EC}, which implies that the stiction problem can be solved by increasing h, while the other parameters are fixed. However, the increment of h is not always a proper option. In the fabrication of a GAA FET, a large gap between the nanowire and substrate increases the subsequent gate etching time, which imposes some difficulties in gate etching without etching damage to the nanowire. N_{EC} is increased as the thickness t is increased,

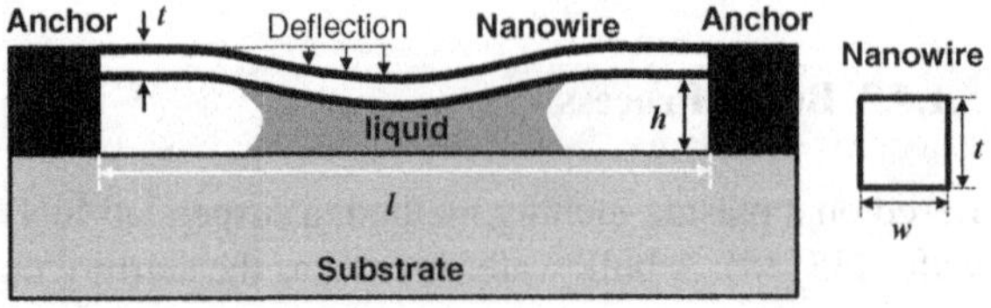

Fig. 5.7 Deflection of a nanowire double clamped by capillary force. The nanowire is pulled toward the substrate as the liquid evaporates

while h is fixed. This indicates that a nanowire with a high aspect ratio tends to be immune to stiction. In addition, the high residual tensile stress σ_r increases N_{EC}, resulting in improved stiction immunity [41].

5.1.5 Applications of Nanowires

5.1.5.1 Bistable Resistor

Bistable current–voltage characteristics are the feature common in various kinds of devices, used for data storage and photodetectors [42–47]. Those applications have been realized in both the metal-oxide-semiconductor field-effect transistor (MOSFET) and the bipolar junction transistor (BJT). In spite of the apparent differences in the operational mechanisms of the MOSFET and the BJT, the channel resistance is modulated by a control terminal such as a gate or a base. Interestingly, bistable current–voltage characteristics can be obtained from a simple nanowire structure by adjusting the doping profile. The bistable resistor (Biristor), also called the biristor, is characterized by the bistability based on a floating-body effect [48–50]. The configuration of a biristor can be realized in a silicon nanowire, composed of three regions of alternating n-type and p-type segments with the form of n-p-n or p-n-p, as shown in Fig. 5.8a. The voltage is applied between two opposite ends while the center region is electrically floated.

For the fabrication of the n-p-n biristor, the top silicon layer on a SOI wafer is initially doped with p-type dopants. The SOI wafer and the doping concentration of the p-type region are important for the bistable characteristics. The BOX allows the silicon nanowire to be electrically floated, and the sufficient doping concentration around 10^{18} cm^{-3} preserves the excess holes. The silicon nanowire is patterned by the photolithography and RIE process. Afterward, the second optical lithography is carried out to mask against ion implants. While the center region is covered with a photoresist, the other regions are counterdoped with n-type dopants. Afterward, the photoresist is stripped, and the dopants are activated by a subsequent thermal annealing process. The operation of the biristor can be understood as a single transistor latch in a floating-body MOSFET or an open-base breakdown in a BJT [51, 52].

A biristor initially remains in a high-resistance state under a low-voltage regime, but it abruptly switches to a low-resistance state via impact ionization under a high-voltage regime. As the applied voltage increases, the impact ionization is triggered, thereby generating excess holes in the floating region. As a result, the potential barrier between the n-type and the p-type regions is lowered. Consequently, the current flowing through the p-type region is amplified by iterative carrier generation, i.e., a positive feedback mechanism. When the current abruptly increases, i.e., when a latch-up occurs, the device remains latched in a low-resistance state for as long as the multiplication process continues. Hence, there are two available latch voltages according to the presence or absence of

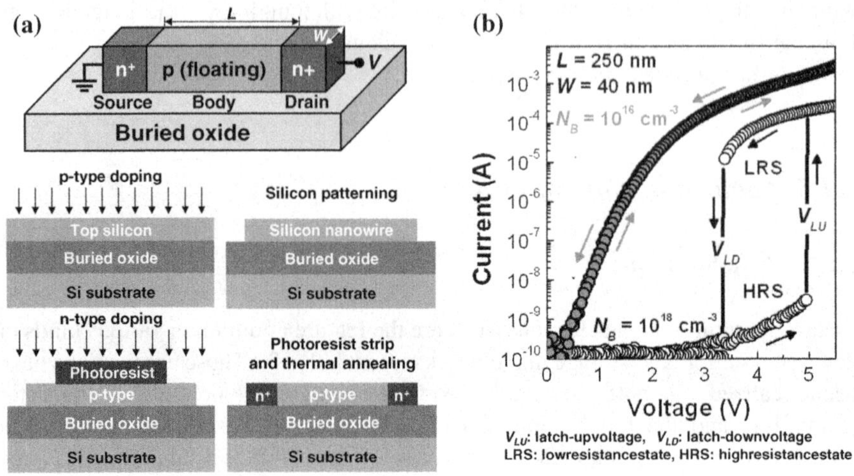

Fig. 5.8 **a** Schematics of a biristor and the process flow. **b** Double-sweep current–voltage characteristics for different body doping concentrations (N_B). In the heavily doped case ($N_B = 10^{18}$ cm^{-3}), bistable states between the *latch-up* and the *latch-down* state can be achieved

excess holes. The experimental data in Fig. 5.8b show counterclockwise hysteresis between a forward and a reverse scan. The latch voltage is greater in the absence of holes than in the presence of holes. Therefore, the device does not turn off until the voltage decreases below the value of the latch-down voltage. That is, the biristor shows bistable resistance between the latch-up and latch-down events. Based on the hysteric characteristics, the biristor can potentially be implemented in various applications including memory, optical, and sensor devices.

5.1.5.2 Fin Flip-Flop Actuated Channel Transistor

Nanoelectromechanical system (NEMS) switches have received much attention due to their low-power consumption by virtue of a steep slope in the current–voltage transfer characteristics. In the NEMS switch, for example, a cantilever beam is suspended over the electrode, and the cantilever moves through the electrostatic force between the gate and the beam [53, 54]. When the beam and the electrode are physically separated, the standby leakage current is ideally zero. In contrast, a considerable amount of current flows when they come into contact.

Conventional MOSFETs have a gate electrode, a solid-state gate dielectric, and an unmovable channel that is mechanically pinned at the substrate and/or the gate dielectric. More functionalities and benefits can be attained if the solid-state gate dielectric of the MOSFET is replaced with air in the same manner as the NEMS switch. As illustrated in Fig. 5.9a, the structure of a fin flip-flop actuated channel transistor (FinFACT) is slightly modified from an independently controlled double-gate FinFET by replacing the gate oxide with an air gap. Thus, the structure has a suspended and movable channel (fin or nanowire) between the two gates on both sides.

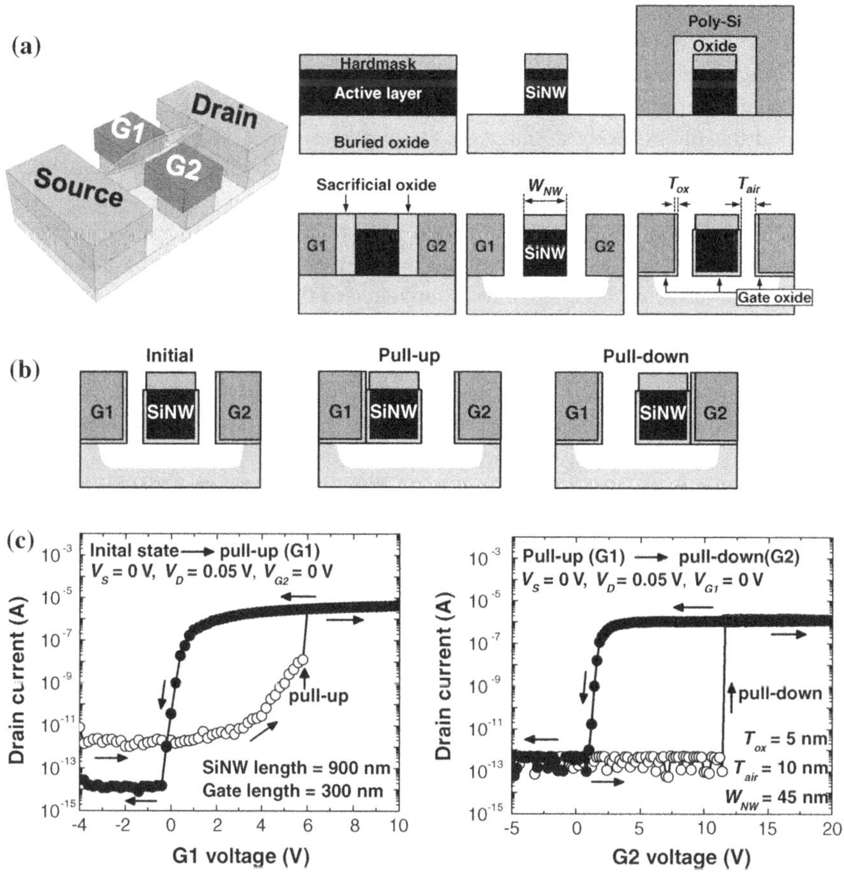

Fig. 5.9 **a** FinFACT and its process flow, **b** mechanical states as determined by the silicon nanowire (SiNW) position, and **c** electrical characteristics of the *pull-up* (flip) and *pull-down* (flop) operation

As the fin is actuated between a flip and a flop state, the device is termed a FinFACT [55–60]. The FinFACT is fabricated on a SOI wafer to create a suspended and movable channel. A fin is patterned with a nitride hardmask. It should be noted that the nanowire fin should be designed in a range of nanoscales to enable mechanical movement. A gate oxide is deposited around the fin as a sacrificial layer, followed by the deposition of the gate. The gate is then separated by chemical–mechanical polishing. Therefore, two gate electrodes are placed at both sides of the silicon fin. After a gate is patterned by a photoresist for the gate, the source and the drain electrode are formed by an ion implantation process. The preexisting sacrificial gate oxide is removed by a wet-etching process, and the BOX beneath the fin is simultaneously recessed. During this process, the suspended fin is formed. Because the fin is floated in air, it is movable and bendable.

Finally, the exposed surfaces of the fin and the gate are re-oxidized in order to form a thinned oxide, which is thinner than the air gap. By delivering a voltage from either gate, the bendable fin can move up (flip) or down (flop); thus, it can adhere to either of the gates according to the electrostatic force between one of the gates and the fin. Although voltage exists at all terminals, the fin remains on the landing electrode due to the adhesion force. By alternating the gate voltage, the fin is attached to gate-1 (G1) or gate-2 (G2); i.e., the electrostatic force attracts the fin from G1 and G2, or vice versa. When the attraction force between G2 and the fin becomes higher than the adhesion force between G1 and the fin, the fin abruptly comes into contact with G2. As the fin attaches to the one of the gates and closes out the air gap, its structure is transformed to a conventional MOSFET because the gate, the gate oxide, and the channel are in contact. Thus, the conventional transfer characteristics of the MOSFETs are achieved, as shown in Fig. 5.9b, c. Moreover, the physical position of the fin represents a binary state; therefore, the FinFACT is applicable for use in transformable logic and digital memory.

5.2 Nanotubes and Nanowires by Bottom-Up Approach

5.2.1 Introduction

Nanotubes and nanowires can be synthesized by catalytic growth based on a bottom-up method. The advantages of the bottom-up method, which is superior to the top-down method, are that nanowires with a small diameter and smooth roughness can be grown, and the chemical composition of the nanowires can be precisely controlled to produce a novel heterogeneous structure. However, controllable methods are necessary to align the nanotubes and nanowires in an array form for large-scale semiconductor device processing. In this section, alignment methods as well as techniques to grow nanotubes and nanowires are introduced.

5.2.2 Carbon Nanotubes

A single-walled carbon nanotube (SWCNT) is a hollow cylinder comprised of carbon. It can be considered as a single graphene sheet rolled up into a seamless tube. Depending on the chirality and diameter, an individual SWCNT can be either a metallic or semiconducting material. Due to its high mobility and ballistic transport characteristics, the SWCNT has been considered as an active channel material for future generation transistors.

There are various methods to synthesize CNTs, including arc discharge [61], laser ablation [62], and CVD [63]. The first two methods utilize a solid-state

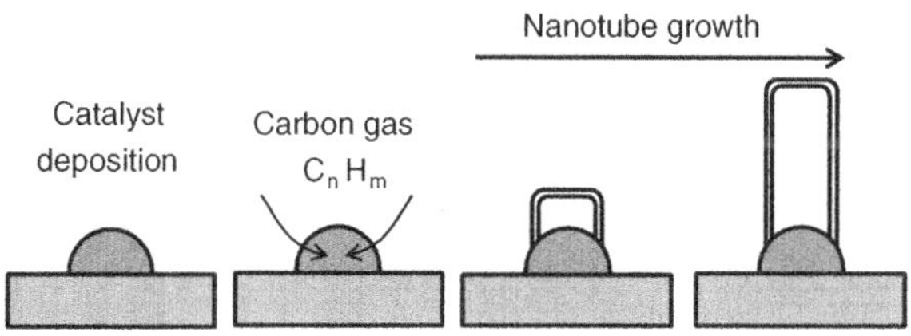

Fig. 5.10 CNT growth from a metal catalyst nanoparticle on a substrate. The catalyst nanoparticle is deposited on the substrate and carbon gas as a carbon source is introduced. The substrate is heated to approximately 700 °C. The carbon gas is decomposed into carbon and hydrogen and diffused toward the surface of the catalyst nanoparticle. The carbon atoms then grow out from the catalyst when the carbon solubility in the metal is beyond the critical limit

carbon precursor (graphite) as the carbon source needed for CNT growth, after which the carbon precursor is evaporated at a high temperature (>1,000 °C). In the arc discharge method, a discharge occurs between two graphite rods in a reaction chamber through which an inert gas flows. As a result, carbonaceous deposition is enabled at one of the rods where CNTs are produced. In the laser ablation method, SWNTs are synthesized in a process in which a mixture of carbon and transition metals is evaporated by a high-intensity laser focusing on a metal–graphite composite target. In the CVD method, hydrocarbon gas containing carbon atoms is utilized as a carbon source. As shown in Fig. 5.10, the hydrocarbon gas is decomposed in a heated furnace (500–1,000 °C) and dissolved into metal catalyst particles where the nanotubes are grown. Thus far, none of the three synthesis methods have been used to obtain CNTs with homogeneous diameters and chirality.

Nanotube assemblies can be formed by deposition of CNTs which are produced from bulk methods and dispersed in a solution. This solution-deposition method makes device fabrication onto a large-scale template feasible and device implementation on diverse substrates possible. CNTs dispersed in a solution can be formed in aligned arrays by applying an external force such as an electric field or mechanical force. As shown in Fig. 5.11a, an electric field attracts and aligns the CNTs along the field lines due to their high polarizability and anisotropic structure [64]. In this case, pre-patterned microelectrodes are needed to generate the electrical field. Figure 5.11b shows that a microfluidic channel utilizes shear force to align CNTs along the flow direction [65]. The density of the CNTs is controlled by concentration of CNTs in the suspension solution and the flow time. As shown in Fig. 5.11c, the Langmuir–Blodgett (LB) technique allows large-scale assembly, which is difficult to achieve by means of an electric field or a microfluidic method [66]. The CNTs in the solution are compressed, and the ordered structure is created. Such aligned CNTs can be transferred from the solution surface onto various substrates. The spacing of the CNT patterns can be controlled through the compression force. By modulating the alignment direction during iterative processes, not only parallel arrays but also crossed arrays can be attained.

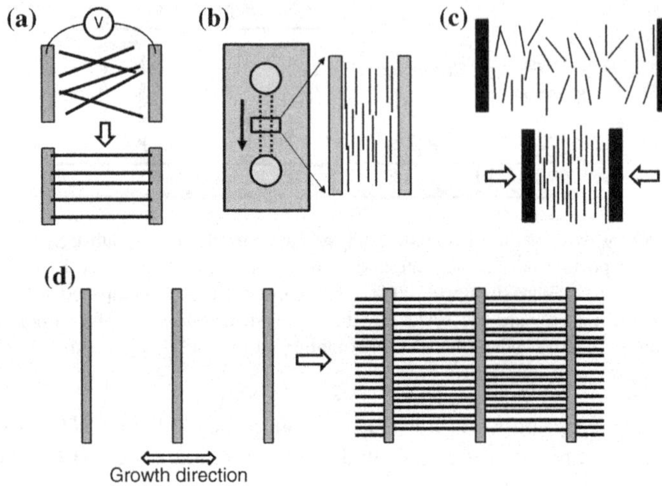

Fig. 5.11 Alignment methods of CNTs and nanotubes from the dispersion solution: **a** Electric-field-assisted method, **b** Microfluidic flow, and **c** the LB technique. **d** The CNTs are directionally grown during the CVD process with catalyst patterns on single-crystalline substrates as guided templates

Because as-synthesized CNTs can be semiconducting or metallic, purification techniques to collect only semiconducting CNTs are required to fabricate electronic high-performance devices. This filtration process relies on the difference in the electrical, chemical, or optical properties between semiconducting and metallic CNTs. Alternating current (AC) dielectrophoresis allows for the separation of metallic CNTs from their semiconducting counterparts in the suspension solution [67]. The difference in the relative dielectric constant between metallic and semiconducting CNTs with respect to the solvent leads to opposite movement of the two species. Chromatography and ultracentrifugation methods can be used to separate the CNTs because DNA or some surfactants which encapsulate CNTs influence the optical properties or buoyant density according to the diameter, bandgap, and/or electronic characteristics [68–70].

In the solution-deposition process, the electrical characteristics are degraded and the lengths of CNTs are shortened due to the high-power sonication and strong acid treatment during the dispersion of bulk CNTs into the solution. To overcome these weaknesses of the "solution" process, a "dry" process was demonstrated to fabricate aligned arrays of CNTs. There are a couple of methods that can be used to align CNTs during the growth process that utilize CVD. The electric-field-directed growth of CNTs has been demonstrated [71, 72]. An electric field is generated with voltage applied to pre-patterned catalyst electrodes. The nanowire is grown along the electric field lines. Due to the high polarizability of CNTs, a large induced dipole moment results in large alignment force, thereby suppressing the randomization of the CNT orientation caused by thermal fluctuation. A single-crystalline substrate such as a sapphire and a quartz wafer can be used as a template for aligned nanotube growth [73–76]. The CNTs are directionally grown along atomic steps of the specific facet in the single-crystalline substrate. This aligned growth is attributed to the strong van der Waals interaction with an increased contact area between the nanotubes and the surface and also to the electrostatic interaction with uncompensated dipoles. High-density SWNT arrays without sacrificing the alignment are attained with spatial or parallel patterns of the catalyst on the single-crystalline substrate [77–80]. As shown in Fig. 5.11d, catalyst patterns are lithographically defined as narrow stripes perpendicular to the preferential growth direction after which the CNTs grow with good alignment. Aligned CNT growth up to the wafer scale has been carried out for VLSI devices [79, 80]. An extra transfer step to implement the aligned CNTs grown on a quartz wafer to a silicon wafer is needed for the subsequent integration onto the silicon wafer [78, 80] for CMOS fabrication.

With current technology, it is difficult to obtain pure semiconducting CNTs among CVD-grown CNTs. Some portion of metallic CNTs is mixed undesirably with the semiconductor CNTs, which degrades the on/off current ratio of devices. The most commonly used technique for the selective removal of metallic CNTs is the electrical breakdown of metallic CNTs. This is achieved by increasing the voltage between the source and drain, whereas the semiconducting CNTs are kept in the off-state, as depicted in Fig. 5.12 [81]. By applying this technique, an on/off current ratio of CNT FETs of less than 5 was exponentially increased to 10^5 [82].

Applications of CNT FETs include future high-performance devices, bio/chemical sensors, nonvolatile memory, and others. Feasible SWCNT FETs with a sub-10 nm gate length have been shown to be able to overcome current Si technology [83]. DNA biosensors have been developed with SWCNTs as sensing materials to detect charges arising from DNA hybridization [84]. A gas sensor has been demonstrated by adapting the electrical characteristics of SWCNT FETs, showing very high sensitive to gases [85]. A crossbar array of SWCNTs can be

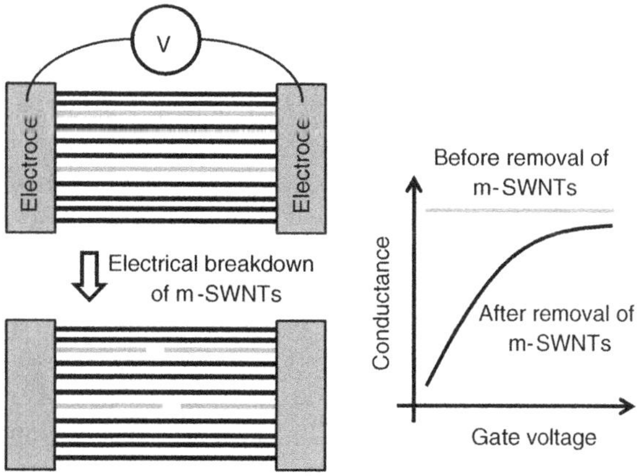

Fig. 5.12 Removal of metallic SWNTs (m-SWNTs) using the electrical breakdown method. By applying a high voltage to the SWNT assembly, m-SWNTs are destroyed and semiconducting SWCNTs are left as they were. This results in a high on/off ratio of the SWNT FETs

used as nonvolatile memory cells based on mechanical actuation [86] as well. However, the uniformity and reliability of CNT devices have yet to be proven.

5.2.3 Silicon Nanowires

The distinctive features of silicon nanowires fabricated by bottom-up methods are their small diameters, the possibility of in situ doping, their heterogeneous composition, and fewer surface dangling bonds. These features cannot be easily achieved with the top-down method.

There are various methods that can be used to fabricate bottom-up silicon nanowires. The vapor–liquid–solid (VLS) mechanism in the CVD process is the most commonly used approach for the growth of silicon nanowires [87]. As shown in Fig. 5.13, the nanoparticle catalyst is deposited as a "seed" material on the silicon wafer. Adsorption of a vapor-phase precursor onto a nanoparticle catalyst in a liquid phase reaches a supersaturation level and the 1-D crystalline nanowire growth occurs from the nucleation site at the liquid–solid interface. Silicon nanowires with a diameter in the range of 3–500 nm can be processed by changing the diameter of the nanoparticle catalyst [88]. Au is the most commonly used catalyst for the nanowire growth in the VLS method. However, Au is incompatible with the standard CMOS process because its diffusivity is unacceptably high and also because it creates deep level traps in the bandgap of Si. These traps act as recombination centers and degrade the performance of the device. Au exhibits high chemical stability, which makes a fine cleaning process difficult. In addition, Au etchants containing alkali ions degrade the device performance.

For those reasons, there have been alternative attempts to develop complementary metal-oxide semiconductor (CMOS) compatible catalysts or catalyst-free growth techniques [89, 90]. Molecular beam epitaxy (MBE) enables the epitaxial growth of silicon nanowires without a catalyst. MBE takes place in an ultrahigh vacuum to generate a directional beam of silicon atoms with long mean-free paths. A highly purified silicon source is evaporated, reached to the wafer, and deposited along the crystal orientation of the wafer. MBE provides good controllability of the gaseous beam by switching the evaporation source. Thus, a heterogeneous structure with different types of doping can be easily produced. However, the achievable diameter of the silicon nanowire remains limited to approximately 40 nm due to the weak supersaturation effect by MBE [91, 92]. Moreover, the low throughput can be problematic because the growth rate of the silicon nanowire is as low as a few nanometers per minute [92].

The bottom-up method utilizes in situ doping by the addition of dopant precursors during the growth process, whereas the top-down method employs ion implantation to inject dopants into the silicon nanowires [93]. A boron-doped and a phosphorous-doped silicon nanowire behave as a p-type and n-type material, respectively, as expected. In addition, other compositions such as Ge can be heterogeneously incorporated into silicon nanowires. Thus, the bottom-up method

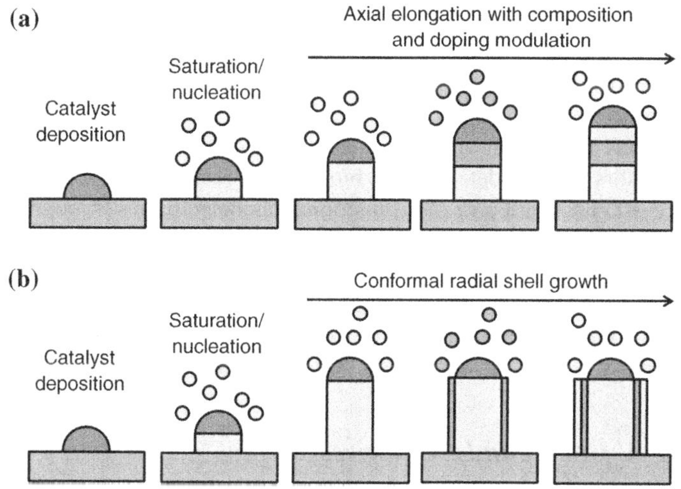

Fig. 5.13 Bottom-up grown nanowire from a metal catalyst nanoparticle on a substrate for **a** an axial and **b** a radial hetero-structure. The axial hetero-structure can be produced by modulating the composition and/or doping precursor during the elongation process. The radial core/shell hetero-structure can be formed by the conformal deposition of a radial shell after the elongation of the core structure

enables the production of a heterogeneous structure, where the composition and/or doping are modulated by control of the precursor components. As shown in Fig. 5.13a, it is possible to grow an axial heterogeneous structure by alternating the flow of different reactants and/or dopants during the elongation of the structure. For example, electronic devices fabricated with an n^+–n–n^+ nanowire structure have been demonstrated. Naturally, these devices could be used as logic device units, e.g., as lithography-independent address decoders [94]. After the elongation of a core nanowire, the condition is altered to increase the conformal deposition of the radial shell and inhibit axial elongation of the nanowire core (Fig. 5.13b). As an example, p-type/intrinsic/n-type (p-i-n) coaxial silicon nanowire solar cells were investigated [95]. Moreover, a heterogeneous structure such as Ge/Si nanowire was also demonstrated for use as a high-performance field-effect transistor [96]. By the iterative control over the nucleation and growth of nanowires, kinked or zigzag nanowires can be grown for use in a novel structure of nanoelectronics, photodetectors, and biological sensors [97].

The alignment of silicon nanowires is necessary when fabricating VLSI devices. Because both silicon nanowires and SWNTs have similar properties in term of their high anisotropy and polarizability, alignment methods of silicon nanowires are very similar to those of the CNTs, as shown in Fig. 5.11a, b, c. These alignment methods are based on an electric field [98] realized by polarizability, on capillary force created by microfluidic channels [99], on the LB technique [100], or on contact printing [101]. Thus, various configurations, including parallel and crossed arrays of silicon nanowires, are achievable. These alignment methods are feasible for horizontal and planar structures. For devices with a vertical structure, the bottom-up method enables the growth of vertically aligned silicon nanowire. Therefore, vertical devices can also be easily fabricated without an additional alignment process [102].

Applications of bottom-up silicon nanowires include nanoelectronic circuits and nanosensors for the detection of biological species. The bottom-up method enables the 3-D integration of multifunctional nanoelectronics through a layer-by-layer process on a flexible substrate [101]. High sensitivity can be achieved in nanowire-based nanosensors due to the high surface-to-volume ratio. In such sensors, the changes in the conductance or resistance monitored through silicon nanowires facilitate the highly sensitive label-free detection of biomolecules [103].

5.2.4 III–V Compound Nanowires

III–V compound semiconductors have been used in high-performance electronics, optoelectronics, and light-emitting diodes. The epitaxial growth of III–V semiconductors on a Si substrate must circumvent a few difficulties, such as lattice mismatch, differences in the crystal structure (a polar zinc blend or a wurtzite structure for III–Vs but a covalent diamond structure for Si), and a large difference in the thermal expansion coefficient [104]. To resolve these difficulties, several techniques for III–V channel formation on a Si substrate have been demonstrated with the use of graded buffer layers for 2-D growth [105] and areal epitaxy through a small open window for 1-D growth [106]. The growth of 1-D nanowire using a bottom-up method is preferred due to the efficient relaxation of the lattice mismatch with narrow diameters and the easy formation of a heterogeneous structure with atomically sharp interface, which cannot easily be realized by a conventional top-down technique. The heterogeneous structure allows the energy bandgap tuning to modulate and enhance the device characteristics.

Similar to the growth of silicon nanowires, metal catalysts such as Au nanoparticles are widely used for the growth of III–V compound nanowires as well. Nanowires of binary III–V materials (GaAs, GaP, InAs, and InP) and ternary III–V materials (GaAs/P and InAs/P) have been produced by the VLS mechanism [107]. Due to the incompatibility of Au with the conventional CMOS process, catalyst-free growth methods have also been developed [106]. III–V compound semiconductor nanowires can be grown on lattice-mismatched substrates; however, there is a critical diameter beyond which well-aligned nanowires cannot be grown [108]. The critical dimension is inversely proportional to the lattice mismatch. Alignment methods of grown III–V compound nanowires are identical as those of the silicon nanowires, as discussed in the previous section.

A superlattice structure of semiconducting nanowire by the iterative stacking of GaAs/GaP can be fabricated for potential applications ranging from nanobarcodes to polarized nanoscale LEDs [109]. InAs/InP radial nanowire hetero-structures have been demonstrated as quantum-confined and high-electron-mobility devices [110].

5.3 Transfer Printing Technique for Electronics on Flexible/Diverse Substrates

5.3.1 Introduction

It is generally accepted that flexible electronics and more advanced electronics on various substrates represent the promising technology of the future for a wide range of applications. Currently, amorphous silicon, poly-crystalline silicon, and organic materials are the dominant semiconductor materials used for thin-film transistors (TFTs) in large-area and/or flexible electronic systems. Moreover, the applications of TFT are rapidly broadened to include the display, solar cell, and X-ray imagers, etc. In these applications, the high-performance devices are not necessarily needed. Rather the large-area and low-cost fabrication methods are more important.

Today's flexible electronic fabrication processes have constraints imposed by materials and/or substrate process limitations. For example, a flexible plastic substrate cannot endure some high-temperature processes due to the low glass-transition temperature of the plastics. However, this has not been a constraint for the conventional wafer-based microelectronic systems. Thus, it has not been possible to grow the single-crystalline semiconductor materials with sufficiently high mobility on a plastic substrate. As a result, only relatively low mobility semiconductor materials had to be used on flexible substrates with existing TFT technology. Moreover, such unconventional substrates as paper, banknotes, vinyl, or fabric can be used but not preferred because of the poor substrate properties such as rough or contaminated surface, vulnerability to environments. Therefore, the development of the fabrication methods of high-performance devices, compatible with flexible and/or soft substrates has become important for multi-faceted applications with a greater variety of functions.

Recently, the transfer printing (TP) technique has been developed to circumvent the aforementioned constraints associated with high-performance electronics on flexible or unconventional substrates. The technique is aptly named, as can be shown by the following examples. As high-quality semiconductors cannot be formed on flexible substrates, high-quality single-crystal semiconductors are alternatively *transferred* from conventional crystalline semiconductor wafers (the *donor*) onto target substrates (the *acceptor*). In addition, harsh fabrication processes such as thermal annealing, which exceed the glass-transition temperature of the acceptor substrate, can be utilized on heat-resistant substrates (the *donor*), and the processed material layer is then *transferred* onto the target substrates (the *acceptor*). Therefore, the main feature of TP is that ideally all types of materials and/or fabrication processes become possible regardless of the materials/substrates that are used. Accordingly, the TP technique has been utilized to realize several high-performance electronic systems on flexible/diverse substrates which were impossible with traditional TFT technology.

In this section, the processes of TP are briefly summarized and a number of high-performance flexible electronic devices created via the TP technique are briefly introduced. Further applications to create high-performance electronic devices on diverse substrates are also provided. Some review papers are available as well with brief overviews [111–113].

5.3.2 Fabrication of the Releasable Nano-/Microstructures on Donor Substrates

This subsection focuses on the TP of the crystalline Si nano-/microstructures that are fabricated from conventional single-crystal Si donor wafers. However, it is important to note that other materials, e.g., Ge, III–V compounds, II–VI compounds, oxides, or releasable functional blocks can also be transferred with similar procedures.

If nano-/microscale structures can be fabricated on high-quality single-crystal silicon wafers that are readily transferrable onto a range of substrates, high-performance electronic systems can be readily realized on any substrate with functional performance comparable to that of conventional wafer-based electronics. Successfully creating releasable nano-/microscale structures on conventional bulk wafers (or donor substrates) is therefore an important step for TP.

A key feature of TP is to make "necessary objects" that *weakly and temporarily* adhere to the donor substrate so that they can be detached easily from this substrate. Figure 5.14a shows typical releasable nano-/microscale structures. A freestanding long beam-type patterned structure is the most frequently transferred structure via the TP method. The ends of the nano-/microbeams are tethered to a mechanically supporting part of the wafer, acting as an anchor. The application and detachment of a stamp on such a donor wafer can at this point release the freestanding structures with fractures induced at the anchors. The released freestanding structures adhere to the stamp, which can be transferred afterward onto the desired target substrates.

There are two main approaches for the generation of these types of releasable structures, as illustrated in Fig. 5.14b, c. One is a straightforward approach that uses the selective etching of the buried sacrificial layer among the multilayers on the wafer, as shown in 5.3.1b. A SOI wafer is a typical example of this type of substrate, which is comprised of multiple layers. Lithographically defined top-silicon patterns (the functional device material for the transfer) are suspended in an ambient environment after the removal of the BOX (sacrificial layer). The very first work using this method demonstrated a successful high-performance TFT with transferred crystalline Si film on a plastic substrate in 2004 [114].

The other approach uses the anisotropic etching properties along certain crystalline planes of crystalline materials. Single-crystal Si shows high etching selectivity along the (110) orientation with the use of wet etchants such as KOH or

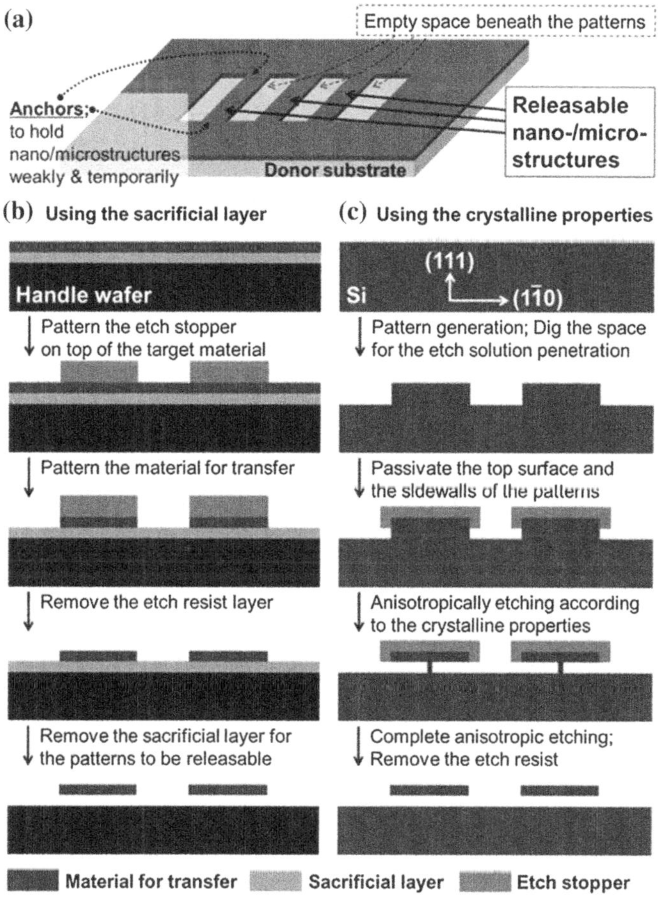

Fig. 5.14 **a** Schematic illustration of typical releasable nano-/microscale structures generated on a donor substrate. Nano-/microstructures are suspended in an ambient environment but weakly adhere to the anchor parts. **b**, **c** Schematics of the fabrication procedures of the releasable nano-/microscale structures from donor substrates using **b** the sacrificial layer removal method, or **c** anisotropic etching according to the crystalline properties

tetra-methyl ammonium hydroxide (TMAH). By using an appropriate etch stopper with a specific shape (from the top view) and the anisotropic etching properties of crystalline materials as mentioned earlier, releasable nano-/microscale patterns can be generated on bulk wafers. One example is depicted in Fig. 5.14c, where Si releasable patterns are generated from a (111) Si wafer [115–117]. Initially, trenches are defined so the sidewalls are terminated on the (110) planes. Then, the designed parts of the Si to be protected from the Si wet etchants are coated with an etch stopper. Afterward, the anisotropic etching by the Si wet etchant produces the freestanding patterns, as depicted. In a similar manner, it is possible to generate the nano-/microscale structures from a (110) Si wafer [118] or other crystalline materials such as GaAs [119, 120] or GaN [121]. However, the fabrication of suspended freestanding patterns by etching sacrificial layer (Fig. 5.14b) is more frequently used because of its straightforward process.

5.3.3 Transfer Printing of the Nano-/Microstructures on Flexible Substrates

After the generation of releasable blocks by using appropriate processes on donor substrates, they can be readily transfer-printed onto the desired target substrates using "stamps" as a temporary holder. Elastomeric polydimethylsiloxane (PDMS) is usually used as the stamp to hold the nano-/microscale patterns by virtue of the soft and conformal contact it affords; however, a thermally released type of tape or a similar type of adhesive tape can also be used as the temporary holder.

Typical TP steps are shown in Fig. 5.15. First, the stamp is attached to the donor wafer, where the nano-/microscale blocks are generated (step 1). Releasing the stamp at a fast rate causes the nano-/microscale patterns to adhere to the stamp and then detach from the donor by the fracturing of the anchor parts (step 2). Making contact with this stamp is then done by attaching it to a desired target substrate that is usually coated with an adhesive layer (step 3), after which the layer is peeled off the stamp, leaving the nano-/microscale patterns on the target substrate (step 4), completing the TP process.

The basic mechanism of the TP is the control of the adhesion between the stamp and releasable patterns at the donor substrate as well as the control of the adhesion between the pattern and the target substrate. The degree of adhesion between the patterns and the elastomeric stamp is controllable by the kinetic control of the peeling rate of the stamp due to the viscoelastic behavior of the elastomer [122, 123]. A fast peeling rate of the stamp generates sufficient adhesion energy between the stamp and the solid objects used in the process. With the fast peeling off of the stamp during the TP processes, nano-/microscale patterns effectively adhere to the stamp and are therefore used to release the patterns from the donor. On the other hand, a slow peeling rate causes insufficient adhesion energy between the patterns and the stamp. Therefore, the slow peeling of the stamp cannot detach the nano-/microscale patterns from the donor. However, the slow peeling of the stamp (which is already *inked* with the nano-/microscale patterns) from an arbitrary surface shows a tendency to leave the patterns on the surface, onto which those patterns come into contact. During the TP process, the slow stamp peel-off process can be utilized to transfer the patterns from the stamp to an arbitrary target substrate, even without adhesives. However, if the target substrate and the transferred patterns are not in direct contact with each other, thin adhesive layers are usually employed for a reliable, complete transfer.

General features of the TP process are described below. The stamps themselves or other adhesive objects can act as desired target substrates to carry the detached

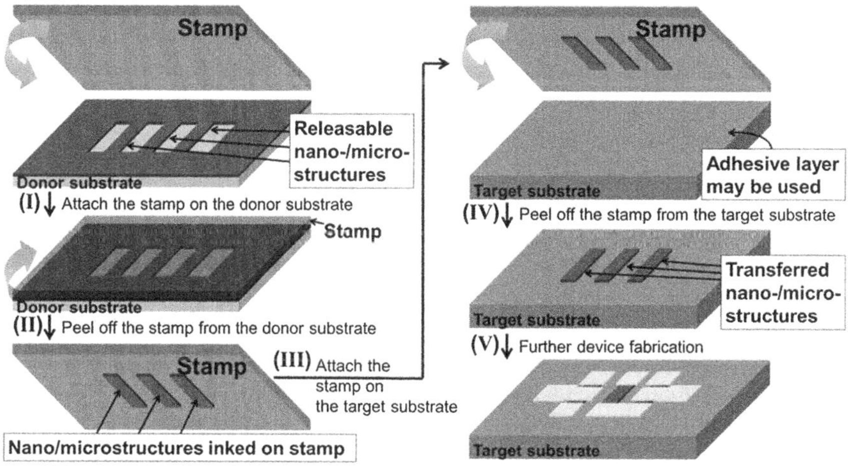

Fig. 5.15 Schematics of the typical transfer steps of nano-/microscale structures from the donor substrate onto the target substrate

blocks from the donor substrates. In addition, because the transferrable nano-/microstructures are fabricated via the traditional top-down procedures on the donor wafer, ordered arrays of the nano-/microstructures can be maintained after the transfer printing process. Moreover, TP can be designed to transfer the patterns selectively from the donor using a patterned PDMS stamp. Hence, an area-expanded printed array can be realized by the multiple repetition of the TP process on a single target substrate. The technique offers an efficient means of implementing the large-area flexible macro-electronics [124].

After the TP process, further fabrication processes can be applied onto a target substrate on which transfer-printed patterns exist (step 5 in Fig. 5.15). Due to substrate limitations, fabrication processes that are compatible only with a certain type of target substrate are available for use. The traditional deposition and lithography processes must be used carefully and only after considering the properties of the target substrates. Processes which are not feasible due to substrate limitations such as a high-temperature process or a chemical treatment should be conducted before the transfer process.

With active channel materials with transfer-printed semiconductor nano-/microscale wires/ribbons on plastic, bendable high-performance single-crystal TFTs have been demonstrated on flexible substrates [114–133]. Low-temperature-processed dielectrics were commonly used as the gate insulator, and the electrodes were formed using conventional metal deposition, lithography, and etching processes [114–126, 128–133]. The observed electrical performances of single-crystal semiconductor TFTs were comparable with conventional FETs with the same channel materials. TFTs could be fabricated with high-temperature processes such as Si doping [125], annealing to make ohmic contacts for GaAs [126], and thermal oxidation for the gate insulator [127] by conducting these processes on donor wafers before transferring onto a plastic substrate. In this manner, flexible substrates are not directly exposed to thermal processes that exceed the substrate glass-transition temperature. TFTs fabricated on plastic materials typically exhibit excellent performance.

The integration of single-crystal semiconductor TFTs on plastic substrate enables bendable high-performance integrated circuits such as inverters, ring oscillators, differential amplifiers, or logic gates such as NOR or NAND [128–133]. Bendable complementary MOS (CMOS) integrated circuits and even conventional wafer-based CMOS chips [130] have been realized by the transfer printing process on both p-type and n-type semiconductor ribbons on a plastic substrate. In a similar manner, dissimilar materials could be transferred sequentially and combined to form hybrid integrated electronic systems [131]. The mechanical bendability of TFTs and ICs fabricated with single-crystal semiconductor nano-/microstructures was evaluated, and satisfactory mechanical stability was confirmed [125, 126, 129–131]. Flexible system-level high-performance integrated electronics fabricated by the TP technique can widen the range of applications of flexible electronics such as the bio-integrated applications for cardiac electrophysiological, brain activity mapping, and other important applications [132, 133].

Other electronic systems such as photovoltaics [134, 135], optoelectronics [135], and light-emitting diodes [136] can also be implemented on flexible plastic materials with high-performance levels via the TP technique. Materials that require high-temperature annealing can be formed on plastic by transferring the pre-annealed layer from the donor wafers to a flexible plastic material. Piezoelectric materials that require high-temperature-driven crystallization are typical examples [137, 138].

5.3.4 Transfer Printing of the Nano-/Microscale Devices on Diverse Substrates

Stretchable structures, more advanced forms beyond flexibility, were demonstrated with the adaptation of thin, "wavy" structures that behave as accordion bellows [139, 140]. Wavy structures were realized by transferring thin nano-/microstructures to *pre-strained* PDMS substrates, where the strain was relieved after the transfer. Stretchable functional devices (active semiconductors with other materials such as dielectrics and electrodes for transistor functions) were fabricated by a dual-transfer process, as illustrated in Fig. 5.16. In brief, the TP was utilized twice. The first instance served to transfer the active materials from a conventional semiconductor wafer, while the second was utilized to transfer fully fabricated functional devices from the second donor to the targeted diverse substrate. Functional devices or integrated-circuit-level systems became stretchable using the dual-transfer process onto *pre-strained* PDMS substrates [139–142]. On the other

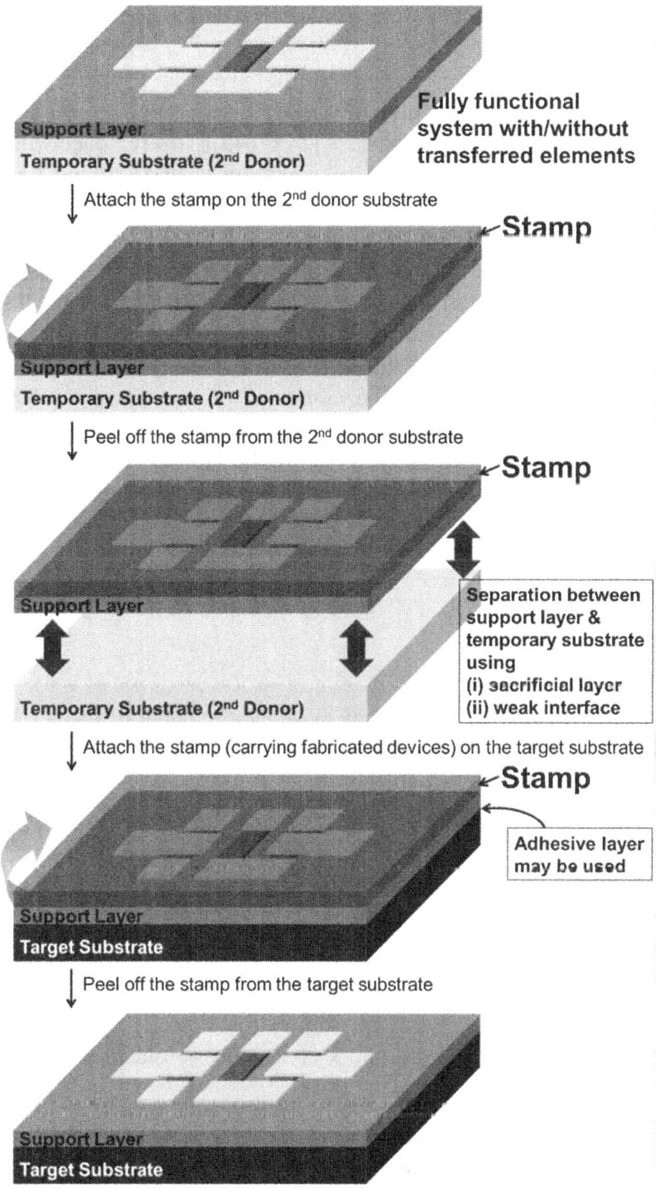

Fig. 5.16 Schematic illustrations of the fabrication procedures for the transfer of nano-/microscale functional devices onto diverse substrates

hand, a non-coplanar "mesh" design for interconnects can provide a wide degree of freedom to enhance the degree of endurance against extreme mechanical deformation of the target substrate [142].

High-performance stretchable electronics realized by the TP technique with a proper wavy/mesh design can be a driving force for developing next-generation electronics. Additionally, considerable efforts have been expended to realize the electronic devices on unconventional substrates in order to make use of the attractive properties of various substrates [143–145]. To keep pace with this trend, the TP technique has been applied to make functional systems on numerous unconventional substrates such as paper, foil, vinyl, fabric, and bio-dissolvable substrates [127, 146–150].

Moreover, functional devices on or within a support layer can be fabricated on diverse substrates by transferring not only the functional device blocks alone but also the entire support layer, as illustrated in Fig. 5.16. Initially, any types of devices, including the devices made up of nanowires processed by chemically synthetic bottom-up approach, can be fabricated on a temporary substrate. Separation between the support layer and the temporary substrate can be accomplished by either etching the sacrificial layer to release the support layer, as is usually done in the dual-transfer process or by using the weak adhesion between the support layer and the temporary substrate [151, 152].

5.4 Problems

5.1 In conventional optical lithography systems, the depth of focus (DOF) is also an important parameter. For a given object distance, the DOF is the range of the image distance for which the object is in focus. Accordingly, the DOF restricts the thickness of the photoresist and the depth of the underlying topography on the wafer. Explain a method that can be used to obtain high-resolution patterns while maintaining the uniformity in terms of the DOF:

$$D_F = \frac{\kappa_2 \lambda}{\mathrm{NA}^2} \tag{5.3}$$

Here, D_F is the depth of focus, and κ_2 is a process-related coefficient.

5.2 Draw a brief process flow to achieve both positive- and negative-toned patterns based on spacer lithography from the given mask. Assume that there are two hardmask layers: the first and second hardmask layers on the active layer (Fig. P 5.2).

Fig. P 5.2 Schematic of the mask

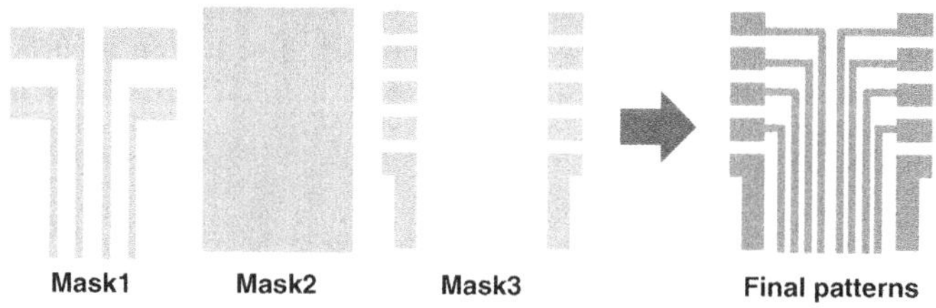

Fig. P 5.3 Schematics of the mask set and final patterns

The positive-toned patterns mean that the spacer layer remains after the first hardmask layer is removed. A negative-toned pattern implies that the spacer layer is removed and the other layers remain.

5.3 Draw a brief process flow to achieve the final patterns from the three masks, as shown in the figure above. Use spacer lithography with conventional optical lithography (i.e., mixed mode patterning) (Fig. P 5.3).

5.4 An advanced MOSFET structures, such as UTB, single-gate SOI, MOSFETs, and (Mug-FETs), have been explored to provide improved controllability of the channel by the gate. Short-channel effects and leakage currents are suitably suppressed as the body thickness becomes thinner. However, the mobility degradation has been reported in thin-body structures. The potentials induced by the fluctuation of the silicon thickness (T_{SOI}) scatter carriers and limit the mobility, as shown in the figure below. The fluctuation of T_{SOI} leads to the spatial fluctuation of the energy level (E_n) (Fig. P 5.4):

$$E_n = \frac{h^2}{8m^* T_{SOI}^2} \tag{5.4}$$

In this equation, h is the Planck constant, and m^* is the effective mass of an electron. This equation was formulated with the assumption that the energy band offset between the conduction band of the oxide and that of the silicon channel was infinite.

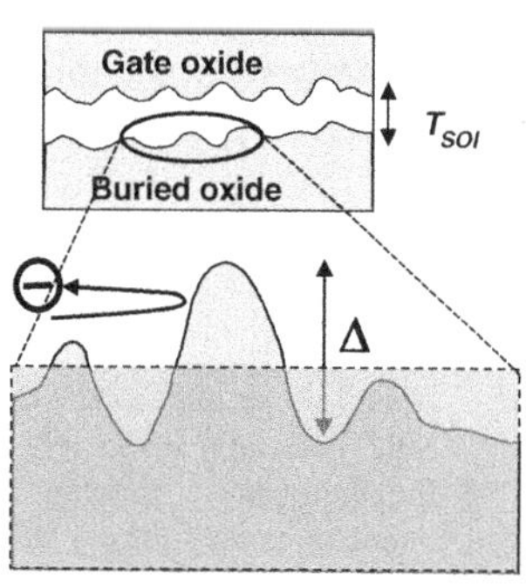

Fig. P 5.4 Schematics of the UTB SOI film. The Si and SiO_2 interface roughness induces the fluctuation of the SOI channel thickness (T_{SOI})

(a) Calculate the potential fluctuation (ΔV) according to the variation of T_{SOI} (Δ).
(b) Derive the relationship between the mobility fluctuation arising from the surface roughness scattering (μ_{SR}) and T_{SOI}. You may need to use the simplified Fermi's golden rule: $\mu_{SR} \propto \left(\frac{1}{\Delta V}\right)^2$ in the time-dependent perturbation theory.

5.5 Strong capillary force cannot be avoided during the fabrication process of a suspended nanowire during the wet etching of a sacrificial layer. In some cases, the magnitude of this force is sufficient to deform and pin the structure to the substrate, which results in device failure. Therefore, the prediction of nanowire stiction is important before the fabrication of the device.

(a) Calculate the maximum allowable length (l_{max}) of the nanowire without stiction from (5.2), where E is 130.2 GPa for Si (100), σ_r is 0.8 GPa for Si/SiO_2, θ_c is 0° for the isopropyl alcohol (IPA) drying, γ_l is 21.7×10^{-3} Nm^{-1} for the IPA drying, h is 100 nm, t ranges from 10 nm to 60 nm, and w is equal to t (aspect ratio (t/w) = 1).
(b) Calculate l_{max} for the aspect ratio (t/w) of 3 as w varies from 10 to 60 nm.
(c) Calculate l_{max} for different sacrificial layers of SiGe, where σ_r is 1.33 GPa for $Si/Si_{0.8}Ge_{0.2}$ and 3.18 GPa for $Si/Si_{0.6}Ge_{0.4}$. The aspect ratio (t/w) is 1.
(d) Deionized wafer (DIW) is commonly used for a final wet cleaning process. Compare and discuss the maximum allowable length (l_{max}) of DIW drying with that of IPA drying.
(e) Describe a strategy for structural optimization of a stiction-immune structure.

5.6 Describe alignment methods of CNTs in both the solution-deposition process and the CVD-based growing process.

5.7 Describe the advantages and disadvantages of bottom-up nanowires compared to top-down nanowires.

5.8 One of the main advantages of the bottom-up method is its ability to produce a hetero-structure, e.g., a Ge/Si core/shell nanowire. Discuss the strength of Ge/Si core/shell nanowire. Refer to the relevant Ref. [96].

5.9 What are the advantages of the transfer process of patterns/devices from a donor substrate (e.g., a single-crystalline silicon wafer or a GaAs wafer) to a target substrate (e.g., flexible substrates or unconventional substrates such as paper)?

References

1. Levinson, H. J., & Arnold, W. H. (1997). Optical lithography. In P. Rai-Choudury (Ed.), *Handbook of microlithography, micromachining, and microfabrication*. Bellingham, WA: SPIE-International Society for Optical Engineering.
2. International Technology Roadmap for Semiconductors. (2011). San Jose, CA: Semiconductor Industry Association.

3. Tritchkov, A., Jeong, S., & Kenyon, C. (2005). Lithography enabling for the 65 nm node gate layer patterning with alternating PSM. *Proceedings of SPIE, 5754*, 215–225.
4. Maenhoudt, M., Versluijs, J., Struyf, H., Olmen, J., & Hove, M. (2005). Double patterning scheme for sub-0.25 k1 single damascene structures at NA = 0.75, $\lambda = 193$ nm. *Proceedings of SPIE, 5754*, 1508–1518.
5. Choi, Y. K., Lindert, N., Xuan, P., Tang, S., & Ha, D., et al. (2001). Sub-20 nm CMOS FinFET Technologies. In *Proceedings of 47th IEEE International Electron Devices Meeting Technical Digest*, (pp. 421–424).
6. Choi, Y. K., King, T. J., & Hu, C. (2002). Nanoscale CMOS spacer FinFET for the terabit era. *IEEE Electron Device Letters, 23*, 25–27.
7. Choi, Y. K., King, T. J., & Hu, C. (2002). A spacer patterning technology for nanoscale CMOS. *IEEE Transactions on Electron Devices, 49*, 436–441.
8. Choi, Y. K., Zhu, J., Grunes, J., Bokor, J., & Somorjai, G. A. (2003). Fabrication of sub-10-nm silicon nanowire arrays by size reduction lithography. *Journal of Physical Chemistry, B107*, 3340–3343.
9. Hua, X., Engelmann, S., Oehrlein, G. S., Jiang, P., Lazzeri, P., et al. (2006). Studies of plasma surface interactions during short time plasma etching of 193 and 248 nm photoresist materials. *Journal of Vacuum Science and Technology B, 24*, 1850–1858.
10. Choi, Y. K., Lee, J. S., Zhu, J., Somorjai, G. A., Lee, L. P., et al. (2003). Sublithographic nanofabrication technology for nanocatalysts and DNA chips. *Journal of Vacuum Science and Technology B, 21*, 2951–2955.
11. Hwang, J., Seo, J., Lee, Y., Park, S., & Leem, J. et al. (2011). A middle-1X nm NAND flash memory cell (M1X-NAND) with highly manufacturable integration technologies. In *Proceedings of 57th IEEE International Electron Devices Meeting Technical Digest*, (pp. 199–202).
12. Bondy, P. K. (1998). Moore's law governs the silicon revolution. *Proceedings of IEEE, 86*, 78–81.
13. Choi, Y. K., Asano, K., Lindert, K., Subramanian, V., & King, T. J., et al. (1999). Ultra-thin body SOI MOSFET for deep-sub-tenth micron era. In *Proceedings of 45th IEEE International Electron Devices Meeting Technical Digest*, (pp. 919–921).
14. Choi, Y. K., Ha, D., & King, T. J., Hu, C. (2001). Ultra-thin body PMOSFETs with selectively deposited Ge source/drain. In *Proceedings of 21th Symposium VLSI Technology*. Digest of Technical Papers, (pp. 19–20).
15. Huang, X., Lee, W. C., Kuo, C., Hisamoto, D., & Chang, L., et al. (1999). Sub 50-nm FinFET: PMOS. In *Proceedings of 45th IEEE International Electron Devices Meeting Technical Digest*, (pp. 67–70).
16. Hisamoto, D., Lee, W. C., Kedzierski, J., Takeuchi, H., Asano, K., et al. (2000). FinFET—a self-aligned double-gate MOSFET scalable to 20 nm. *IEEE Transactions on Electron Devices, 47*, 2320–2325.
17. Yu, B., Chang, L., Ahmed, S., Wang, H., & Bell, S., et al. (2002) FinFET scaling to 10 nm gate length. In *Proceedings of 48th IEEE International Electron Devices Meeting Technical Digest*, (pp. 251–254).
18. Yang, F. L., Lee, D. H., Chen, H. Y., Chang, C. Y., & Liu, S. D., et al. (2004) 5 nm-Gate Nanowire FinFET. In *Proceedings of 24th Symposium VLSI Technology*. Digest of Technical Papers, (pp. 196–197).
19. Frank, D. J., Taur, Y., & Wong, H. S. P. (1998). Generalized scale length for two-dimensional effects in MOSFET's. *IEEE Electron Device Letters, 19*, 385–387.
20. Chang, L., Choi, Y. K., Ha, D., Ranade, P., Xiong, S., et al. (2003). Extremely scaled silicon nano-CMOS devices. *Proceedings of IEEE, 91*, 1860–1873.
21. Doyle, B. S., Datta, S., Doczy, M., Hareland, S., Jin, B., et al. (2003). High performance fully-depleted tri-gate CMOS transistors. *IEEE Electron Device Letters, 24*, 263–265.
22. Park, J. T., Colinge, J. P., & Diaz, C. H. (2001). Pi-gate SOI MOSFET. *IEEE Electron Device Letters, 22*, 405–406.

23. Yang, F. L., Chen, H. Y., Chcn, F. C., Huang, C. C., & Chang, C. Y., et al. (2002). 25 nm CMOS omega FETs. In *Proceedings of 48th IEEE International Electron Devices Meeting Technical Digest*, (pp. 255–258).
24. Colinge, J. P., Gao, M. H., Romano, A., Maes, H., & Claeys, C., (1990). Silicon-on-insulator "gate-all-around device". In *Proceedings of 36th IEEE International Electron Devices Meeting Technical Digest*, (pp. 595–598).
25. Singh, N., Agarwal, A., Bera, L. K., Liow, T. Y., Yang, R., et al. (2006). High-performance fully depleted silicon nanowire (diameter ≤ 5 nm) gate-all-around CMOS devices. *IEEE Electron Device Letters, 27*, 383–386.
26. Suk, S. D., Lee, S. Y., Kim, S. M., Yoon, E. J., & Kim, M. S., et al. (2005). High performance 5 nm radius twin silicon nanowire MOSFET (TSNWFET): Fabrication on bulk Si wafer, characteristics, and reliability. In *Proceedings of 51th IEEE International Electron Devices Meeting Technical Digest*, (pp. 552–555).
27. Liu, H. I., Biegelsen, D. K., Johnson, N. M., Ponce, F. A., & Pease, R. F. W. (1993). Self-limiting oxidation of Si nanowires. *Journal of Vacuum Science and Technology B, 11*, 2532–2537.
28. Kedzierski, J., Bokor, J., & Anderson, E. (1999). Novel method for silicon quantum wire transistor fabrication. *Journal of Vacuum Science and Technology B, 17*, 3244–3247.
29. Tezuka, T., Toyoda, E., Nakaharai, S., Irisawa, T., & Hirashita, N., et al. (2007). Observation of mobility enhancement in strained Si and SiGe tri-gate MOSFETs with multi-nanowire channels trimmed by hydrogen thermal etching. In *Proceedings of 53th IEEE International Electron Devices Meeting Technical Digest*, (pp. 887–890).
30. Bangsaruntip, S., Cohen, G. M., Majumdar, A., Zhang, Y., & Engelmann, S. U., et al. (2009). High performance and highly uniform gate-all-around silicon nanowire MOSFETs with wire size dependent scaling. In *Proceedings of 55th IEEE International Electron Devices Meeting Technical Digest*, (pp. 297–300).
31. Monfray, S., Skotnicki, T., Morand, Y., Descombes, S., & Coronel, P., et al. (2002). 50 nm—gate all around (GAA)—silicon on nothing (SON)—devices: A simple way to co-integration of GAA transistors within bulk MOSFET process. In *Proceedings of 22th Symposium VLSI Technology*. Digest of Technical Papers, (pp. 108–109).
32. Lee, S. Y., Kim, S. M., Yoon, E. J., Oh, C. W., Chung, I., et al. (2003). A novel multibridge-channel MOSFET (MBCFET): Fabrication technologies and characterization. *IEEE Transactions on Nanotechnology, 2*, 253–257.
33. Bera, L. K., Nguyen, H. S., Singh, N., Liow, T. Y., & Huang, D. X., et al. (2006). Three dimensionally stacked SiGe nanowire array and gate-all-around p-MOSFET. In *Proceedings of 52th IEEE International Electron Devices Meeting Technical Digest*, (pp. 551–554).
34. Hubert, A., Nowak, E., Tachi, K., Maffini-Alvaro, V., & Vizioz, C., et al. (2009). A stacked SONOS technology, up to 4 levels and 6 nm crystalline nanowires, with gate-all-around or independent gates (Φ-flash), suitable for full 3D integration. In *Proceedings of 55th IEEE International Electron Devices Meeting Technical Digest*, (pp. 637–640).
35. Ng, R. M. Y., Wang, T., Liu, F., Zuo, X., He, J., et al. (2009). Vertically stacked silicon nanowire transistors fabricated by inductive plasma etching and stress-limited oxidation. *IEEE Electron Device Letters, 30*, 520–522.
36. Pott, V., Moselund, K. E., Bouvet, D., Michielis, L., & Ionescu, A. M. (2008). Fabrication and characterization of gate-all-around silicon nanowires on bulk silicon. *IEEE Transactions on Nanotechnology, 7*, 733–744.
37. Moon, D. I., Choi, S. J., Kim, C. J., Kim J. Y., & Lee, J. S., et al. (2010). Ultimately scaled 20 nm unified-RAM. In *Proceedings of 56th IEEE International Electron Devices Meeting Technical Digest*, (pp. 284–287).
38. Moon, D. I., Choi, S. J., Kim, C. J., Kim, J. Y., Lee, J. S., et al. (2011). Silicon nanowire all-around gate MOSFETs built on a bulk substrate by all plasma-etching routes. *IEEE Electron Device Letters, 32*, 452–454.

39. Mastrangelo, C. H., & Hsu, C. H. (1993). Mechanical stability and adhesion of microstructures under capillary forces—part I: Basic theory. *Journal of Microelectromechanical Systems, 2*, 33–43.
40. Kovacs, G. T. A. (1998). *Micromachined transducers sourcebook*. NewYork: McGraw-Hill.
41. Han, J. W., Moon, D. I., & Choi, Y. K. (2009). High aspect ratio silicon nanowire for stiction immune gate-all-around MOSFETs. *IEEE Electron Device Letters, 30*, 864–866.
42. Okhonin, S., Nagoga, M., Sallese, J. M., & Fazan, P. (2002). A capacitorless 1T-DRAM cell. *IEEE Electron Device Letters, 23*, 85–87.
43. Sakui, K., Hasegawa, T., Fuse, T., Watanabe, S., Ohuchi, K., et al. (1989). A new static memory cell based on the reverse base current effect of bipolar transistors. *IEEE Transactions on Electron Devices, 36*, 1215–1217.
44. Hetherington, D. L., Klem, J. F., & Weaver, H. T. (1992). An integrated GaAs n-p-n-p thyristor/JFET memory cell exhibiting nondestructive read. *IEEE Electron Device Letters, 13*, 476–478.
45. Fossum, E. R. (1997). CMOS image sensors: Electronic camera-on-a-chip. *IEEE Transactions on Electron Devices, 44*, 1689–1698.
46. Tada, K., & Okada, Y. (1986). Bipolar transistor carrier-injected optical modular/switch: Proposal and analysis. *IEEE Electron Device Letters, 7*, 605–606.
47. Jaecklin, A. A. (1982). Turn-on phenomena in optically and electrically fired thyristors. *IEEE Transactions on Electron Devices, 29*, 1552–1560.
48. Han, J. W., & Choi, Y. K., (2010). Bistable resistor—gateless silicon nanowire memory. In *Proceedings of 30th Symposium VLSI Technology*. Digest of Technical Papers, (pp. 171–172).
49. Han, J. W., & Choi, Y. K. (2010). Biristor-bistable resistor based on a silicon nanowire. *IEEE Electron Device Letters, 31*, 797–799.
50. Moon, D. I., Choi, S. J., Kim, J. Y., Ko, S. W., & Kim, M. S., et al. (2012). Highly endurable floating body cell memory: Vertical biristor. In *Proceedings of 58th IEEE International Electron Devices Meeting Technical Digest*, (pp. 749–752).
51. Chen, C. E. D., Matloubian, M., Sundaresan, R., Mao, B. Y., Wei, C. C., et al. (1988). Single-transistor latch in SOI MOSFETs. *IEEE Electron Device Letters, 9*, 636–639.
52. Reisch, M. (1992). On bistable behavior and open-base breakdown of bipolar transistors in the avalanche regime—modeling and applications. *IEEE Transactions on Electron Devices, 39*, 1398–1409.
53. Jang, W. W., Lee, J. O., Yang, H. H., & Yoon, J. B. (2008). Mechanically operated random access memory (MORAM) based on an electrostatic microswitch for nonvolatile memory applications. *IEEE Transactions on Electron Devices, 55*, 2785–2789.
54. Lee, J. O., Kim, M. W., Ko, S. D., Kang, H. O., & Bae, W. H., et al. (2009). 3-terminal nanoelectromechanical switching device in insulating liquid media for low voltage operation and reliability improvement. In *Proceedings of 55th IEEE International Electron Devices Meeting Technical Digest*, (pp. 621–624).
55. Han, J. W., Ahn, J. H., Kim, M. W., Yoon, J. B., & Choi, Y. K. (2009) Monolithic integration of NEMS-CMOS with a fin flip-flop actuated channel transistor (FinFACT). In *Proceedings of 55th IEEE International Electron Devices Meeting Technical Digest*, (pp. 621–624).
56. Kwon, S. G., Han, J. W., & Choi, Y. K. (2010). A bendable-channel FinFET for logic application. *IEEE Electron Device Letters, 31*, 624–626.
57. Han, J. W., Ahn, J. H., Kim, M. W., Lee, J. O., Yoon, J. B., et al. (2010). Nanowire mechanical switch with a built-in diode. *Small, 6*, 1197–1200.
58. Han, J. W., Ahn, J. H., & Choi, Y. K. (2010). FinFACT—fin Flip-FlopActuated channel transistor. *IEEE Electron Device Letters, 31*, 764–766.
59. Choi, S. J., Ahn, J. H., Han, J. W., Seol, M. L., Moon, D. I., et al. (2011). Transformable functional nanoscale building blocks with wafer-scale silicon nanowires. *Nano Letters, 11*, 854–859.

60. Baek, D. J., Choi, S. J., Ahn, J. H., Kim, J. Y., & Choi, Y. K. (2012). Addressable nanowire field-effect-transistor biosensors with local backgates. *IEEE Transactions on Electron Devices, 59*, 2507–2511.
61. Ebbesen, T. W., & Ajayan, P. M. (1992). Large-scale synthesis of carbon nanotubes. *Nature, 358*, 220–222.
62. Guo, T., Nikolaev, P., Thess, A., Colbert, D. T., & Smalley, R. E. (1995). Catalytic growth of single-walled nanotubes by laser vaporization. *Chemical Physics Letters, 234*, 49–54.
63. Li, W. Z., Xie, S. S., Qian, L. X., Chang, B. H., Zou, B. S., et al. (1996). Large-scale synthesis of aligned carbon nanotubes. *Science, 274*, 1701–1703.
64. Chen, X. Q., Saito, T., Yamada, H., & Matsushige, K. (2001). Aligning single-wall carbon nanotubes with an alternating-current electric field. *Applied Physics Letters, 78*, 3714.
65. Xiong, X., Jaberansari, L., Hahm, M. G., Busnaina, A., & Jung, Y. J. (2007). Building highly organized single-walled-carbon-nanotube networks using template-guided fluidic assembly. *Small, 3*, 2006–2010.
66. Li, X., Zhang, L., Wang, X., Shimoyama, I., Sun, X., et al. (2007). Langmuir-blodgett assembly of densely aligned single-walled carbon nanotubes from bulk materials. *Journal of the American Chemical Society, 129*, 4890–4891.
67. Krupke, R., Hennrich, F., Löhneysen, H., & Kappes, M. M. (2003). Separation of metallic from semiconducting single-walled carbon nanotubes. *Science, 301*, 344–347.
68. Zheng, M., Jagota, A., Semke, E. D., Diner, B. A., McLean, R. S., et al. (2003). DNA-assisted dispersion and separation of carbon nanotubes. *Nature Materials, 2*, 338–342.
69. Zheng, M., Jagota, A., Strano, M. S., Santos, A. P., Barone, P., et al. (2003). Structure-based carbon nanotube sorting by sequence-dependent DNA assembly. *Science, 302*, 1545–1548.
70. Arnold, M. S., Green, A. A., Hulvat, J. F., Stupp, S. I., & Hersam, M. C. (2006). Sorting carbon nanotubes by electronic structure using density differentiation. *Nature Nanotechnology, 2*, 60–65.
71. Zhang, Y., Chang, A., Cao, J., Wang, Q., Kim, W., et al. (2001). Electric-field-directed growth of aligned single-walled carbon nanotubes. *Applied Physics Letters, 79*, 3155.
72. Joselevich, E., & Lieber, C. M. (2002). Vectorial growth of metallic and semiconducting single-wall carbon nanotubes. *Nano Letters, 2*, 1137–1141.
73. Ismach, A., Segev, L., Wachtel, E., & Joselevich, E. (2004). Atomic-step-templated formation of single wall carbon nanotube patterns. *Angewandte Chemie International Edition, 116*, 6266–6269.
74. Kocabas, C., Hur, S. H., Gaur, A., Meitl, M. A., & Shim, M., et al (2005). Guided growth of large-scale, horizontally aligned arrays of single-walled carbon nanotubes and their use in thin-film transistors. *Small, 1*, 1110–1116.
75. Ismach, A., Kantorovich, D., & Joselevich, E. (2005). Carbon nanotube graphoepitaxy: highly oriented growth by faceted nanosteps. *Journal of the American Chemical Society, 127*, 11554–11555.
76. Han, S., Liu, X., & Zhou, C. (2005). Template-free directional growth of single-walled carbon nanotubes on a- and r-plane sapphire. *Journal of the American Chemical Society, 127*, 5294–5295.
77. Kocabas, C., Shim, M., & Rogers, J. A. (2006). Spatially selective guided growth of high-coverage arrays and random networks of single-walled carbon nanotubes and their integration into electronic devices. *Journal of the American Chemical Society, 128*, 4540–4541.
78. Kang, S. J., Kocabas, C., Ozel, T., Shim, M., Pimparkar, N., et al. (2007). High-performance electronics using dense, perfectly aligned arrays of single-walled carbon nanotubes. *Nature Nanotechnology, 2*, 230–236.
79. Zhou, W., Rutherglen, C., & Burke, P. J. (2008). Wafer scale synthesis of dense aligned arrays of single-walled carbon nanotubes. *Nano Research, 1*, 158–165.
80. Patil, N., Lin, A., Myers, E. R., Ryu, K., Badmaev, A., et al. (2009). Wafer-scale growth and transfer of aligned single-walled carbon nanotubes. *IEEE Transactions on Nanotechnology, 8*, 498–504.

81. Collins, P. G., Arnold, M. S., & Avouris, P. (2001). Engineering carbon nanotubes and nanotube circuits using electrical breakdown. *Science, 27*, 706–709.
82. Patil, N., Lin, A., Zhang, J., Wei, H., & Anderson, K., et al. (2009) VMR: VLSI-compatible metallic carbon nanotube removal for imperfection-immune cascaded multi-stage digital logic circuits using carbon nanotube FETs. In *Proceedings of 55th IEEE International Electron Devices Meeting Technical Digest*, (pp. 573–576).
83. Franklin, A. D., & Chen, Z. (2010). Length scaling of carbon nanotube transistors. *Nature Nanotechnology, 5*, 858–862.
84. Star, A., Tu, E., Niemann, J., Gabriel, J. C. P., Joiner, C. S., et al. (2006). Label-free detection of DNA hybridization using carbon nanotube network field-effect transistor. *Proceedings of the National Academy of Sciences of the United States of America, 103*, 921–926.
85. Kong, J., Franklin, N. R., Zhou, C., Ghapline, M. G., Peng, S., et al. (2000). Nanotube molecular wires as chemical sensors. *Science, 287*, 622–625.
86. Rueckes, T., Kim, K., Joselevich, E., Tseng, G. Y., Cheung, C. L., et al. (2000). Carbon nanotube-based nonvolatile random access memory for molecular computing. *Science, 289*, 94–97.
87. Wagner, R. S., & Ellis, W. C. (1964). Vapor-liquid-solid mechanism of single crystal growth. *Applied Physics Letters, 4*, 89.
88. Morales, A. M., & Liber, C. M. (1998). A laser ablation method for the synthesis of crystalline semiconductor nanowires. *Science, 279*, 208–211.
89. Molnar, W., Lugstein, A., Pongratz, P., Auner, N., Bauch, C., et al. (2010). Subeutectic synthesis of epitaxial Si-NWs with diverse catalysts using a novel Si precursor. *Nano Letters, 10*, 3957–3961.
90. Schmidt, V., Wittenmann, J. V., Senz, S., & Cösele, U. (2009). Silicon nanowires: a review on aspects of their growth and their electrical properties. *Advanced Materials, 21*, 2681–2702.
91. Werner, P., Zakharov, N. D., Gerth, G., Schubert, L., & Cösele, U. (2009). On the formation of Si nanowire by molecular beam epitaxy. *International Journal of Materials Research, 97*, 1008–1015.
92. Schubert, L., Werner, P., Zakaharov, N. D., Gerth, G., Kolb, F. M., et al. (2004). Silicon nanowhiskers grown on <111> Si substrates by molecular-beam epitaxy. *Applied Physics Letter, 84*, 4968.
93. Cui, Y., Duan, X., Hu, J., & Liber, C. M. (2000). Doping and electrical transport in silicon nanowires. *Journal of Physics Chemistry B, 104*, 5213–5216.
94. Yang, C., Zhong, Z., & Lieber, C. M. (2005). Encoding electronic properties by synthesis of axial modulation-doped silicon nanowires. *Science, 310*, 1304–1307.
95. Tian, B., Zheng, X., Kempa, T. J., Fang, Y., Yu, N., et al. (2007). Coaxial silicon nanowires as solar cells and nanoelectronic power sources. *Nature, 449*, 885–890.
96. Xiang, J., Lu, W., Hu, Y., Yan, H., & Lieber, C. M. (2006). Ge/Si nanowire heterostructures as high-performance field-effect transistors. *Nature, 441*, 489–493.
97. Tian, B., Xie, P., Kempa, T. J., Bell, D. C., & Lieber, C. M. (2009). Single-crystalline kinked semiconductor nanowire superstructures. *Nature Nanotechnology, 4*, 824–829.
98. Duan, X., Huang, Y., Cui, Y., Wang, J., & Lieber, C. M. (2001). Indium phosphide nanowires as building blocks for nanoscale electronic and optoelectronic devices. *Nature, 409*, 66–69.
99. Huang, Y., Duan, X., Wei, Q., & Lieber, C. M. (2001). Directed assembly of one-dimensional nanostructures into functional networks. *Science, 26*, 630–633.
100. Whang, D., Jin, S., Wu, Y., & Lieber, C. M. (2003). Large-scale hierarchical organization of nanowire arrays for integrated nanosystems. *Nano Letters, 3*, 1255–1259.
101. Javey, A., Nam, S., Friedman, R. S., Yan, H., & Lieber, C. M. (2007). Layer-by-layer assembly of nanowires for three-dimensional, multifunctional electronics. *Nano Letters, 7*, 773–777.

102. Goldberger, J., Hochbaum, A. I., Fan, R., & Yang, P. (2006). Silicon vertically integrated nanowire field effect transistors. *Nano Letters, 6*, 973–977.
103. Cui, Y., Wei, Q., Park, H., & Lieber, C. M. (2001). Nanowire nanosensors for highly sensitive and selective detection of biological and chemical species. *Science, 293*, 1289–1292.
104. Mårtensson, T., Svensson, C. P. T., Wacaser, B. A., Larsson, M. W., & Seifert, W. (2004). Epitaxial III–V nanowires on silicon. *Nano Letters, 4*, 1987–1990.
105. Fang, S. F., Adomi, K., Iyer, S., Morkoç, H., Zabel, H., et al. (1990). Gallium arsenide and other compound semiconductors on silicon. *Applied Physics Letters, 68*, R31.
106. Mohan, P., Motohisa, J., & Fukui, T. (2005). Controlled growth of highly uniform, axial/radial direction-defined, individually addressable InP nanowire arrays. *Nanotechnology, 16*, 2903–2907.
107. Duan, X., & Lieber, C. M. (2000). General synthesis of compound semiconductor nanowires. *Advanced Materials, 12*, 298–302.
108. Chuang, L. C., Moewe, M., Chase, C., Kobayashi, N. P., Chang-Hasnain, C., et al. (2007). Critical diameter for III-V nanowires grown on lattice-mismatched substrates. *Applied Physics Letters, 90*, 043115.
109. Gudiksen, M. S., Lauhon, L. J., Wang, J., Smith, D. C., & Lieber, C. M. (2002). Growth of nanowire superlattice structures for nanoscale photonics and electronics. *Nature, 415*, 617–620.
110. Jiang, X., Qihua, Xiong, Nam, S., Qian, F., Li, Y., et al. (2007). InAs/InP radial nanowire heterostructures as high electron mobility devices. *Nano Letters, 7*, 3214–3218.
111. Sun, Y., & Rogers, J. A. (2007). Inorganic semiconductors for flexible electronics. *Advanced Materials, 19*, 1897–1916.
112. Baca, A. J., Ahn, J. H., Sun, Y., Meitl, M. A., Menard, E., et al. (2008). Semiconductor wires and ribbons for high-performance flexible electronics. *Angewandte Chemie International Edition, 47*, 5524–5542.
113. Carlson, A., Bowen, A. M., Huang, Y., Nuzzo, R. G., & Rogers, J. A. (2012). Transfer printing techniques for materials assembly and micro/nanodevice fabrication. *Advanced Materials, 24*, 5284–5318.
114. Menard, E., Lee, K. J., Khang, D. Y., Rogers, J. A., & Nuzzo, R. G. (2004). A printable form of silicon for high performance thin film transistors on plastic substrates. *Applied Physics Letters, 84*, 5398–5400.
115. Mack, S., Meitl, M., Baca, A., Zhu, Z. T., & Rogers, J. A. (2006). Mechanically flexible thin-film transistors that use ultrathin ribbons of silicon derived from bulk wafers. *Applied Physics Letters, 88*, 213101.
116. Ko, H. C., Baca, A., & Rogers, J. A. (2006). Bulk quantities of single-crystal silicon micro/nanoribbons generated from bulk wafers. *Nano Letters, 6*, 2318–2324.
117. Baca, A. J., Meitl, M. A., Ko, H. C., Mack, S., Kim, H. S., et al. (2007). Printable single-crystal silicon micro/nanoscale ribbons, platelets and bars generated from bulk wafers. *Advanced Functional Materials, 17*, 3051–3062.
118. Lee, K. J., Ahn, H., Motala, M. J., Nuzzo, R. G., Menard, E., et al. (2010). Fabrication of microstructured silicon (μs-Si) from a bulk Si wafer and its use in the printing of high-performance thin-film transistors on plastic substrates. *Journal of Micromechanics and Microengineering, 20*, 075018.
119. Sun, Y., & Rogers, J. A. (2004). Fabricating semiconductor nano/microwires and transfer printing ordered arrays of them onto plastic substrates. *Nano Letters, 4*, 1953–1959.
120. Sun, Y., Khang, D. Y., Hurley, K., Nuzzo, R. G., & Rogers, J. A. (2005). Photolithographic route to the fabrication of micro/nanowires of III-V semiconductors. *Advanced Functional Materials, 15*, 30–40.
121. Lee, K., Lee, J., Hwang, H., Reitmeier, Z., Davis, R. F., et al. (2005). A printable form of single crystal gallium nitride for flexible optoelectronic systems. *Small, 1*, 1164–1168.
122. Meitl, M. A., Zhu, Z. T., Kumar, V., Lee, K. J., Feng, X., et al. (2006). Transfer printing by kinetic control of adhesion to an elastomeric stamp. *Natural Material, 5*, 33–38.

123. Feng, X., Meitl, M. A., Bowen, A. M., Huang, Y., Nuzzo, R. G., et al. (2007). Competing fracture in kinetically controlled transfer printing. *Langmuir, 23*, 12555–12560.
124. Lee, K., Motala, M. J., Meitl, M. A., Childs, W. R., Menard, E., et al. (2005). Large area, selective transfer of microstructured silicon (us-Si): A printing-based approach to high performance thin film transistors supported on flexible substrates. *Advanced Materials, 17*, 2332–2336.
125. Zhu, Z. T., Menard, E., Hurley, K., Rogers, J. A., & Nuzzo, R. G. (2005). Spin on dopants for high-performance single-crystal silicon transistors on flexible plastic substrates. *Applied Physics Letters, 86*, 133507.
126. Sun, Y., Kim, S., Adesida, I., & Rogers, J. A. (2005). Bendable GaAs metal-semiconductor field effect transistors formed with printed GaAs wire arrays on plastic substrates. *Applied Physics Letter, 87*, 083501.
127. Chung, H. J., Kim, T. I., Kim, H. S., Wells, S. A., Jo, S., et al. (2011). Fabrication of releasable single-crystal silicon-metal oxide field-effect devices and their deterministic assembly on foreign substrates. *Advanced Functional Materials, 21*, 3029–3036.
128. Sun, Y., Kim, H. S., Menard, E., Kim, S., Adesida, I., et al. (2006). Printed arrays of aligned GaAs wires for flexible transistors, diodes and circuits on plastic substrates. *Small, 2*, 1330–1334.
129. Ahn, J. H., Kim, H. S., Menard, E., Lee, K. J., Zhu, Z., et al. (2007). Bendable integrated circuits on plastic substrates by use of printed ribbons of single-crystalline silicon. *Applied Physics Letters, 90*, 213501.
130. Kim, D. H., Ahn, J. H., Kim, H. S., Lee, K. J., Kim, T. H., et al. (2008). Complementary logic gates and ring oscillators on plastic substrates by use of printed ribbons of single-crystalline silicon. *IEEE Electron Device Letters, 29*, 73–76.
131. Ahn, J. H., Kim, H. S., Lee, K. J., Jeon, S., Kang, S. J., et al. (2006). Heterogeneous three dimensional electronics using printed semiconductor nanomaterials. *Science, 314*, 1754–1757.
132. Viventi, J., Kim, D. H., Moss, J. D., Kim, Y. S., Blanco, J. A., et al. (2010). A conformal, bio-interfaced class of silicon electronics for mapping cardiac electrophysiology. *Science Translated Medicine, 2*, 24ra22.
133. Viventi, J., Kim, D. H., Vigeland, L., Frechette, E. S., Blanco, J. A., et al. (2011). Flexible, foldable, actively multiplexed, high-density electrode array for mapping brain activity in vivo. *Natural Neuroscience, 14*, 1599–1605.
134. Yoon, J., Baca, A. J., Park, S. I., Elvikis, P., Geddes, J. B., et al. (2008). Ultrathin silicon solar microcells for semitransparent, mechanically flexible and microconcentrator module designs. *Nature Materials, 7*, 907–915.
135. Yoon, J., Jo, S., Chun, I. S., Jung, I., Kim, H. S., et al. (2010). GaAs photovoltaics and optoelectronics using releasable multilayer epitaxial assemblies. *Nature, 465*, 329–333.
136. Park, S. I., Xiong, Y., Kim, R. H., Elvikis, P., Meitl, M., et al. (2009). Printed assemblies of inorganic light-emitting diodes for deformable and semitransparent displays. *Science, 325*, 977–981.
137. Qi, Y., Jafferis, N. T., Lyons, K, Jr, Lee, C. M., Ahmad, H., et al. (2010). Piezoelectric ribbons printed onto rubber for flexible energy conversion. *Nano Letters, 10*, 524–528.
138. Park, K. I., Xu, S., Liu, Y., Hwang, G. T., Kang, S. J. L., et al. (2010). Piezoelectric BaTiO3 thin film nanogenerator on plastic substrates. *Nano Letters, 10*, 4939–4943.
139. Khang, D. Y., Jiang, H., Huang, Y., & Rogers, J. A. (2006). A stretchable form of single crystal silicon for high performance electronics on rubber substrates. *Science, 311*, 208–212.
140. Sun, Y., Choi, W. M., Jiang, H., Huang, Y. Y., & Rogers, J. A. (2006). Controlled buckling of semiconductor nanoribbons for stretchable electronics. *Nature Nanotechnology, 1*, 201–207.
141. Kim, D. H., Ahn, J. H., Choi, W. M., Kim, H. S., Kim, T. H., et al. (2008). Stretchable and foldable silicon integrated circuits. *Science, 320*, 507–511.
142. Kim, D. H., Song, J., Choi, W. M., Kim, H. S., Kim, R. H., et al. (2008). Materials and noncoplanar mesh designs for integrated circuits with linear elastic responses to extreme

mechanical deformations. *Proceedings of the National Academy of Sciences of the United States of America, 105*, 18675–18680.
143. Irimia-Vladu, M., Głowacki, E. D., Voss, G., Bauer, S., & Sariciftci, N. S. (2012). Green and biodegradable electronics. *Material Today, 15*, 340–346.
144. Berggren, M., Nilsson, D., & Robinson, N. D. (2007). Organic materials for printed electronics. *Nature Material, 6*, 3–5.
145. Tobjörk, D., & Österbacka, R. (2011). Paper electronics. *Advanced Materials, 23*, 1935–1961.
146. Kim, D. H., Kim, Y. S., Wu, J., Liu, Z., Song, J., et al. (2009). Ultrathin silicon circuits with strain-isolation layers and mesh layouts for high-performance electronics on fabric, vinyl, leather, and paper. *Advanced Materials, 21*, 3703–3707.
147. Kim, D. H., Kim, Y. S., Amsden, J., Panilaitis, B., Kaplan, D. L., et al. (2009). Silicon electronics on silk as a path to bioresorbable implantable devices. *Applied Physics Letter, 95*, 133701.
148. Kim, D. H., Lu, N., Ma, R., Kim, Y. S., Kim, R. H., et al. (2011). Epidermal electronics. *Science, 333*, 838–843.
149. Hwang, S. W., Tao, H., Kim, D. H., Cheng, H., Song, J. K., et al. (2012). A physically transient form of silicon electronics. *Science, 337*, 1640–1644.
150. Kim, D. H., Lu, N., Ghaffari, R., Kim, Y. S., Lee, S. P., et al. (2011). Materials for multifunctional balloon catheters with capabilities in cardiac electrophysiological mapping and ablation therapy. *Nature Materials, 10*, 316–323.
151. Lee, C. H., Kim, D. R., & Zheng, X. (2010). Fabricating nanowire devices on diverse substrates by simple transfer-printing methods. *Proceedings of the National Academy of Sciences of the United States of America, 107*, 9950–9955.
152. Lee, C. H., Kim, D. R., & Zheng, X. (2011). Fabrication of nanowire electronics on nonconventional substrates by water-assisted transfer printing method. *Nano Letters, 11*, 3435–3439.

Chapter 6
Characterization of Nanowire Devices Under Electrostatic Discharge Stress Conditions

Wen Liu and Juin J. Liou

Abstract Discussed in this chapter is the electrostatic discharge (ESD) as the pervasive threat from fabrication, packaging to assembly operation of IC. The ESD robustness of the gate-all-around Silicon nanowire FETs and poly-Si nanowire thin-film FETs is characterized and compared with other types of FETs including bulk/SOI FinFETs and MOSFETs. The ESD performance is specified in terms of the figures of merit such as the failure current, trigger voltage, on-state resistance, leakage current, and failure current density, etc. By using these figures of merits, the ESD performance of nanowire FETs is characterized as a function of gate length, channel shape and material, operation modes (bipolar or diode), process variation, and layout topologies. Moreover, the failure mechanism of nanowire FETs subject to ESD stresses is investigated by means of electrical characterization, optical microscopic observation, and failure analysis. Finally, the optimal nanowire structure and design window are proposed in the light of the ESD performance evaluations presented.

Abbreviation

ESD	Electrostatic discharge
IC	Integrated circuits
SOI	Silicon-on-insulator
HBM	Human body model
TLP	Transmission line pulsing
NWTFT	Nanowire thin-film transistor
NW	Nanowire

W. Liu (✉) · J. J. Liou
School of Electrical Engineering and Computer Science, University of Central Florida, Orlando, USA
e-mail: wenliu876@gmail.com

J. J. Liou
e-mail: liou@eecs.ucf.edu

D. M. Kim and Y.-H. Jeong (eds.), *Nanowire Field Effect Transistors: Principles and Applications*, DOI: 10.1007/978-1-4614-8124-9_6,

GAA NWFET	Gate-all-around silicon nanowire field-effect transistor
FOM	Figure of merit
SCS	Semiconductor characterization system

6.1 Introduction

Electrostatic discharge (ESD) is one of the most prevalent threats to electronic components. It is an event that transfers a large amount of charges between two objects with different potentials. ESD occurs throughout the lifespan of integrated circuits (IC), from fabrication, package, to assembly operation, and finally ending at the user's sites. According to several studies conducted in the past two decades, ESD failure percentages as total failure modes range from 10 to 90 % depending on the various device structures and materials [1–5]. As a result, the ability of developing effective on- and off-chip ESD protection structures is a must-have in the design and manufacturing of modern and emerging ICs.

There is no "one size fits all" strategy for the realization of ESD protection solutions. Every technology or each circuit application requires a customized ESD consideration that includes the devices' operating voltage, trigger voltage, leakage current, breakdown voltage, and footprint. The clamping and shunting capabilities of ESD protection elements depend significantly on fabrication processes, circuit topologies, and layout parameters. At the same time, considerable efforts are also needed to meet the ESD turn-on speed requirement while minimizing the layout area and possible adverse impact of protection elements on the core circuit's performance. Designs of ESD protection solutions become increasingly challenging as the CMOS technology continues to scale toward the 32 nm node and beyond, and the thinner gate oxides, shallower junction depths, and smaller channel lengths all make the next-generation IC even more vulnerable to the ESD stresses.

The Si nanowire transistor has been deemed as a promising alternative CMOS technology in the beyond-Moore era due to its superior electrostatic channel control, low standby leakage current, higher carrier mobility, better noise performance, and ease of fabrication in the existing Si processing. However, the extremely scaled dimension, silicon-on-insulator (SOI) structure, and silicided diffusion also lead to localized high current density, high electric field, and heat generation in such a device, making the device highly susceptible to ESD stress. Moreover, multiple parallel nanowires may be needed to achieve a satisfactory on-current level, and unevenly distributed currents among nanowires may further degrade the device's ESD tolerance and raise challenges to the design and commercialization of the nanowire-based electronics.

Recently, the Human Body Model (HBM) ESD robustness of two promising Si nanowire transistors has been investigated for the first time [6–8]. The ESD figures of merit, including failure current (I_{t2}), leakage current (I_{leakage}), trigger voltage (V_{t1}), and on-state resistance (R_{on}) were characterized as a function of major design parameters, such as gate length, nanowire diameter, and nanowire counts, by transmission line pulsing (TLP) system. Failure analysis was performed to investigate the failure mechanism subject to ESD. The ESD robustness of nanowire transistors was compared with that of other advanced CMOS devices including sub-90 nm bulk/SOI MOSFETs and FinFETs. The main findings of these works will be presented and summarized in this chapter.

6.2 ESD Performance of Poly-Si Nanowire Thin-Film Transistor

6.2.1 Device Structure and DC Characteristic

The first device under study is called the poly-Si nanowire thin-film transistor (NWTFT) fabricated by National ChiaoTung University, Taiwan [9]. Such a device offers advantages of simple fabrication flow, reliable contact, low cost, and precise alignment of nanowires [10]. The 3D structure schematics of pre- and post-top gate formation are illustrated in Fig. 6.1. The simplified layout top view and cross-section views along two different nanowire (NW) positions are presented in Fig. 6.2.

The steady-state (DC) transfer characteristics in the logarithm and linear scales of three N-type NWTFTs with the following makeups are shown in Fig. 6.3: nanowires with a rectangular cross section of 60 × 18 nm, three channel lengths of 0.4, 1.0, and 2.0 μm, and a gate oxide thickness of 20 nm.

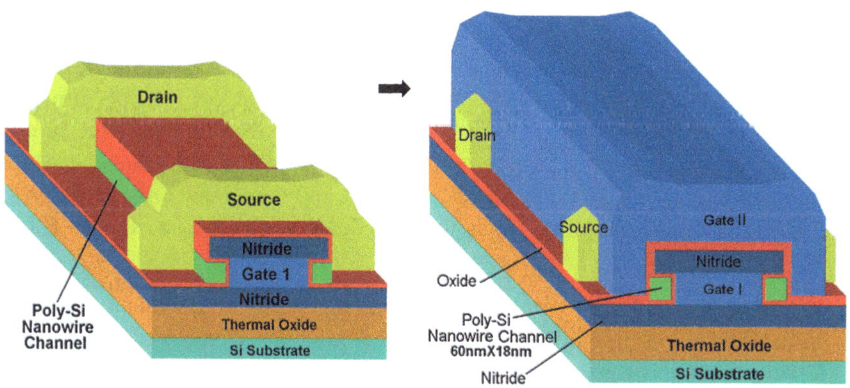

Fig. 6.1 NWTFT before and after top gate (gate II) formation

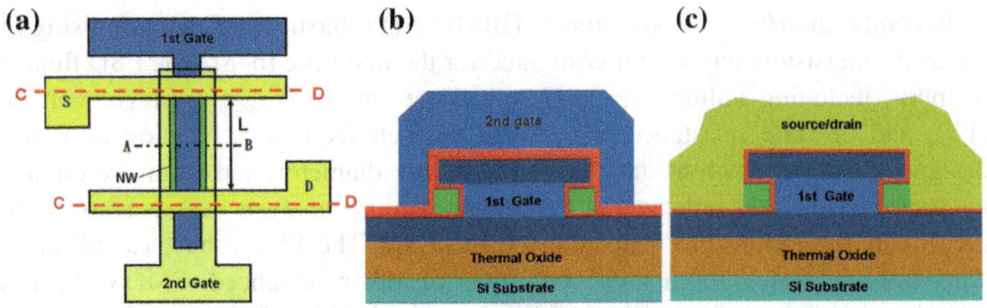

Fig. 6.2 (**a**) Layout top view and cross-section views along black dashed line *A–B* (**b**) and *red dashed line C–D* (**c**)

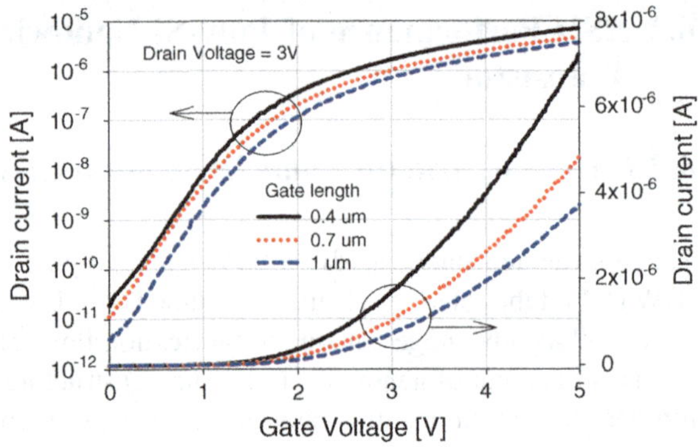

Fig. 6.3 Transfer characteristic of N-type NWTFTs

6.2.2 ESD Performance at a Glance

N-type NWTFTs having various channel numbers (1, 9, 25, 50, 100), channel lengths (0.4, 0.7, 1, 4 μm), nanowire widths (43 and 18 nm), and a fixed nanowire spacing of 500 nm in both the bipolar mode (gate grounded) and diode mode (gate-drain connected) were characterized under the HBM ESD condition. HBM pulses with a 100 ns pulse width and 10 ns rise time were generated using the TLP system. Increasing pulse magnitudes were applied to the devices until hard failure occurs. The TLP I–V curve provides important information for ESD performance evaluation. Such information includes the failure current (I_{t2}) defined as the current at which the device fails, the trigger voltage (V_{t1}) defined as the voltage at which the device starts to conduct considerable current, and the on-state resistance (R_{on}) that is related to the TLP I–V slope after the device is turned on. The TLP I–V results shown in Fig. 6.4 can be summarized as follows:

1. In the bipolar mode, the NWTFTs turn on around 10 to 15 V without snapback;
2. In the diode mode, the NWTFTs trigger at around 0.6 V and exhibit a linear I–V behavior after triggering. Its failure current is about four times higher than that of the bipolar mode;

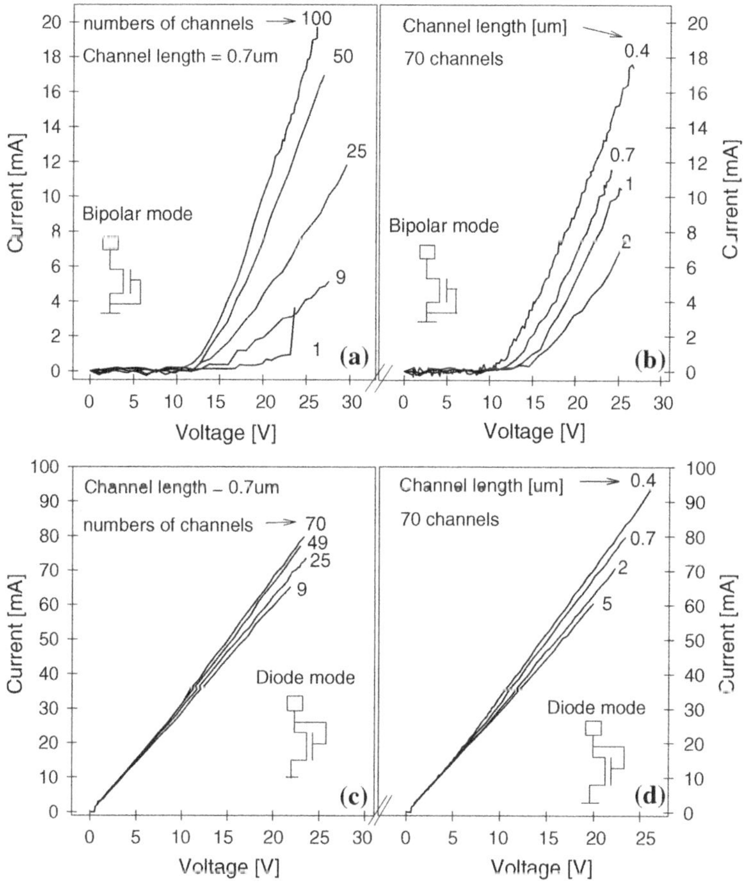

Fig. 6.4 TLP I–V characteristics of N-type NWTFTs in bipolar mode (**a** and **b**) and diode mode (**c** and **d**)

3. In both the bipolar and diode modes, shorter channel devices possess better ESD performance. Also, I_{t2} and R_{on} increase and decrease with increasing channel number, respectively, but they do not scale linearly with the number of channels.

The above-mentioned characteristics stem from the floating body effect in SOI and grain structures in poly-Si material [6]. More discussions will be given later.

6.2.3 Effect of Nanowire Dimension

To further investigate the ESD robustness's dependency on nanowire dimensions, N-type NWTFTs having wide and narrow nanowires ($W = 43$ nm and $W = 18$ nm) were measured and compared in terms of I_{t2} and failure current density J_{t2} as a function of channel length, as shown in Fig. 6.5, where J_{t2} is calculated using the following equation:

$$J_{t2} = \frac{I_{t2}}{\text{nanowire width} \times \text{number of nanowires}} \tag{6.1}$$

According to Fig. 6.5, a shorter channel length leads to higher I_{t2} and J_{t2}. Thanks to the reduced numbers of grains within the shorter nanowires, current conduction takes place deeper in the junction at a lowered source barrier potential [11], thus giving rise to a higher I_{t2}. NWTFT with narrower nanowires exhibits a higher J_{t2} due to the presence of the relatively high resistivity in these narrow channels, a mechanism similar to that of planar GGNMOS imbedded with ballast resistors [12], in which the increased voltage drop across the spreading resistance minimizes the likelihood of filamentation.

The total width of NWTFT does not only depend on the nanowire width, but also on the number of nanowire channels in parallel. As illustrated in Fig. 6.6, a larger number of channels increases the overall I_{t2} but to a lesser extent decreases J_{t2}, a trend confirming that I_{t2} does not scale linearly with the channel number. Devices having narrow nanowires offer a higher J_{t2} as shown in Fig. 6.6b, further demonstrating that a higher percentage of nanowires are conducting current in a narrow nanowire device than that of the wide counterpart. However, as shown in Fig. 6.6c, the narrow nanowire device with 100 channels has a leakage current that is too high for practical ESD protection applications.

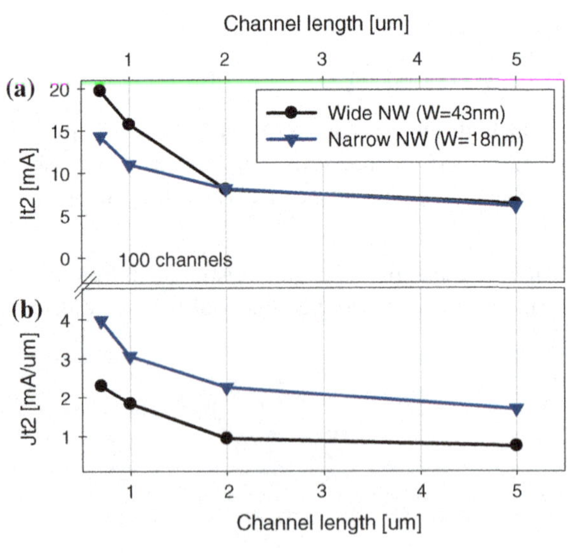

Fig. 6.5 Comparisons of (**a**) I_{t2}, and (**b**) J_{t2} of the bipolar mode NWTFTs having different channel lengths and nanowire widths

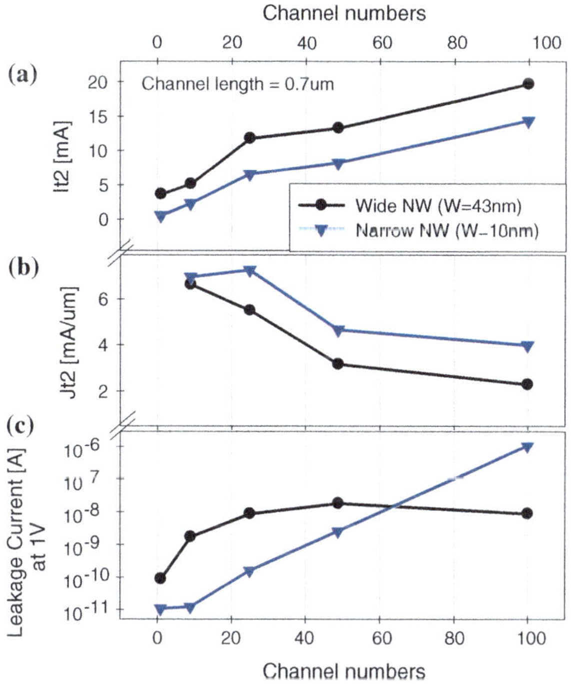

Fig. 6.6 Comparisons of (**a**) I_{t2}, (**b**) J_{t2}, and (**c**) leakage current of N-type NWTFTs having various channel numbers

6.2.4 Effect of Plasma Treatment

Plasma treatment is typically used to passivate the inherent inter-intra-grain-boundary defects located in the poly-Si to reduce performance fluctuation, DC leakage current, and threshold voltage [13]. Therefore, 3 hours' NH_3 plasma treatment at 300 °C was applied to all the NWTFTs [14]. However, ESD failure current of N-type NWTFT does not benefit from the plasma treatment, except for the reduced leakage current. It is believed that the nanowires become more fragile after the plasma treatment, because the hot-carrier endurance and thermal stability degrade after passivation [13]. As the avalanche breakdown and impact ionization are main mechanisms underlying the ESD performance, the plasma-induced degradation reduces the ESD robustness of plasma-treated NWTFTs.

6.2.5 Layout Optimization

The decreasing J_{t2} versus increasing channel number (Fig. 6.6b) suggests that there is a non-uniform current distribution among the multiple channels of NWTFTs. This undesirable effect can be minimized by optimizing the drain and source layout topologies. Figure 6.7 presents NWTFTs having three different drain

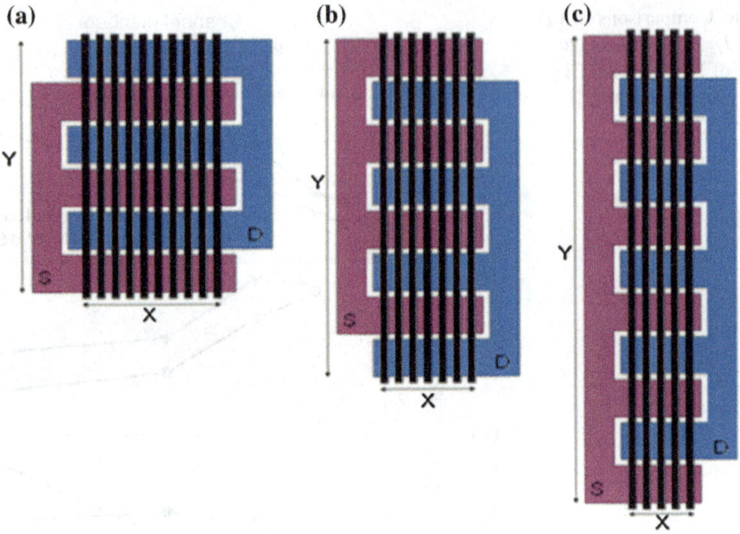

Fig. 6.7 Layout topologies of N-type NWTFTs having total channel numbers of 50 but different drain/source fingers and nanowires in parallel

Table 6.1 Comparison of multiple channel NWTFTs having the different layouts shown in Fig. 6.7

Layout topology	(a)	(b)	(c)
Channel number	50	49	50
Drain/source fingers	5	7	10
Number of nanowire in parallel	10	7	5
Failure current I_{t2} (mA)	12.5	13.3	16.9
Effective layout area (X × Y) (μm^2)	653	606	589
Leakage current (at 1 V) (nA)	3.84	1.81	0.71

and source layouts but the same total channel number of 50, and their TLP measurement results are compared in Table 6.1. In the direction from (a) to (c) in Fig. 6.7, while the drain and source finger number is increased from 5, 7 to 10, the channel number for each drain and source pair is decreased from 10, 7 to 5. Note that the black lines denote the nanowire channels, and the effective layout area is calculated by X × Y.

It is evident that the layout in Fig. 6.7c yields the highest I_{t2}, lowest leakage current, and smallest effective layout area, followed by the layout in Fig. 6.7b, and the layout in Fig. 6.7a renders the worst ESD performance. The layout topology shown in Fig. 6.7c has least number of channels in parallel for each pair of drain and source. Therefore, uneven distribution of current is less likely to occur in such a layout. Moreover, the higher I_{t2} in Fig. 6.7c can be attributed to the longer and less nanowires in parallel that offers a better heat dissipation than other topologies.

Compared to the conventional planar MOSFET, the NWTFT allows for a larger degree of freedom to design the layout for performance optimization and size reduction. In addition to raising the total channel number, increasing the drain and source finger numbers with reduced numbers of nanowires on each drain/source finger is an effective approach to enhance the ESD robustness of multi-channel NWTFTs.

6.3 ESD Performance of Gate-All-Around Silicon Nanowire Field-Effect Transistors

6.3.1 Device Structure and DC Performance

The second type of nanowire FET under study is called the gate-all-around silicon nanowire field-effect transistors (GAA NWFET), fabricated by the Institute of Microelectronics, Singapore. Top-down approach synthesis was implemented to obtain a precise control over the nanowire dimension and spacing. The process starts from nanowire patterning and etching from SOI wafers and further trimmed by thermal oxidation. Poly-Si gate is then deposited and trimmed before drain/source implantation and activation. Both gate and drain/source extension regions went through salicidation process to optimize threshold voltage and reduce contact resistance. More detailed process information can be found in [15, 16].

Off-scale 3D and 2D schematics of the NWFET are shown in Fig. 6.8. The cylinder-shaped nanowires are surrounded by a SiO_2 layer of 5 nm thickness and a NiSi fully silicide metal gate. The number of parallel nanowires in a single transistor ranges from 1 to 1,000 with a fixed spacing of 400 nm. Figure 6.9 shows the DC transfer characteristics (drain voltage = 1.2 V) of the single-nanowire NWFETs having NW diameters (D) of 5 and 10 nm, and gate lengths of 250, 750, and 1,750 nm, in both logarithmic scale (left Y-axis) and linear scale (right Y-axis).

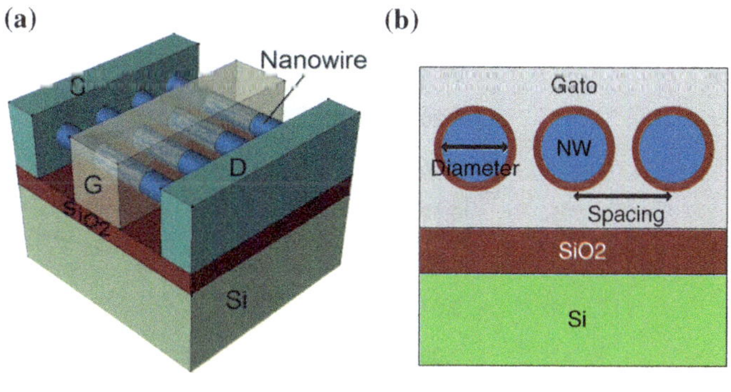

Fig. 6.8 **a** 3D schematic and **b** cross section of a multi-channel NWFET

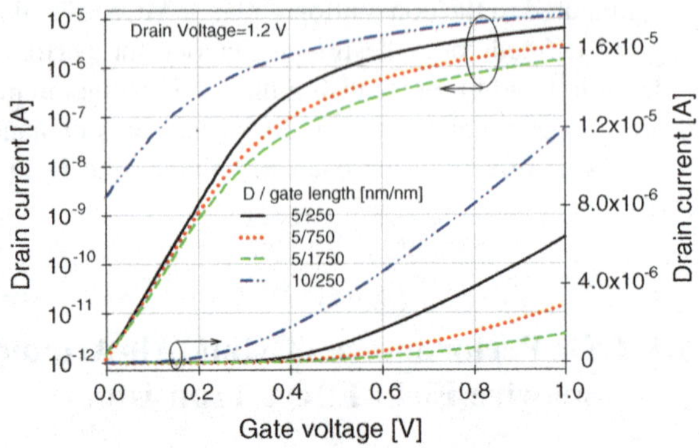

Fig. 6.9 Transfer characteristic of N-type NWFETs

6.3.2 ESD Performance Evaluation

The TLP I–V characteristics of N-type and P-type NWFETs, having various gate lengths, 1,000 nanowires in parallel with diameters of 10 and 5 nm, and operating in the bipolar and diode modes are shown in Fig. 6.10. To provide a fair comparison between the different types of devices, the same TLP system and stressing setup applied to the NWTFT were used for NWFET.

As illustrated in Fig. 6.10, snapback behavior is also absent in the NWFETs due to the floating base of parasitic BJT. The devices turn on at around 3–10 V in the bipolar mode and 0.6 V in the diode mode. Highest I_{t2} is observed at a gate length of 750 nm.

Based on the TLP and DC results, the ESD design window of NWFET can be determined and is shown in Fig. 6.11. The DC operation voltage of NWFET is typically below 1.5 V. A gate oxide of 5 nm should have a transient gate oxide

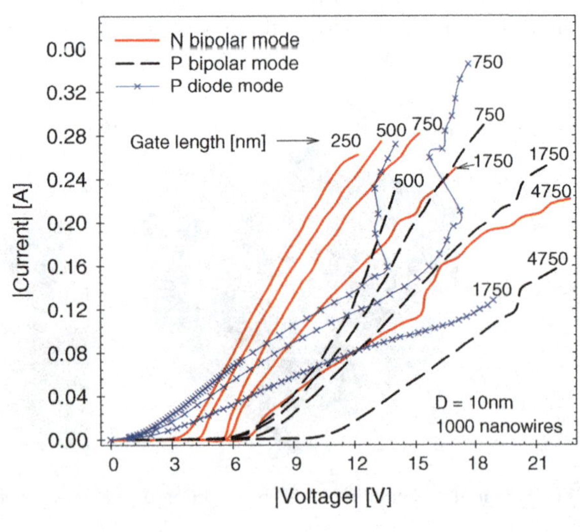

Fig. 6.10 TLP I–V curves for the N-type and P-type NWFETs operating in the bipolar and diode modes

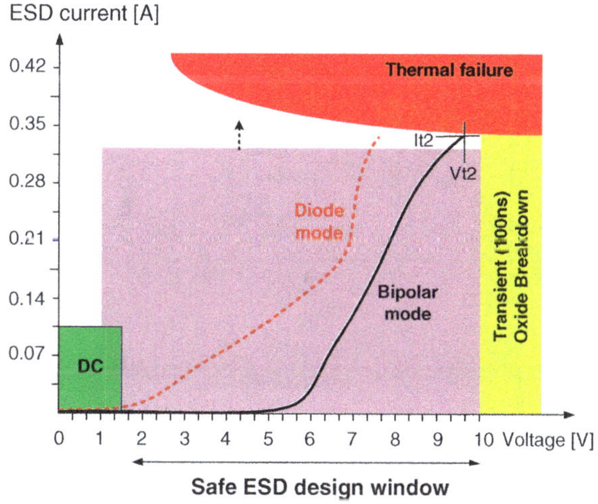

Fig. 6.11 ESD design window of NWFETs

breakdown voltage of about 10 V under the HBM ESD stress [1, 17, 18]. These two voltages constitute the low and upper bounds, respectively, of the ESD window. The bipolar-mode NWFET triggers within the ESD design window, whereas several diode-mode NWFET may need to be connected in series to achieve a higher trigger voltage needed in certain applications. Further improvements are desired to enhance the failure current and reduce the on-state resistance of NWFET.

6.3.3 *Effect of Gate Length*

Gate length is one of the major parameters that significantly impact the NWFET's ESD performances, including trigger voltage (V_{t1}), failure current (I_{t2}), on-state resistance (R_{on}), and leakage current ($I_{leakage}$). In this section, such ESD figures of merit are extracted from TLP I–V curves and post-stress leakage currents for both N-type and P-type NWFETs having nanowire diameters (D) of 5 and 10 nm. The results are summarized and explained as follows.

As shown in Fig. 6.12a, V_{t1} increases with increasing gate length, and N-type devices possess a smaller trigger voltage than P-type ones. The trigger mechanism is determined by the turn on of the parasitic BJT by impact ionization [19]:

$$\beta \times (M - 1) \geq 1 \tag{6.2}$$

where β is the common-emitter current gain and M is the multiplication factor. Since the gate length governs the base width of parasitic BJT, increasing the gate length reduces β. The lower carrier mobility of P-type NWFET also gives rise to a

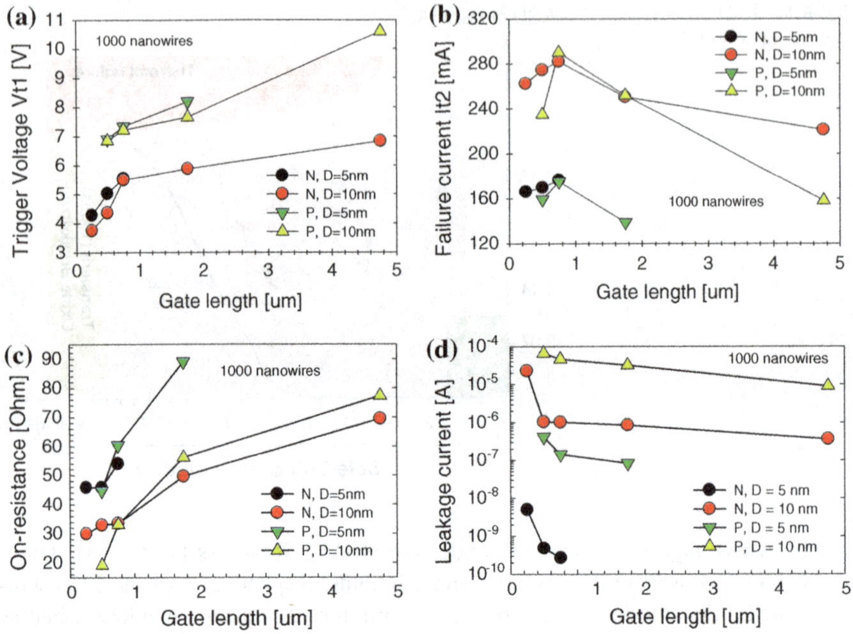

Fig. 6.12 Results of **a** V_{t1}, **b** I_{t2}, **c** R_{on}, and **d** $I_{leakage}$ versus gate length of N-type and P-type NWFETs

lower β, which in turn requires a larger avalanche multiplication factor M, and thus a larger drain voltage, to turn on the BJT.

As can be seen in Fig. 6.12b, I_{t2} increases with increasing gate length initially but drops when the gate length exceeds 750 nm, a trend similar to that of the FinFETs [20]. As explained in Table 6.2, a longer gate can impose both positive and negative impacts on I_{t2}, and the best tradeoff is achieved by using medium gate length so that there is a balance between the heat dissipation and power consumption. As expected, a smaller D gives rise to a lower failure current because a narrow nanowire is more likely to be impaired by localized heating generated from the ESD stress.

Table 6.2 Impacts of gate length on failure current I_{t2}

Why a larger gate length improves I_{t2}?	Why a larger gate length reduces I_{t2}?
Harder to induce drain-to-source filamentation [39]	Higher voltage drop and larger energy consumption [18, 20, 31]
Higher ballistic resistance and better current distribution [31]	Higher degree of dimension fluctuation of the nanowire cross-section [20]
Larger volume associated with the gate improves heat dissipation [18, 20]	More defects [40]

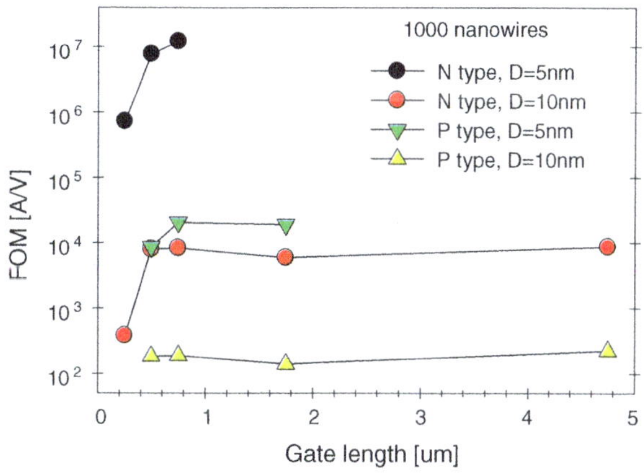

Fig. 6.13 FOM versus gate length of N-type and P-type NWFETs

The on-state resistance R_{on} and leakage current $I_{leakage}$ (measured at 1 V for N type and −1 V for P-type devices) of the same devices are compared in Fig. 6.12c and d, respectively. Narrower nanowire ($D = 5$ nm) and longer gate length lead to a higher R_{on}, but lower $I_{leakage}$, because the resistance of a nanowire is directly proportional to its length and inversely proportional to its cross-section area. For devices having the same gate length and nanowire diameter, N-type NWFETs have a lower $I_{leakage}$ than its P-type counterpart. The P-type devices with a diameter of 10 nm are unsuitable for ESD protection applications due to the relatively high $I_{leakage}$.

The best candidacies for ESD protection applications should possess a high I_{t2}, small R_{on}, and low leakage current. This can be represented by the following figure of merit (FOM):

$$\mathrm{FOM} = \frac{I_{t2}}{R_{on} \times I_{leakage}} \tag{6.3}$$

As indicated in Fig. 6.13, based on the measured data, an N-type NWFET with a medium gate length and narrow nanowire has the highest FOM. Note that the FOM is a simplified evaluation and dominated by high values of $1/I_{leakage}$. A thorough consideration of each ESD parameter is suggested for any specific ESD protection application where a certain parameter may have a higher priority than others.

6.3.4 Channel Number Scaling Effect

Multi-channel NWFETs are often fabricated to offer scalable on-state current without changing the dimension of nanowire. The simplified multi-channel layout is shown in Fig. 6.14, and N-type devices with a 5 nm diameter and 500 nm gate length, 100, 500, and 1,000 nanowires were characterized and shown in Fig. 6.15.

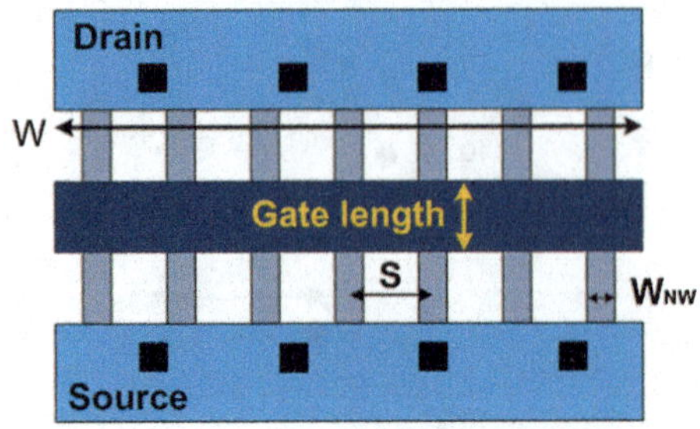

Fig. 6.14 Simplified layout of multi-channel NWFET

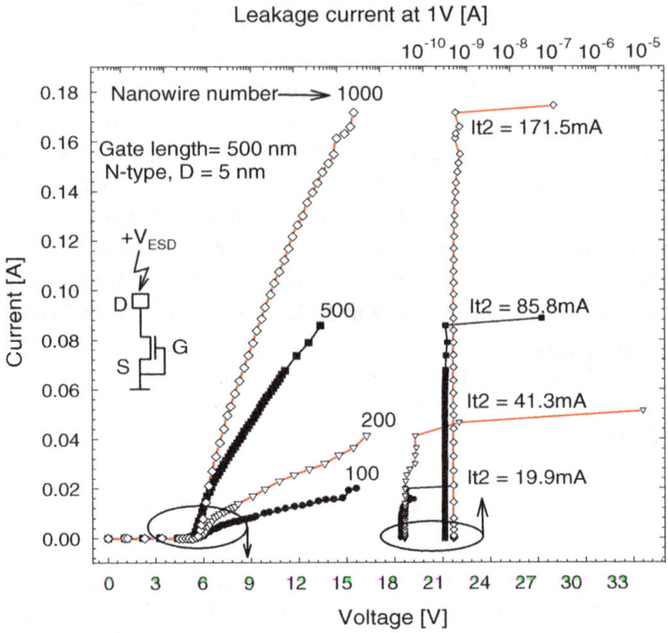

Fig. 6.15 TLP *curves* and leakage currents (measured at 1 V) for the bipolar-mode N-type NWFETs with different nanowire numbers

Here the curves on the right-hand side (top abscissa) are the leakage currents measured after each corresponding I–V points plotted on the left-hand side (bottom abscissa). The results suggest that I_{t2} of these devices scales almost linearly with the number of nanowires, at the cost of increasing the leakage current when the nanowire number goes up. The linearity relationship of nanowire number (N) and I_{t2} can be fitted using

$$I_{t2} = 0.166 \times N + 4.9 \tag{6.4}$$

The dependency of FOM (calculated by Eq. 6.3) on channel number indicates that adding nanowires provides an alternative solution to enhance the FOM of the NWFETs.

6.3.5 Failure Analysis

Different ESD-induced failure mechanisms occurring in conventional MOSFET devices, such as interconnect burn-out [21], metal-contact rupture [21], and junction melting [22], have been studied and are well known. Recent research work on FinFETs revealed a new failure mechanism of crystal structure change/fusing due to the ESD stress [23]. Hence a thorough study and understanding of ESD-induced failures in NWFETs is needed.

P-type NWFETs fabricated at the Institute of Microelectronics, Singapore, were considered in this study [15]. The devices consist of 1,000 nanowires in parallel, each with a diameter of 5 nm and gate length of 250 nm. The devices were stressed with HBM equivalent pulses generated by the Barth 4002 TLP tester with the gate of NWFET floating, a configuration frequently used for ESD protection applications [24, 25]. Keithley semiconductor characterization system (SCS) was also used for measuring the pre- and post-stress DC current–voltage (I–V) characteristics. Several identical NWFETs were characterized. While some variations on the measured data were observed, quite consistent trends were obtained.

Figure 6.16a shows the pulsed I–V curve and DC leakage current (the curve on the left) of a P-type NWFET. The four stressing conditions undertaken, with increasing stress level, are indicated by the N1, N2, N3, and N4 points. N1 is the pre-stress condition, N2 and N3 are the post-stress conditions where the device is still functional, and N4 is the post-stress condition where the pulsed I–V curve ends and device is completely damaged. The device failure is evidenced by the increase in the leakage current by 5 orders of magnitude after N3. For easier description of the experimental procedure, I_{t2} was taken as the current causing the device to fail.

The complete experimental procedure is as follows: (1) at N1, a fresh device was first measured using Keithley 4200-SCS to obtain the pre-stress DC I–V curves; (2) TLP stresses were then applied to the device and gradually increased until the pulsed current reached about 1/3 of I_{t2} (i.e., N2); (3) conducted post-stress DC I–V measurements at N2; (4) applied TLP pulses until the pulsed current reached around 2/3 of I_{t2} (i.e., N3); (5) conducted post-stress DC I–V measurements at N3; (6) applied TLP pulsing again until hard failure occurred (i.e., N4); and (7) conducted post-damage I–V measurements at N4.

According to Fig. 6.16c and d, the saturation current at N2, Idsat(N2), can be up to 11 % higher than that at N1, Idsat(N1), while the gate at N2 still exerts an effective control on the channel conduction. The relatively weak ESD stress at N2 induces avalanche breakdown that leads to electron trapping in the oxide–Si

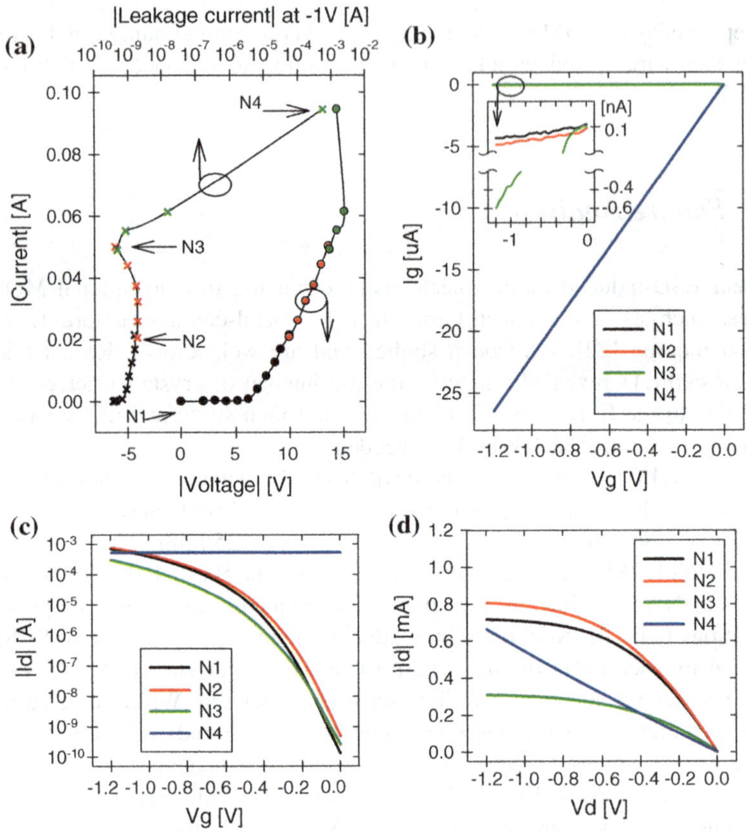

Fig. 6.16 Measured results of a P-type NWFET subject to the TLP stress: **a** I–V (*bottom x-axis*) and the corresponding leakage current (*top x-axis*) curves, **b** gate leakage Ig vs. Vg (drain and source grounded) at N1, N2, N3, and N4, and zoom in of Ig–Vg characteristics at N1, N2, and N3 (*insert*), **c** transfer characteristic Id–Vg with drain voltage Vd = −1.2 V at N1, N2, N3, and N4, and **d** output characteristic Id–Vd with Vg = −1.2 V at N1, N2, N3, and N4

interface and a decrease in the threshold voltage |*V*th| [26]. Therefore, under the same bias conditions, drain current is increased after the device is stressed modestly. The difference in the post-stress Ig–Vg curves at N1 and N2 (Fig. 6.16b) stems from the charge trapping/detrapping in association with the gate oxide defect accumulation and trap-assisted tunneling [27]. At N3, Idsat(N3) reduces to 43 % of Idsat(N1) (Fig. 6.16c and d), after the device is subjected to the relatively severe stress. Minor gate oxide degradation is observed from the Ig–Vg characteristics in the inset of Fig. 6.16b. This observation, together with the increased threshold voltage $|V_{th}|$ and reduced transconductance, suggest the occurrence of partial nanowire damage and modest gate oxide degradation [28–30]. At N4 where the catastrophic failure takes place, the device behaves like a resistor, having linear Ig versus Vg and linear output characteristic as shown in Fig. 6.16b and d. This

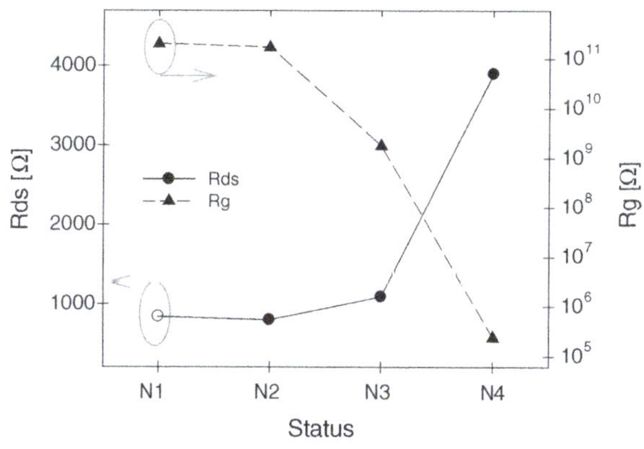

Fig. 6.17 Gate resistance Rg and drain–source resistance Rds at N1, N2, N3, and N4

implies that the current in the NWFET is conducted through the impaired drain–nanowire–source junction and gate oxide.

Calculating the slopes of the Ig–Vg curve (Fig. 6.16b) and the linear region of Id–Vd curve (Fig. 6.16d) yields the gate resistance Rg and drain–source resistance R_{ds}, and the resulting R_{ds} and R_g values at N1, N2, N3, and N4 are plotted in Fig. 6.17.

SEM and TEM were used to analyze the destructive failure at N4. Fig. 6.18 illustrates the TEM cross-section image and SEM surface image of the damaged device, respectively, indicating that the fusing of a large percentage of nanowires as well as the surrounding gate oxide is the main mechanism underlying the ESD-induced failure of the NWFET. The severity of the fusing varies considerably among the damaged areas (see circled areas in Fig. 6.18a and arrow pointing regions in Fig. 6.18b). This is due to the non-uniform distribution of ESD current among the multiple nanowires. It is worth mentioning that no physical defects could be observed at the N2 and N3 status.

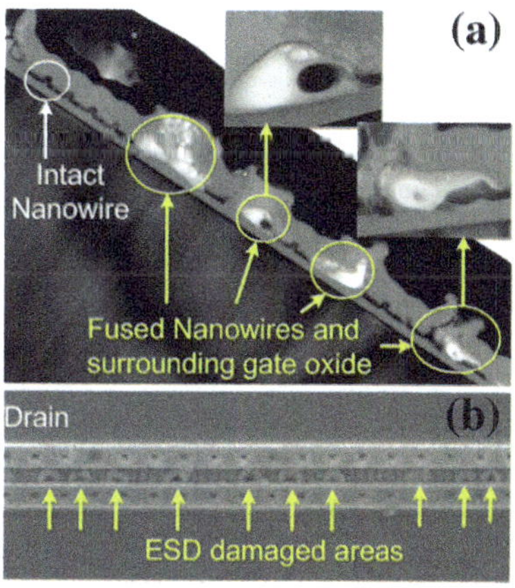

Fig. 6.18 **a** TEM cross-section image and **b** SEM surface image at N4

6.4 Comparison of ESD Performances of CMOS, FinFET, and Nanowire Technologies

6.4.1 GAA Si NWFET versus Poly-Si NWTFT

First of all, the ESD performances of the two different types of nanowire FETs under study are compared. To obtain a consistent assessment, the devices considered have similar dimensions. Because of the different nanowire sizes and geometries, the failure current densities J_{t2}, J_{t2}', and J_{t2}'' normalized by the effective channel width, effective silicon area, and total width, respectively, are also examined (see Table 6.3). The definitions of J_{t2}, J_{t2}', and J_{t2}'' are given below:

$$J_{t2} = \frac{I_{t2}}{\text{nanowire width} \times \text{number of nanowires}} \tag{6.5}$$

$$J_{t2}' = \frac{I_{t2}}{\text{nanowire cross} - \text{section area} \times \text{number of nanowires}} \tag{6.6}$$

$$J_{t2}'' = \frac{I_{t2}}{\text{total layout width}} \tag{6.7}$$

As summarized in Table 6.3, the GAA Si NWFET possesses a higher failure current as well as failure current densities than the poly-Si NWTFT, owing in part to the circular nanowire cross section. Higher electric field and current density exist in the corner regions of the rectangular nanowires within the poly-Si NWTFT, making such a device more susceptible to the ESD stress. The different trigger voltages found in poly-Si NWTFT and Si NWFET result from the different nanowire materials (poly-silicon vs. silicon), gate structures (GAA vs. two-sided gates), and nanowire geometries. Overall, the GAA Si NWFET is a better candidate for ESD protection application thanks to its lower trigger voltage and leakage current, higher failure current, and smaller area.

Table 6.3 ESD performances of N-type GAA Si NWFET and poly-Si NWTFT having similar gate lengths and channel numbers

Devices	GAA Si NWFET		Poly-Si NWTFT
NW Dimension (nm)	$D = 5$	$D = 10$	$W = 18$
Gate length (nm)	750		700
Channel numbers	100		
I_{t2} (mA)	25.2	37.8	14.3
J_{t2} (mA/μm)	50.4	37.8	7.9
J_{t2}' (A/μm^2)	3.21	1.20	0.13
J_{t2}'' (mA/μm)	0.634	0.951	0.73
Trigger voltage (V)	6.2	5.28	12.9
$I_{leakage}$ (at 1 V) (nA)	0.014	2.19	1,040
Layout size (μm^2)	121.8		1,209
R_{on} (Ω)	689.4	377.6	720.2

6.4.2 GAA Si NWFET versus FinFETs and Planar MOSFETs

We now compare the ESD performances of the NWFETs with those of the devices fabricated using other modern technology nodes, such as FinFET, 32, 45, and 65 nm SOI/bulk CMOS technologies [20, 31–38], and the results are summarized in Table 6.4. To maintain a consistent assessment, the devices considered have similar makeups. Moreover, because of the different device sizes and geometries, the failure current densities J_{t2}, J_{t2}', and J_{t2}'' are normalized by the effective silicon width, total layout width, and layout area, respectively. For NWFET and FinFET:

$$J_{t2} = \frac{I_{t2}}{D \times N} \tag{6.8}$$

$$J_{t2}' = \frac{I_{t2}}{D \times N + (N-1) \times S} \tag{6.9}$$

For planar MOSFET:

$$J_{t2} = J_{t2}' = \frac{I_{t2}}{W} \tag{6.10}$$

where N is the number of nanowires or fins, S is the spacing between adjacent nanowires or fins, and W is the width of planar MOSFET.

The data in Table 6.4 suggest that the NWFETs possess a higher J_{t2} than all other devices due to the miniature size and circular cross section of conduction channels. However, J_{t2}' and $J_{t2}''$of the NWFETs are smaller than that of the planar MOSFETs. This can be ascribed to the relatively large nanowire spacing (400 nm), which gives rise to a larger NWFET layout width and area. Area efficiency of NWFET can be improved by multi-finger-multi-channel layout shown in Fig. 6.19.

Table 6.4 Comparison of failure current densities in various technologies

Technology node	Gate length (nm)	D or W (μm)	J_{t2} (mA/μm)	J_{t2}' (mA/μm)	J_{t2}'' (mA/μm^2)
NWFET	750	0.01	29	0.73	0.24
	250	0.005	33.2	0.42	0.17
	250	0.01	26.2	0.66	0.26
32 nm bulk MOSFET	150 [32]	NA	7 ~ 10		2.9 ~ 5.4
45 nm bulk MOSFET	260 [33]	240	7.2 ~ 11.8		3.7
	240 [34]	240	6.5 ~ 11		2.7 ~ 4.9 [32, 34]
45 nm SOI MOSFET	40 [35]	468	1.88 ~ 4.91		1.4 ~ 1.76
65 nm bulk MOSFET	260 [36]	100 ~ 400	4 ~ 11.8		1.27 ~ 3.76
	270 [37]	20	8 ~ 15		2.36 ~ 3.75
SOI FinFET	250 [20]	0.03	7.5	1.2	0.054 ~ 0.108 [38]
	250 [31]	0.02	11.5	1.15	0.938
Bulk FinFET	250 [31]	0.02	6.5 ~ 26	0.65 ~ 2.61	0.056 ~ 0.224

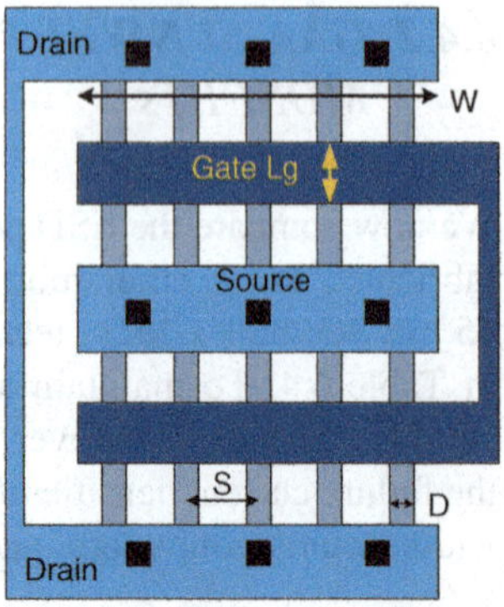

Fig. 6.19 Multi-finger–multi-channel layout of NWFET

For instance, NWFET with 100 nanowires, 750 nm gate length, and 5 nm diameter is capable of reducing more than 21 % of effective layout area and facilitating a uniform turn on of the multiple channels. The ESD performances of the NWFET and SOI FinFET, on the other hand, are quite comparable to each other.

6.5 Conclusion

This chapter presented experimental study pertinent to the ESD robustness of two promising nanowire devices called the poly-Si NWTFT and gate-all-around Si nanowire field-effect transistor (GAA NWFET).

For the NWTFT, a higher ESD robustness can be obtained by decreasing the channel length and increasing the number of channels. Plasma treatment commonly used to enhance the electrical performance of poly-Si-based devices does not seem to improve the NWTFT's ESD current-handling capability. Nonetheless, since the NWTFT-based integrated circuits typically operate at relatively low operating voltages (i.e., around or below 1–3 V), the high trigger voltage and low robustness of the gate-grounded (bipolar-mode) NWTFT can impose a great difficulty in implementing ESD protection for these nanowire-based integrated circuits. More research work is definitely needed to address these issues, and the diode-mode NWTFT having a trigger voltage of about 0.6 V appears to be more attractive for ESD applications.

The measurement results of the GAA NWFETs suggested that these devices are in general suitable for serving as ESD protection elements. It was found that adjusting the gate length, nanowire diameter, and nanowire number is an efficient and simple way to design NWFET with optimal ESD performances to meet the ESD design constrains and targets. By experimentally investigating the reliability and failure mechanisms of NWFETs subject to the HBM ESD event, it was found that nondestructive nanowire burn-out and gate oxide degradation occurred under a modest ESD stress, and destructive fusing of nanowires and surrounding gate oxides gave rise to the catastrophic failure under a severe ESD stress.

The comparison between NWTFT and GAA NWFET indicated that the latter is a better candidate for ESD protection applications with regard to the overall ESD performance per effective area. Compared with the FinFET and planar MOSFET, NWFET offered the highest failure current density normalized by the effective silicon width, thanks to the circular cross section and miniature size of the nanowire channels. Among all the NWFETs characterized, the one with 1,000 nanowires, 10 nm diameter, and 750 nm gate length was the most robust with a failure current of 0.29 A under the bipolar mode. This corresponds to an HBM ESD tolerance of 435 V, a figure lower than the typical industry standard of 500–2,000 V for consumer ICs. Multi-finger and multi-channel layout topology was found to be an effective solution to improve the area efficiency and ESD robustness without process changes.

In summary, the nanowire devices possess several favorable features for ESD protection applications: (1) the floating body under the gate-grounded configuration enables no-snapback I–V characteristic; (2) the multi-finger (drain and source) and multi-nanowire layout improves area efficiency; and (3) the ESD performances of gate-all-round nanowire FETs are superior to the SOI FinFETs in terms of the failure current density and trigger voltage. With the continuous development of nanowire technology and deeper understanding of its physical mechanism, it is to be expected that effective and robust ESD protection solutions for such an emerging technology will be developed and become available in the near future.

References

1. Semenov, O., Sarbishaei, H., & Sachdev, M. (2008). *ESD protection device and circuit design for advanced CMOS technologies*. Berlin: Springer.
2. Wagner, R., Hawkins, C., & Soden, J. (1993). Extent and cost of EOS/ESD damage in an IC manufacturing process. In *Proceedings EOS/ESD Symposium*, (pp. 49–49).
3. Green, W. D. T. (1988). A review of EOS/ESD field failures in military equipment. In *Proceedings EOS/ESD Symposium*, (pp. 7–13).
4. Russ, C. (1999) *ESD Protection devices for CMOS technologies: Processing impact, modeling, and testing issues*, PhD, Technical University Munich.
5. Huang, J. B., Wang, G. (2001). ESD protection design for advanced CMOS. In *International Symposium on Optoelectonics and Microelectronics* (pp. 123–131).
6. Liu, W., Liou, J. J., Chung, A., Jeong, Y. H., Chen, W. C., & Lin, H. C. (2009). Electrostatic discharge robustness of Si nanowire field-effect transistors. *Electron Device Letters, IEEE, 30*, 969–971.
7. Liu, W., Liou, J., Jiang, Y., Singh, N., Lo, G., Chung, J., et al. (2010). Investigation of sub-10-nm diameter, gate-all-around nanowire field-effect transistors for electrostatic discharge applications. *Nanotechnology, IEEE Transactions on, 9*, 352–354.
8. Liu, W., Liou, J. J., Jiang, Y., Singh, N., Lo, G., Chung, J., et al. (2010). Failure analysis of Si nanowire field-effect transistors subject to electrostatic discharge stresses. *Electron Device Letters, IEEE, 31*, 915–917.
9. Lin, H. C., Lee, M. H., Su, C. J., Huang, T. Y., Lee, C., & Yang, Y. S. (2005). A simple and low-cost method to fabricate TFTs with poly-Si nanowire channel. *Electron Device Letters, IEEE, 26*, 643–645.

10. Su, C. J., Lin, H. C., Tsai, H. H., Huang, T. Y., & Ni, W. X. (2007) Fabrication and characterization of poly-Si nanowire devices with performance enhancement techniques. In *International Symposium on VLSI Technology, Systems and Applications* (pp. 1–2).
11. Amerasekera, A., & Duvvury, C. (1995). The impact of technology scaling on ESD robustness and protection circuit design. *Components, Packaging, and Manufacturing Technology, Part A, IEEE Transactions on, 18*, 314–320.
12. Okushima, M., Shinzawa, T., & Morishita, Y. (2007). Layout technique to alleviate soft failure for short pitch multi finger ESD protection devices. in *Proceedings EOS/ESD Symposium* (pp. 1A. 5-1-1A. 5–10).
13. Wang, F. S., Tsai, M. J., & Cheng, H. C. (1995). The effects of NH_3 plasma passivation on polysilicon thin-film transistors. *Electron Device Letters, IEEE, 16*, 503–505.
14. Hsu, H. H., Lin, H. C., Chan, L., & Huang, T. Y. (2009). Threshold-voltage fluctuation of double-gated Poly-Si nanowire field-effect transistor. *Electron Device Letters, IEEE, 30*, 243–245.
15. Jiang, Y., Liow, T., Singh, N., Tan, L., Lo, G., Chan, D., & Kwong, D. (2008). Nanowire FETs for low power CMOS applications featuring novel gate-all-around single metal FUSI gates with dual Φm and VT tune-ability. In *International Electron Devices Meeting* (*IEDM*), (pp. 1–4).
16. Singh, N., Agarwal, A., Bera, L., Liow, T., Yang, R., Rustagi, S., et al. (2006). High-performance fully depleted silicon nanowire (diameter $\leq$5 nm) gate-all-around CMOS devices. *Electron Device Letters, IEEE, 27*, 383–386.
17. Gossner, H., (2004) ESD protection for the deep sub micron regime-a challenge for design methodology. In *Proceedings of the International Conference on VLSI Design* (pp. 809–818).
18. Russ, C. (2008). ESD issues in advanced CMOS bulk and FinFET technologies: Processing, protection devices and circuit strategies. *Microelectronics Reliability, 48*, 1403–1411.
19. Amerasekera, A., & Verwey, J. (1992). ESD in integrated circuits. *Quality and Reliability Engineering International, 8*, 259–272.
20. Trémouilles, D., Thijs, S., Russ, C., Schneider, J., Duvvury, C., Collaert, N., Linten, D., Scholz, M., Jurczak, M., & Gossner, H. (2007). Understanding the optimization of sub-45 nm FinFET devices for ESD applications. In *Proceedings EOS/ESD Symposium*, (pp. 7A. 5-1-7A. 5–8).
21. Anderson, W. R., Eppes, D., & Beebe, S. (2009). Metal and silicon burnout failures from CDM ESD testing. In *Proceedings EOS/ESD Symposium*, (pp. 1–8).
22. Duvvury, C., & Amerasekera, A. (1996). State-of-the-art issues for technology and circuit design of ESD protection in CMOS ICs. *Semiconductor Science and Technology, 11*, 833.
23. Gossner, H., Russ, C., Siegelin, F., Schneider, J., Schruefer, K., Schulz, T., Duvvury, C., Cleavelin, C. R., & Xiong, W. (2006). Unique ESD failure mechanism in a MuGFET technology. In *International Electron Devices Meeting* (*IEDM*), (pp. 1–4).
24. Chou, H. M., Lee, J. W., & Li, Y. (2007). A floating gate design for electrostatic discharge protection circuits. *Integration, the VLSI Journal, 40*, 161–166.
25. Chang, H. H., & Ker, M. D. (1998). Improved output ESD protection by dynamic gate floating design. *Electron Devices, IEEE Transactions on, 45*, 2076–2078.
26. Wu, E. Y., Vollertsen, R. P., Sune, J., & La Rosa, G. (2009). *Reliability wearout mechanisms in advanced CMOS technologies* (vol. 12). Wiley-IEEE Press.
27. Zhang, L., Wang, R., Zhuge, J., Huang, R., Kim, D. W., Park, D., & Wang, Y. (2008) Impacts of non-negligible electron trapping/detrapping on the NBTI characteristics in silicon nanowire transistors with tin metal gates. In *International Electron Devices Meeting (IEDM)* (pp. 1–4).
28. Pompl, T., Wurzer, H., Kerber, M., Wilkins, R., & Eisele, I. (1999) Influence of soft breakdown on NMOSFET device characteristics. In *Proceedings International Reliability Physics Symposium (IRPS)*, pp. 82–87.
29. Ille, A., Stadler, W., Pompl, T., Gossner, H., Brodbeck, T., Esmark, K., et al. (2009). Reliability aspects of gate oxide under ESD pulse stress. *Microelectronics Reliability, 49*, 1407–1416.

30. Park, Y. B., & Schroder, D. K. (1998). Degradation of thin tunnel gate oxide under constant Fowler-Nordheim current stress for a flash EEPROM. *Electron Devices, IEEE Transactions on, 45*, 1361–1368.
31. Griffoni, A., Thijs, S., Russ, C., Trémouilles, D., Linten, D., Scholz, M., Collaert, N., Witters, L., Meneghesso, G., & Groeseneken, G. (2009). Next generation bulk FinFET devices and their benefits for ESD robustness. In *Proceedings EOS/ESD Symposium* (pp. 1–10).
32. Li, J., Chatty, K., Gauthier, R., Mishra, R., & Russ, C. (2009). Technology scaling of advanced bulk CMOS on-chip ESD protection down to the 32 nm node. In *Proceedings EOS/ ESD Symposium* (pp. 1–7).
33. Chatty, K., Alvarez, D., Abou-Khalil, M., Russ, C., Li, J., & Gauthier, R. (2008) Investigation of ESD performance of silicide-blocked stacked NMOSFETs in a 45 nm bulk CMOS technology. In *Proceedings EOS/ESD Symposium* (pp. 304–312).
34. Alvarez, D., Chatty, K., Russ, C., Abou-Khalil, M. J., Li, J., Gauthier, R., et al. (2009). Design optimization of gate-silicided ESD NMOSFETs in a 45 nm bulk CMOS technology. *Microelectronics Reliability, 49*, 1417–1423.
35. Mitra, S., Gauthier, R., Li, J., Abou-Khalil, M., Putnam, C. S., Halbach, R., & Seguin, C. (2008). ESD protection using grounded gate, gate non-silicided (GG-GNS) ESD NFETs in 45 nm SOI technology. In *Proceedings EOS/ESD Symposium* (pp. 312–316).
36. Li, J., Alvarez, D., Chatty, K., Abou-Khalil, M. J., Gauthier, R., Russ, C., Seguin, C., & Halbach, R. (2006). Analysis of failure mechanism on gate-silicided and gate-non-silicided, drain/source silicide-blocked ESD NMOSFETs in a 65 nm bulk CMOS technology. In *International Symposium on the Physical and Failure Analysis of Integrated Circuits* (pp. 276–280).
37. Chatty, K., Alvarez, D., Gauthier, R., Russ, C., Abou-Khalil, M., & Kwon, B. (2007) Process and design optimization of a protection scheme based on NMOSFETs with ESD implant in 65 nm and 45 nm CMOS technologies. in *Proceedings EOS/ESD Symposium*, (pp. 7A. 2-1-7A. 2–10).
38. Russ, C., Gossner, H., Schulz, T., Chaudhary, N., Xiong, W., Marshall, A., et al. (2007). ESD evaluation of the emerging MUGFET technology. *Device and Materials Reliability, IEEE Transactions on, 7*, 152–161.
39. Bock, K., Russ, C., Badenes, G., Groeseneken, G., & Deferm, L. (1998). Influence of well profile and gate length on the ESD performance of a fully silicided 0.25 μm CMOS technology. *Components, Packaging, and Manufacturing Technology, Part C, IEEE Transactions on, 21*, 286–294.
40. Vinson, J. E., & Liou, J. J. (1998). Electrostatic discharge in semiconductor devices: An overview. *Proceedings of the IEEE,* (vol. 86, pp. 399–420).

Chapter 7
Green Energy Devices

Byung-Gook Park

Abstract This chapter treats nanowire FETs as applied to the green energy. The discussion is carried out from two perspectives, namely the renewable energy and the efficiency of energy consumption. In this context, the nanowire tunneling field effect transistor (NTFET) is considered as the low power device and its ON and OFF states are elaborated, based upon the voltage controlled energy band. Moreover, the operation of NTFETs is analyzed in terms of the size and shape of the nanowire, doping, multi-junction structure, and the materials used. In addition, nanowire solar cell is elaborated based upon the photo-generation, separation, and collection of e-h pairs. In particular, the enhanced absorption of solar radiation is highlighted via decreased reflectance, light trapping, and the resonance effect. Additional, the collection efficiency of photo-generated e-h pairs in nanowires is discussed based upon the cell structure, decoupling of the absorption and collection processes, and the types of junctions. Finally the process flows for producing vertical nanowires are presented, together with the fabrication of the nanowire photo-cells therein.

Abbreviation

CMOS	Complementary MOS
FET	Field-effect transistor
SS	Subthreshold swing
IMOS	Impact-ionization MOS
TFET	Tunneling field-effect transistor
SOI	Silicon-on-Insulator
SON	Silicon-on-nothing
LMR	Leaky-mode resonance
VLS	Vapor–liquid–solid
AAO	Anodic aluminum oxide

B.-G. Park (✉)
Seoul National University, Seoul, Korea
e-mail: bgpark@snu.ac.kr

D. M. Kim and Y.-H. Jeong (eds.), *Nanowire Field Effect Transistors: Principles and Applications*, DOI: 10.1007/978-1-4614-8124-9_7,

CIGS	Copper indium gallium selenide
VSS	Vapor–solid–solid
PECVD	Plasma-enhanced chemical vapor deposition
ITO	Indium tin oxide
EQE	External quantum efficiency
MOCVD	Metal–organic chemical vapor deposition
MBE	Molecular beam epitaxy
CBE	Chemical beam epitaxy
CVD	Chemical vapor deposition
PCBM	Phenyl-C_{61}-butyric acid methyl ester

7.1 Introduction

The global climate change is one of the most imminent challenges to the human race. Figure 7.1 shows temperature changes during the last 2,000 years in various reconstructions and actual measurement [1]. The single value for year 2004 is also shown for comparison. What is clear in this graph is that the mean surface temperature of the earth is increasing rapidly in a very short time period ($\sim$100 years). Scientists are almost certain that the global warming is mainly caused by increasing concentrations of greenhouse gases, such as carbon dioxide (CO_2), methane (CH_4), and ozone (O_3), that are generated by human activities. In order to reduce climate change, we have to reduce greenhouse gas emissions. Policies for reduced emissions include increased use of renewable energy and increased energy efficiency.

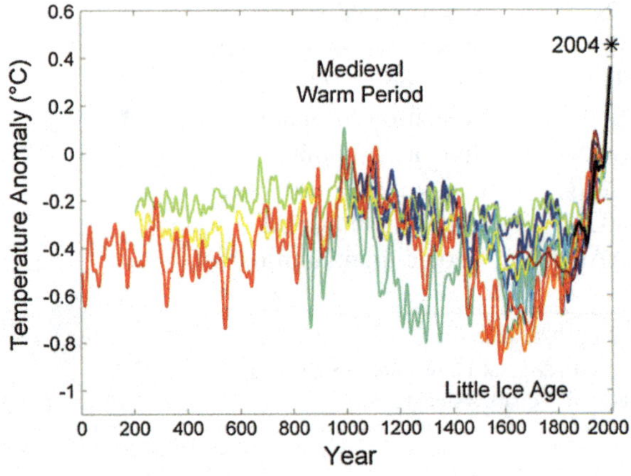

Fig. 7.1 2,000 years of mean surface temperatures according to different reconstructions from climate proxies, each smoothed on a decadal scale. The instrumental temperature record is overlaid in black. From Wikipedia, "Global warming," http://en.wikipedia.org/wiki/Global_warming

In this chapter, we discuss two approaches for reducing the power consumption and the climate change. The first approach is related to the efficient use of energy. By making electronic devices use less power, we can save electricity which is one of the major sources of carbon emission due to the use of fossil fuels. Thus, as an ultralow-power nanowire device, tunneling field-effect transistors (TFETs) are discussed in depth. The second approach is related to the use of renewable energy. Converting solar energy into electricity is one of the most attractive solutions to modern energy issues because the solar energy is produced with almost zero carbon emission. Nanowire solar cells will be on the focus of discussion in the last section.

7.2 Nanowire Tunneling Field-Effect Transistors

7.2.1 Low-Power Operation and Subthreshold Swing

Let us consider the power consumption in a complementary MOS (CMOS) inverter composed of an n-channel MOSFET and a p-channel MOSFET (Fig. 7.2). Assuming that the input of the CMOS inverter is a pulse train (called a clock) with a frequency *f*, we first evaluate the energy consumed during one complete clock cycle. Multiplying it by the clock frequency, we can obtain the power consumption for the switching operation. Let us first consider a high–low transition (discharging process) at the output node. During this transition, the capacitance, *C*, which represents the total parasitic capacitance connected to the output node, will be discharged from V_{DD} to 0. The instantaneous power, *p*, that the n-channel MOSFET consumes is given by

$$p = v_O i_D = -v_O C \frac{dv_O}{dt} \tag{7.1}$$

The energy loss during the high–low transition can be obtained by integrating *p* with respect to time for the complete transition.

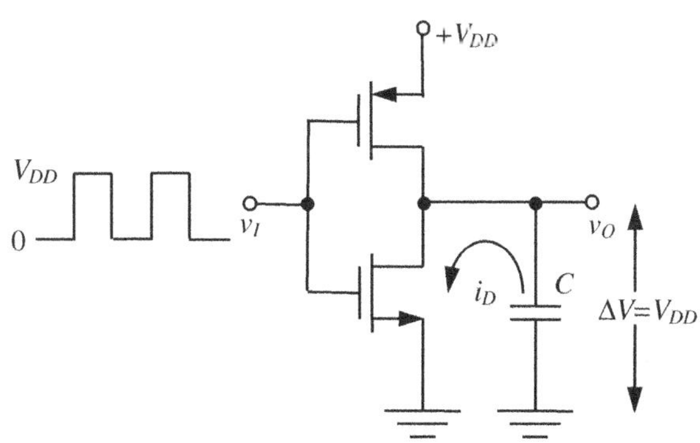

Fig. 7.2 CMOS inverter circuit for the calculation of power consumption: v_I is the input voltage, v_O is the output voltage, V_{DD} is the power supply voltage, and i_D is the drain current of a MOSFET

$$W_{\mathrm{HL}} = \frac{1}{2} C V_{\mathrm{DD}}^2 \tag{7.2}$$

The instantaneous power that the p-channel MOSFET consumes during the low–high transition (charging process) is given as

$$p = (V_{\mathrm{DD}} - v_O) i_D = (V_{\mathrm{DD}} - v_O) C \frac{dv_O}{dt} \tag{7.3}$$

The energy loss during the low–high transition is

$$W_{\mathrm{LH}} = \frac{1}{2} C V_{\mathrm{DD}}^2 \tag{7.4}$$

One clock cycle includes both the high–low and low–high transitions, so that we need to add up Eqs. (7.2) and (7.4) for the total energy loss per clock. Now, if we multiply it by the clock frequency *f*, we obtain the switching power consumption.

$$P_{\mathrm{switching}} = f C V_{\mathrm{DD}}^2 \tag{7.5}$$

In this equation, we can see that the power supply voltage (V_{DD}) plays an important role in power consumption. Thus, for low-power operation, V_{DD} reduction is essential. But if we want to reduce V_{DD}, the threshold voltage of MOSFETs should also be reduced. Otherwise, the performance of the circuit will be degraded significantly. The reduced threshold voltage, however, brings in the exponential increase in the OFF current as shown in Fig. 7.2. The large OFF current is responsible for the high leakage power consumption, since the leakage power is given by the product of the supply voltage and the OFF current, I_{OFF}, i.e.,

$$P_{\mathrm{leakage}} = V_{\mathrm{DD}} I_{OFF} \tag{7.6}$$

This leakage power is consumed not only during the period of switching but also during the standby time. For the applications in which the equipment stays mostly in the standby mode, high leakage power consumption can be fatal.

Figure 7.3 shows the tight linkage between the threshold voltage and the OFF current in field-effect transistors (FETs). Since log I_D is linearly dependent on the gate bias voltage in the subthreshold region, the OFF current increases exponentially as the threshold voltage decreases linearly. The subthreshold swing (SS) is defined as the inverse of the slope of the log I_D versus V_G curve in the subthreshold region. The SS means the amount of gate voltage required for the one-decade change in the drain current. At room temperature, the typical value of SS is 70–90 mV/decade in MOSFETs and 60 mV/decade in bipolar transistors. For the ON/OFF current ratio (I_{ON}/I_{OFF}) of 10^5, the threshold voltage of a MOSFET should be at least 0.3 V. This is the reason why it is difficult to reduce the power supply voltage of MOSFETs below 1 V.

In order to reduce the switching power of CMOS circuits, the SS of MOSFETs should be reduced. Unfortunately, there has been a strict limit in reducing the SS

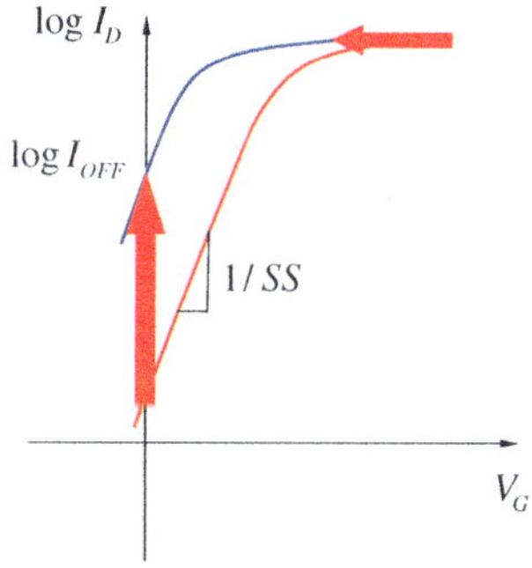

Fig. 7.3 Correlation of the threshold voltage with ON, OFF, and subthreshold currents. Shown are the exponential increase in the subthreshold current with the gate voltage, the subthreshold swing (SS, the inverse of the slope of the log I_D versus V_G curve in the subthreshold region), and the tight correlation of the threshold voltage with the exponential increase in OFF current

of conventional FETs. The lowest value of the SS achievable in a conventional FET at room temperature is 60 mV/decade. The origin of such an SS limit lies in the thermal carrier injection from the source to the channel as shown in Fig. 7.4. Here, the gate bias controls the energy barrier height between the source and the channel, and carriers in the source have energy distributed by the Boltzmann law. Since the carriers can be injected into the channel only when they have a kinetic energy higher than the barrier height, the current injected into the channel is proportional to the Boltzmann factor, $e^{-q(V_0-V)/k_BT}$, i.e.,

$$I = Ae^{-q(V_0-V)/k_BT} \tag{7.7}$$

where V_0 is the barrier height without the bias, V. If we take the logarithm from both sides and differentiate, we obtain

$$\frac{dV}{d(\log I)} = \frac{k_BT}{q\log e} \tag{7.9}$$

Fig. 7.4 Origin of the 60 mV/decade subthreshold swing limit in (**a**) a p–n junction and (**b**) a Schottky junction. The bias voltage across these junctions is shown to decrease the energy barrier height, inducing the exponential injection of carriers into the channel following the Boltzmann probability factor, $\exp(-qV/k_BT)$

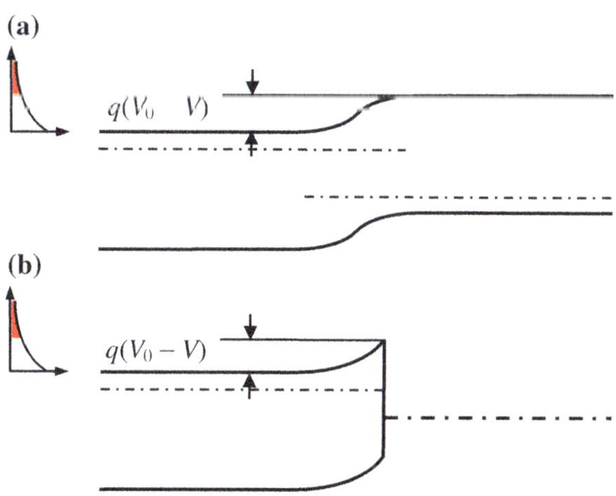

At room temperature (300 K), this value is about 60 mV/decade. Due to the voltage drop in the gate oxide, only a portion of the gate voltage can be used to lower the barrier between the source and the channel. Considering the voltage division between the channel and the gate oxide in a field-effect transistor, we obtain

$$SS = \frac{dV_G}{d(\log I)} = \frac{k_B T}{q \log e}\left(1 + \frac{C_{ch}}{C_{ox}}\right) \tag{7.10}$$

where C_{ch} is the channel capacitance per unit area and C_{ox} is the oxide capacitance per unit area. Thus, the SS of conventional FETs is larger than 60 mV/decade at room temperature. The only way of reducing the SS below 60 mV/decade in conventional FETs is to reduce the temperature, but the power required for cooling exceeds the power saving from the lower supply voltage.

To overcome the 60 mV/decade limit in the SS, a few mechanisms that do not depend on thermal carrier injection have been proposed. One of such mechanisms is injection by impact ionization, which was proposed by Gopalakrishnan [2]. Impact-ionization MOS (IMOS) devices based on this mechanism use modulation of avalanche breakdown voltage of a gated p-i-n diode. Another mechanism is injection by inter-band tunneling [3], and the device that utilizes such a mechanism is called a TFET. The third mechanism is injection by mechanical contact, which is implemented by a *mechanical switch*. In this section, we will discuss the TFETs, since they appear to be the most promising and CMOS-compatible devices.

7.2.2 Concept of Tunneling Field-Effect Transistor

Let us consider a reverse-biased p^+–n^+ junction shown in Fig. 7.5. With the Fermi level pinned within the valence band in the p^+-source electrode, the number of electrons in the conduction band is negligible. Thus, we may consider only the tunneling of valence band electrons into the n^+-channel region.

The injection of electrons into the channel via tunneling is a non-thermal process for the reasons as follows. First, the Fermi occupation factor for electrons in the valence band of the p^+-source is practically equal to unity. Also, the current density of inter-band tunneling is given by the following equation [4],

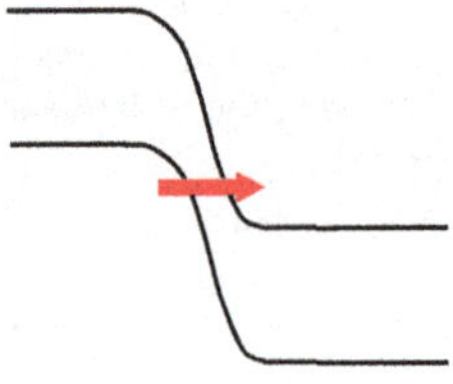

Fig. 7.5 Inter-band tunneling in a p^+–n^+ junction. The valence band electrons tunnel through the triangular barrier formed by the bandgap in the depletion region of a p^+–n^+ junction

$$J_{\mathrm{IB}} = \frac{\sqrt{2m^*}q^3 E V_d}{4\pi^3 \hbar^2 E_g^{1/2}} \exp\left(-\frac{4\sqrt{2m^*}E_g^{3/2}}{3qE\hbar}\right) \tag{7.11}$$

where the electric field, E, in the junction region is given by

$$E = \sqrt{\frac{2qN_a(V_d + V_0)}{\varepsilon_{\mathrm{si}}}} \tag{7.12}$$

and E_g denotes the bandgap. Clearly, the overall expression of the current is independent of T, confirming that the injection process is non-thermal.

If we replace the n^+-source with a p^+-source in an n-channel MOSFET as in Fig. 7.6, we obtain a TFET. In this device, the electron injection from the source occurs through a tunnel junction between the source and the channel. Unlike normal MOSFETs, this structure has an asymmetric structure.

Figure 7.7 shows the band diagrams of a TFET for its OFF and ON states. When the gate bias voltage is below the threshold voltage, electron tunneling from the valence band of the source to the conduction band of the channel is blocked. Holes can be injected into the channel because of the lowered barrier, but they are blocked by the large potential barrier formed at the reverse-biased channel–drain junction. If the gate bias voltage is above the threshold voltage, electrons tunnel from the valence band of the source to the conduction band of the channel. Once injected into the channel, the electrons are transported to the drain as in a conventional MOSFET.

The advantage of the TFET lies in its carrier injection mechanism. Since the tunneling rate is independent of temperature, that is, the carrier injection mechanism is non-thermal, the SS can be smaller than the thermal injection limit (60 mV/decade at room temperature). Figure 7.8 shows such an example in a fabricated TFET [3]. The TFET is fabricated on a silicon-on-insulator (SOI) structure, and both the gate oxide and the body are relatively thin. Such a structure enhances the gate control over the channel, so that the intensity of the electric field in the tunneling region is maintained high. In addition, the buried oxide in the SOI structure cuts off the possible leakage path from the source to the body. Such a blockage is automatically provided in most of the three-dimensional device structures with a floating body.

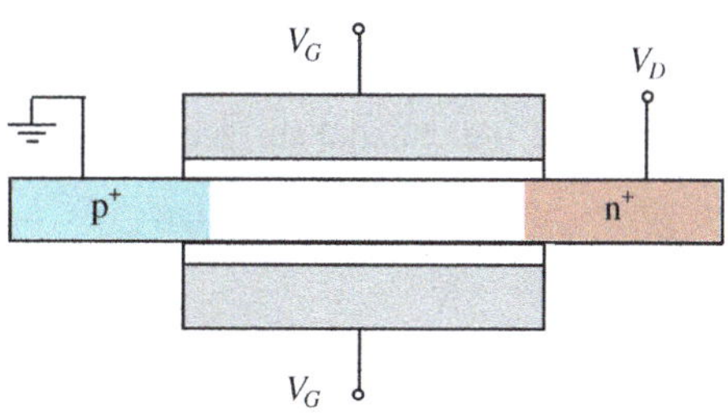

Fig. 7.6 Tunneling field-effect transistor (TFET). A p^+-source is used instead of the n^+-source in order to form a tunnel junction between the source and the channel

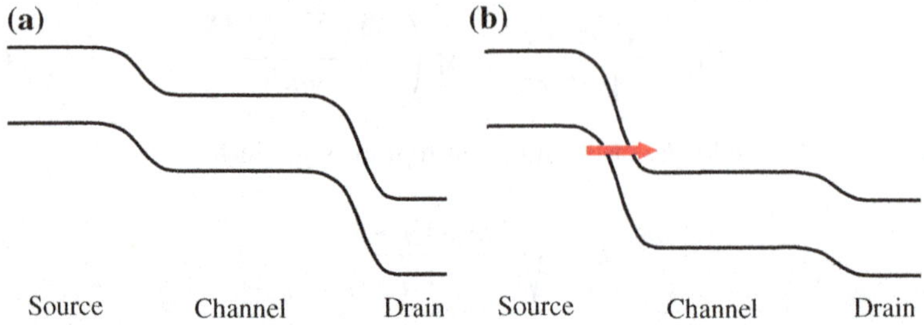

Fig. 7.7 Band diagram of a TFET: **a** OFF-state: electron tunneling from the valence band of the source to the conduction band of the channel is blocked. **b** ON-state: electrons tunnel from the valence band of the source to the conduction band of the channel

7.2.3 Silicon Nanowire Tunneling Field-Effect Transistors

One major issue with silicon (Si) TFETs is the relatively low current density compared with that of MOSFETs. As can be seen in Fig. 7.7, the typical current density of a TFET is more than an order of magnitude smaller than that of a typical MOSFET. The reason behind such a low current is the relatively large bandgap of silicon. The bandgap of silicon is about 1.1 eV, and this value is significantly larger than that of germanium (0.67 eV). This is the reason why germanium has been used for tunnel diodes, instead of silicon.

A possible way to increase the current density of Si TFETs is to enhance the electric field, in which case the effect of large bandgap can be compensated. In Eq. (7.11), it is clear that the ratio of $E_g^{3/2}$ to the electric field determines the exponential factor and we can increase this factor by increasing the electric field. In order to increase the electric field at the source–channel junction, the gate should maintain a tighter control over the channel. The nanowires offer a most efficient structure for enhancing the gate control over the channel, since the gate surrounds the channel and the electric field is concentrated around and inside the channel.

Figure 7.9 shows the schematic diagram of a nanowire TFET and its cross-section at the channel region. The cross-sectional shape of the channel is chosen to be circular, not rectangular, because the channel with a rectangular cross-section suffers from the *corner effect*. In rectangular channels, the field is crowded in the corners, while, in cylindrical channels, the field is uniformly distributed around the circle. Thus, cylindrical channels do not suffer from corner effect.

The circular cross-section of a cylindrical channel is shown in Fig. 7.9b. The gate oxide thickness is t_{ox}, and the radius of the channel is R_{ch}. The capacitance per unit area of the channel surface is given as

$$C_{ox} = \frac{\varepsilon_{\text{OX}}}{R\ln\left(1 + \frac{t_{\text{OX}}}{R}\right)} \tag{7.13}$$

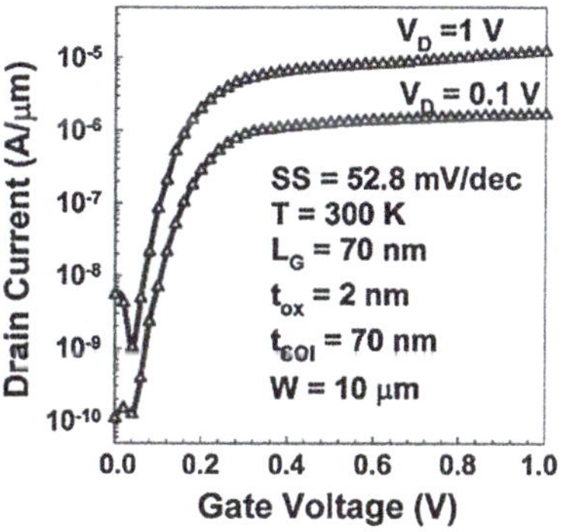

Fig. 7.8 Subthreshold characteristics of a TFET. The subthreshold swing of this device is less than 60 mV/decade at room temperature. Reproduced with permission from [3]. ©2007 IEEE

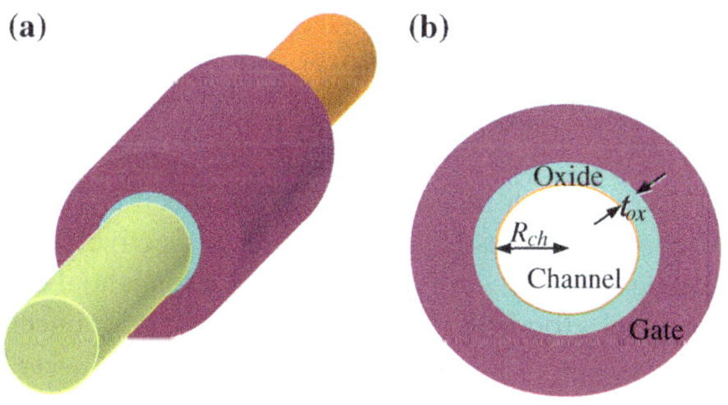

Fig. 7.9 Nanowire tunneling field-effect transistor: **a** schematic diagram and **b** the cross-section of the channel. The gate oxide thickness is t_{ox}, and the radius of the channel is R_{ch}

In this equation, the equivalent oxide thickness is $R\ln(1 + t_{ox}/R)$. Let us divide it by t_{ox} to normalize it and define $x = R/t_{ox}$. We can easily show that the normalized equivalent oxide thickness is smaller than 1 for all $x > 0$. Thus, the equivalent oxide thickness is always smaller than t_{ox}. Such a small effective oxide thickness reflects the effects of enhanced field in the cylindrical channel and also results in increased capacitance. Using a high dielectric constant material, we can further decrease the equivalent oxide thickness and increase the gate capacitance.

The increased capacitance tightens the gate control over the channel, and the tighter gate control enhances the electric field at the junction between the source and the channel. In addition to the increased gate capacitance, the cylindrical channel structure further enhances the electric field at the junction between the source and the channel because of the three-dimensional field enhancement effect within the channel. Figure 7.10 shows the comparison of a double-gate structure with a cylindrical channel structure, in terms of the electric field in the channel. In the comparison, the effect of the smaller equivalent oxide thickness on the cylindrical channel has been compensated by introducing thicker oxide, so that both oxides have the same equivalent thickness. As can be seen in this figure, the high-field region in the cylindrical channel is extended toward the center ($y = 0$) further than that in the double-gate structure.

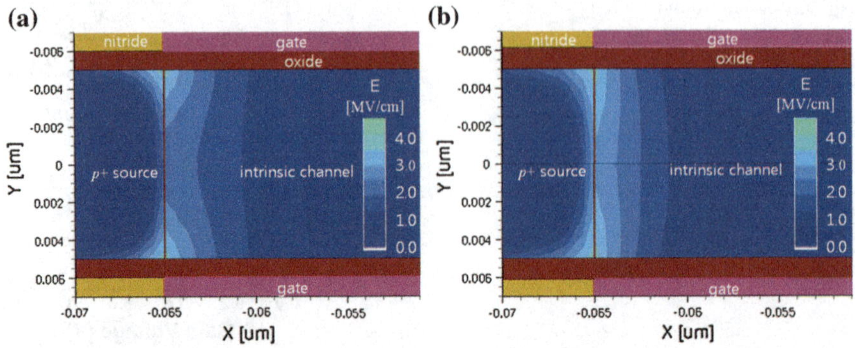

Fig. 7.10 Comparison of electric field in the channel: **a** double-gate structure and **b** cylindrical channel structure. To eliminate the effect of the smaller equivalent oxide thickness in the cylindrical channel device, a thicker oxide with the same equivalent oxide thickness (1 nm) is used. In addition, the channel diameter (10 nm) of the cylindrical channel structure is the same as the channel thickness of the double-gate structure

Even though the difference in the electric field of double-gate and cylindrical channel structures is not drastic, the respective carrier injection rates by tunneling can be drastically different due to the exponential dependence of the tunneling rate on the electric field. The comparison of carrier injection rate via tunneling in these two structures is shown in Fig. 7.11. As expected, the region of high carrier injection is confined to the small area near the oxide–silicon interface in the case of the double-gate TFET, whereas the region of high carrier injection in the cylindrical channel TFET is much larger and extended toward the center ($y = 0$) of the cylinder. Thus, the cylindrical channel is shown to be a superior structure for TFETs.

Another method of enhancing the electric field is to insert a thin n(p)-layer between the $p^{+}(n^{+})$-source and the channel in an n(p)-channel TFET. Such a

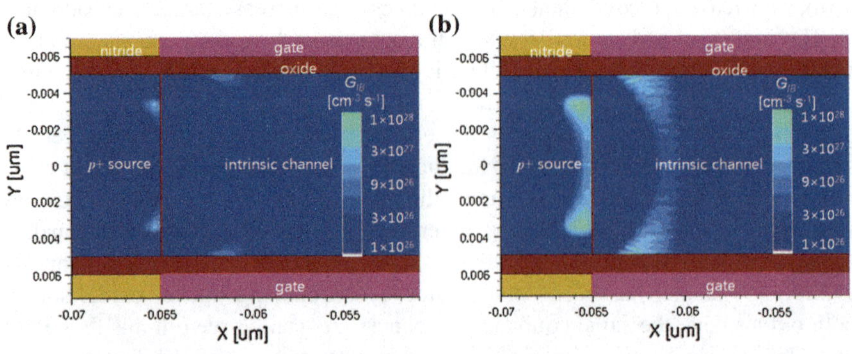

Fig. 7.11 Comparison of carrier generation rate (G_{IB}) by tunneling in the channel: **a** double-gate structure and **b** cylindrical channel structure. All the dimensions of the devices are maintained the same as those in Fig. 7.10

structure is depicted in Fig. 7.12. Figure 7.12a shows source, channel, and drain doping in a conventional n-channel TFET. If we incorporate a thin n-doped layer between p^+-region and i-region, we obtain a cylindrical channel with an n-doped layer. This n-doped layer enhances the electric field by providing the space charge in the depletion region.

We can understand such a field enhancement effect by considering a simplified model. To illustrate the basic principle of field enhancement, we just consider the one-dimensional effect in this model. We further assume that the source/drain doping is high enough to be treated as infinite concentration. In addition, the n-doped layer is assumed to be completely depleted. The same bias voltage is applied to both structures under comparison. Such a model is illustrated in Fig. 7.13. If we consider the charge densities in the entire device structure, the respective charge profiles are shown in Fig. 7.13a and b. Due to the infinite doping concentration, there should be delta function charge densities at the source/drain edge of the channel. In the n-channel TFET without an n-doped layer, there should be no charge in the channel region. In the n-channel TFET with an n-doped layer, there should be depletion charge in the n-doped layer. The electric fields for the n-channel TFETs without and with the n-doped layer are shown in Fig. 7.13c and d, respectively. Since the integration of electric field gives the potential difference across the channel, the area marked by slanting lines in the electric field plots should be the same, for given bias. This condition leads to the following relationship between the applied reverse bias, V_R, and the maximum electric field, E_{max}.

$$V_R + V_{bi} = \frac{qN_D x_n}{2\varepsilon_{\text{si}}} + \left(E_{max} - \frac{\rho_n}{\varepsilon_{\text{si}}}\right) x_{\text{ch}} \tag{7.14}$$

Hence, we can obtain the maximum electric field as

$$E_{max} = \frac{V_R + V_{bi}}{x_{\text{ch}}} + \frac{qN_D}{\varepsilon_{\text{si}}}\left(1 - \frac{x_n}{2x_{\text{ch}}}\right) \tag{7.15}$$

Since $x_n < x_{ch}$ and the electric field in the structure without the n-doped layer is $(V_R + V_{bi})/x_{ch}$, the maximum electric field with the n-doped layer is higher than that without the n-doped layer.

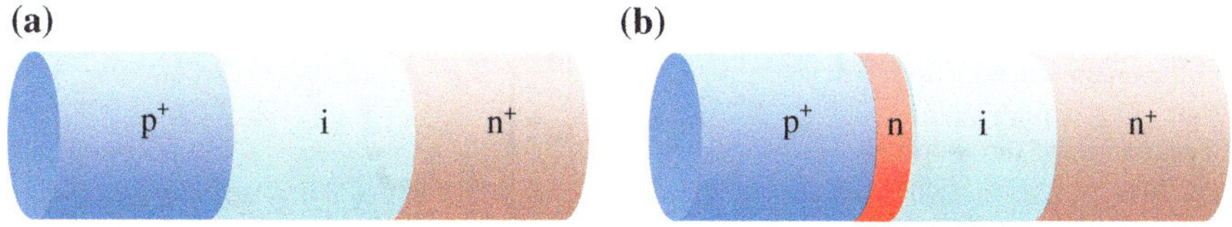

Fig. 7.12 Insertion of n-doped layer between the p^+-source and the channel in an n-channel TFET: **a** without an n-doped layer and **b** with an n-doped layer. The purpose of the n-doped layer is to enhance the electric field at the junction between the source and the channel

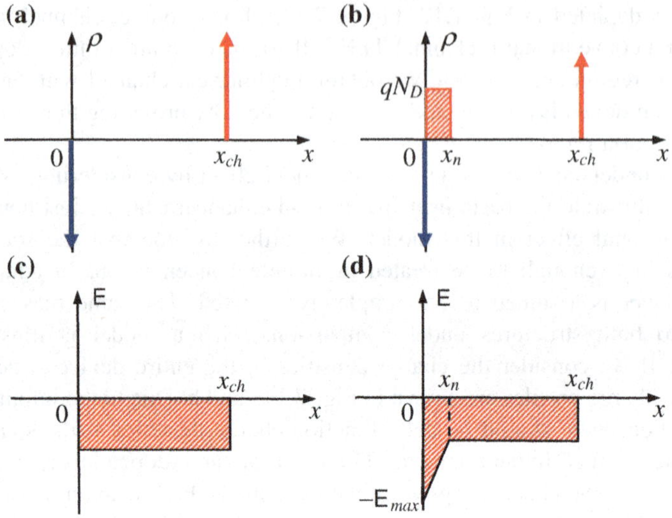

Fig. 7.13 Space charge and electric fields in the TFET structures shown in Fig. 7.12: **a** charge density in the device without n-doped layer, **b** charge density in the device with n-doped layer, **c** electric field in the device without n-doped layer, and **d** electric field in the device with n-doped layer

In order to confirm the effect of the n-doped layer on the drain current in a cylindrical channel TFET, device simulation has been carried out and the results are shown in Fig. 7.14. In this device, the channel diameter (d_{ch}) is 10 nm, the oxide thickness is 1.1 nm, and the channel length is 22 nm. The doping concentration of the n-doped layer (N_D) is 3×10^{19} cm^{-3}, and its thickness (x_n) is 4 nm. The device with the n-doped layer shows a dramatic improvement in the transfer characteristics. The average SS decreases significantly, and the ON current is improved at least by an order of magnitude. The I–V characteristics can be improved further by decreasing the channel diameter.

Fig. 7.14 Effect of a thin n-doped layer on the drain current in a cylindrical channel TFET. The channel diameter (d_{ch}) is 10 nm, the oxide thickness is 1.1 nm, and the channel length is 22 nm. The doping concentration of the n-doped layer (N_D) is 3×10^{19} cm^{-3}, and its thickness (x_n) is 4 nm

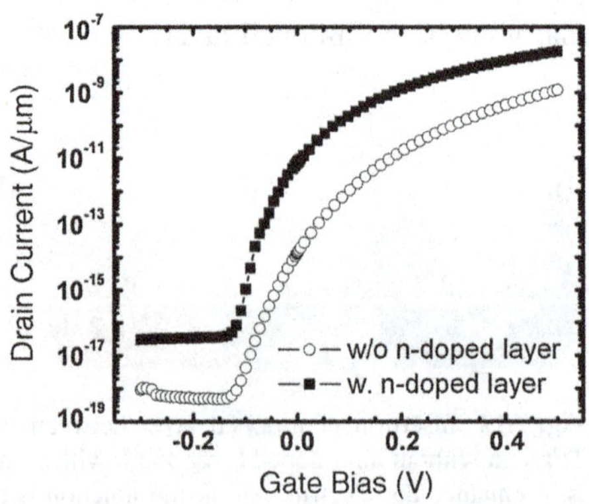

Experimentally, both vertical and horizontal channel nanowire TFETs can be fabricated. The vertical channel TFETs are relatively easy to fabricate. Vertical nanowires with a cylindrical shape can be grown with a metal nanoparticle as a catalyst [5], or etched (and oxidized) by using a patterned mask [6, 7]. Figure 7.15 shows vertical nanowire arrays formed by etching silicon with resist mask and converting silicon pillars into nanowires by oxidation [8]. The array has a pitch of 500 nm, the diameter of nanowires is about 20 nm, and their height is 1 μm. The vertical nanowires can be turned into nanowire TFETs by forming the drain, gate electrode, and source. Figure 7.16 shows a schematic cross-section of a TFET structure. The gate is formed around the nanowire, and the drain, the channel, and the source are formed in sequence vertically. In order to implement a steep doping profile at the source junction, they have used the dopant-segregated silicidation process that causes dopants to pile up at the silicide edge.

Horizontal channel formation is usually more difficult than vertical channel formation. To form a horizontal channel, a nanowire should be patterned by lithography and etch [9]. For the isolation of channel from the substrate, a SOI or silicon-on-nothing (SON) structure is often used. Figure 7.17 shows a typical shape of a horizontal channel nanowire TFET. The main advantage of horizontal channel devices over vertical channel devices is the ease of gate and contact formation. As can be seen in Fig. 7.17, the arrangement of source, channel, and drain in the horizontal nanowire device is the same as that of a planar device, facilitating the gate and contact formation.

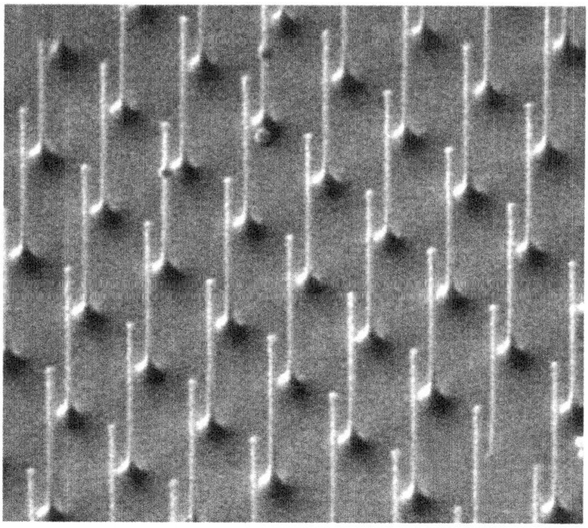

Fig. 7.15 Vertical nanowire arrays with pitch of 500 nm. The diameter of nanowires is about 20 nm, and their height is 1 μm. Reproduced with permission from [8]. ©2008 IEEE

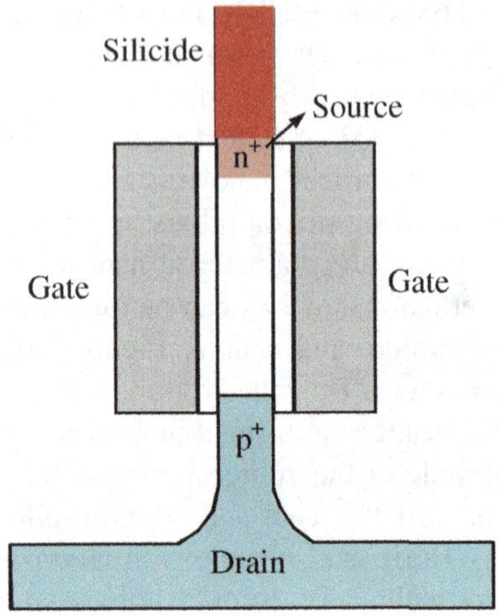

Fig. 7.16 Schematic cross-section of a vertical nanowire TFET structure

Fig. 7.17 Schematic diagram of a horizontal nanowire TFET structure

7.2.4 Bandgap Engineering for Nanowire Tunneling Field-Effect Transistors

We can boost the current density in a silicon-based TFET by changing the source material with a narrow bandgap material such as germanium. The advantage of using a narrow bandgap material can be easily seen by examining the inter-band tunneling current density given in Eq. (7.11). As clear from the equation, the dominant factor for tunneling is the exponential function whose argument is proportional to the 3/2 power of the bandgap. Hence, the current density should be a sensitive function of the bandgap and should in fact increase exponentially with decreasing bandgap. Thus, using a narrow bandgap material as the channel material is an efficient method of boosting the current density.

There is, however, a problem associated with the small bandgap channel material, i.e., the ambipolar behavior due to inter-band tunneling under the condition of low gate and high drain bias. The ambipolar behavior is originated from the tunneling current at the drain junction. When the bandgap of the channel material is small, the normal ON current under the condition of high gate and drain bias voltage is increased as shown in Fig. 7.18a, since both the width and the height of the tunnel barrier are reduced, compared with those of the large bandgap material. On the other hand, the narrow bandgap material increases the OFF

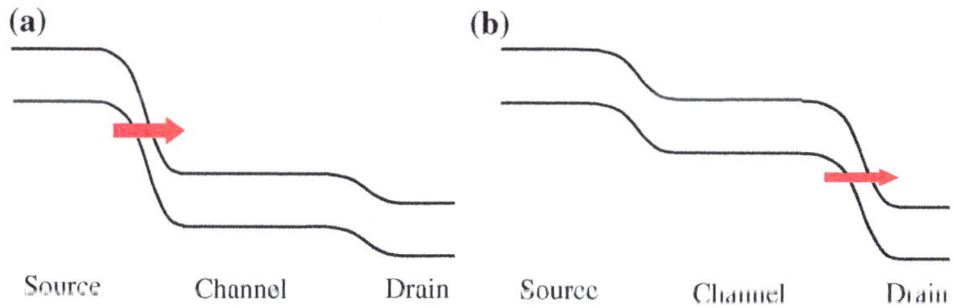

Fig. 7.18 Ambipolar behavior of a nanowire TFET with a narrow bandgap channel material: **a** under high gate and high drain bias voltage and **b** under low gate and high drain bias voltage

current under the condition of low gate and high drain bias. As shown in Fig. 7.18b, the narrow bandgap in the drain edge of the channel may form a low and thin tunnel barrier, so that the tunneling current at the drain junction would not be negligible. This current will increase exponentially, as the gate bias voltage decreases.

One can mitigate the ambipolar behavior by using the narrow bandgap material only at the source region while maintaining a large bandgap material near the drain region. Figure 7.19 shows a bandgap-engineered TFET structure which is designed to have a large ON current, while the ambipolar behavior is suppressed. Under the high gate and high drain bias voltage (Fig. 7.19a), the source junction is reverse-biased and the current density increases due to the narrow bandgap of the source material. Under the low gate and high drain bias voltage (Fig. 7.19b), the drain junction is reverse-biased, but not much current can flow due to the thick tunnel barrier formed by the wide bandgap material.

Incorporation of a narrow bandgap material near the source region is relatively easy in a vertical channel nanowire TFETs, since the lattice mismatch between two materials does not matter much in the vertical nanowire structure due to the small cross-sectional area. Nanowire TFETs with a germanium source have been the most sought-after structure up to now [10, 11]. Germanium (Ge) is a group IV material just like silicon (Si), so that the material compatibility with silicon is good

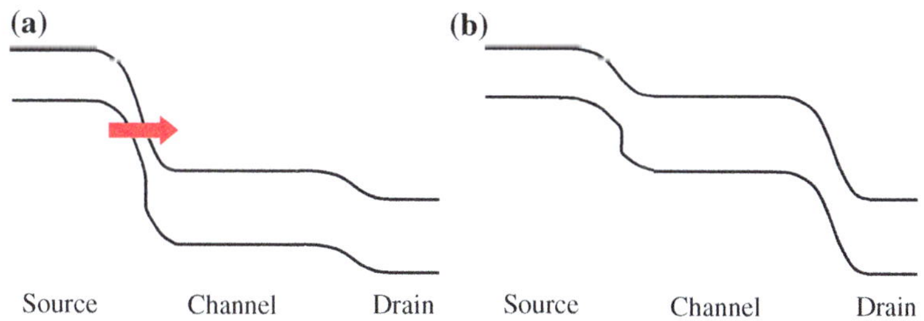

Fig. 7.19 Bandgap-engineered n-channel TFET with a *narrow* bandgap channel material at the source region and a *wide* bandgap material in the channel and the drain region: **a** under high gate and high drain bias voltage and **b** under low gate and high drain bias voltage

except for the lattice mismatch. In a vertical nanowire structure grown with a metal nanoparticle as a catalyst, the formation of germanium source on top of silicon can be easily done by switching the source gas. Simulated subthreshold characteristics for all-Si, all-Ge, and Ge-source Si TFETs are shown in Fig. 7.20 [10]. The all-Si TFET shows about two orders of magnitude lower current than the all-Ge and Ge-source Si TFET. In the all-Ge TFET, however, ambipolar behavior is clearly observed. The Ge-source Si TFET exhibits the highest ON current for the same OFF current.

What is important in determining the current density is not only the bandgap but also the band offset of the heterojunction. In Fig. 7.19, the band offset occurs only in the valence band, and the conduction band has no offset at the heterointerface. This type of band alignment is not favorable to the p-channel TFET. As shown in Fig. 7.21a, there would be an additional potential barrier caused by the band offset. The effect of the narrow bandgap on the source is compensated by the additional potential barrier. In order to utilize the advantage of the narrow bandgap in the source, the band offset should occur only in the conduction band as shown in Fig. 7.21b. Note that the band diagram in Fig. 7.21b is the mirror image of that in Fig. 7.19a. Thus, by considering hole injection from the source to the channel, we can easily explain how the ON current is increased and the ambipolar behavior is suppressed in this bandgap-engineered device.

To implement the band diagram in Fig. 7.21b, indium arsenide (InAs) is often used as a source material [11, 12]. InAs has a bandgap of 0.36 eV and an electron affinity of 4.9 eV. Thus, InAs forms a staggered gap heterojunction with Si as shown in Fig. 7.22. The conduction band offset is 0.85 eV, and the valence band offset is −0.11 eV. The staggered gap heterojunction can provide a very low and thin tunnel barrier near the heterojunction. Such a band profile is very useful in boosting current. In addition to the small bandgap of InAs, the favorable band

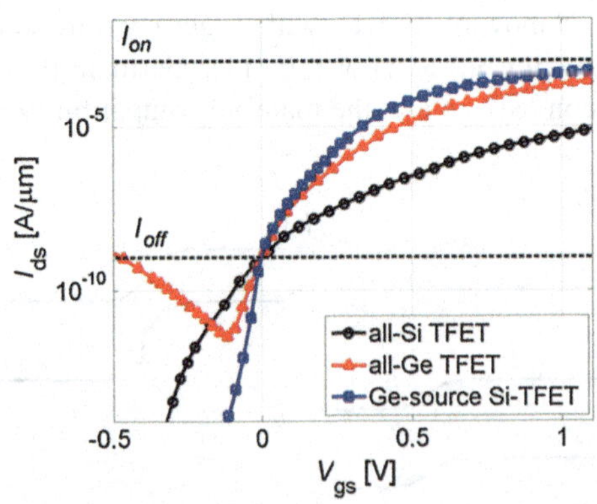

Fig. 7.20 Comparison between the homostructure and heterostructure TFETs by simulation. The dashed lines represent the ON current (10^{-3} A/μm) and OFF current (10^{-9} A/μm) of the ITRS 2003 device specifications. The subthreshold characteristics are shifted horizontally to achieve the required off current at $V_{gs} = 0$ V. Reproduced with permission from [10]. ©2008 AIP

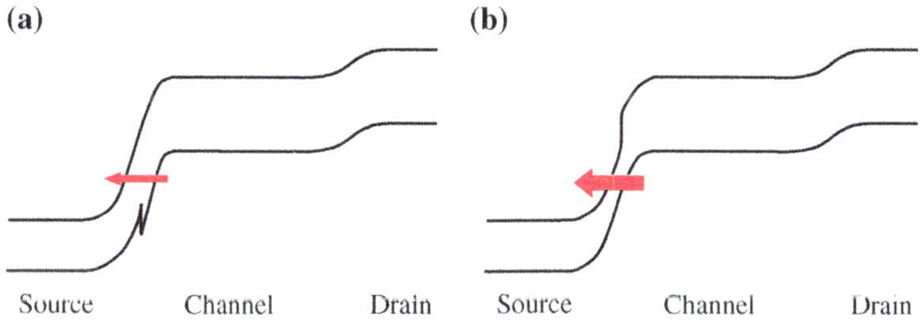

Fig. 7.21 Effect of band alignment on the tunneling current of p-channel TFETs: **a** when the band offset occurs in the valence band and **b** when the band offset occurs in the conduction band

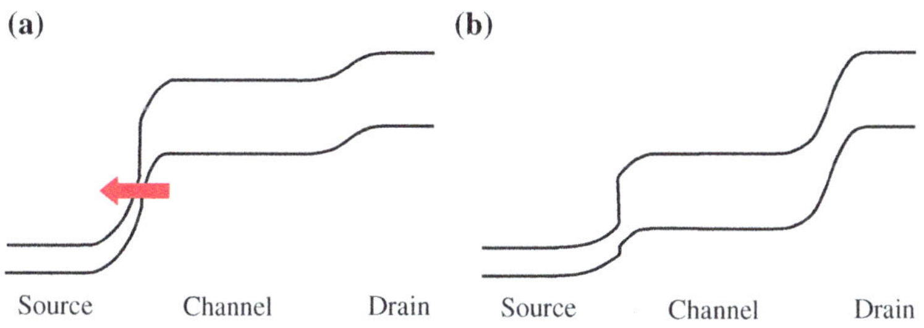

Fig. 7.22 Bandgap-engineered p-channel TFET with a narrow bandgap channel material (InAs) at the source region and a wide bandgap material (Si) at the channel and the drain regions: **a** under high gate and high drain bias voltage and **b** under low gate and high drain bias voltage

offset in the InAs–Si heterojunction can enable a large ON current in the p-channel TFET.

7.3 Nanowire Solar Cells

7.3.1 Backgrounds

A tremendous amount of energy is pouring down on the earth during the day. Figure 7.23 shows the solar radiation spectrum [13]. Even though various gases (O_3, O_2, H_2O, CO_2) in the atmosphere absorb photons with certain wavelengths, the wavelength of photons at the sea level is still distributed in a wide range including the infrared region. As shown in Fig. 7.24, however, the absorption coefficient of light in silicon is high only for short wavelengths [14]. Thus, on the average, photons should travel quite a long distance through silicon without absorption, and many carriers are generated deep within a silicon solar cell.

In order to increase the amount of carrier generation in a solar cell, the cell thickness should be increased. In planar solar cells, however, a thicker absorbing

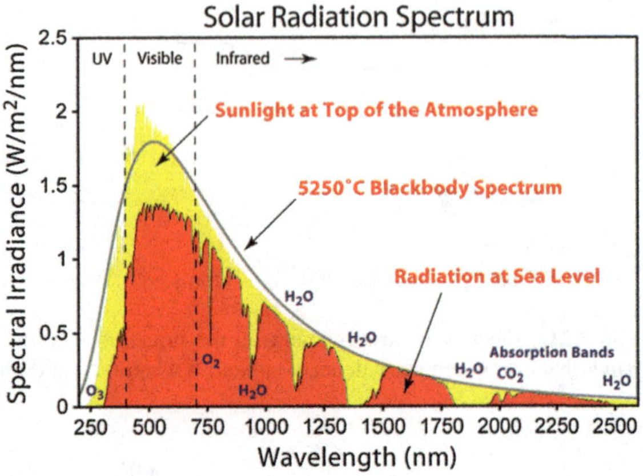

Fig. 7.23 Solar radiation spectra at the top of atmosphere and at sea level. The difference between the two is due to the absorption occurring in the atmosphere. Also shown is the blackbody spectrum at 5,250 °C. http://en.wikipedia.org/wiki/Sunlight

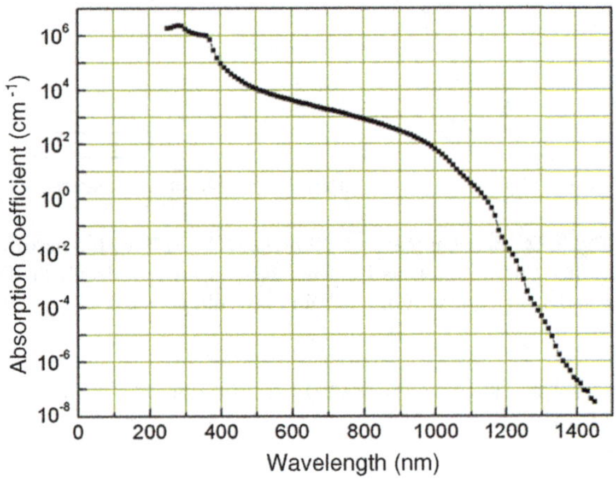

Fig. 7.24 Absorption in silicon. Many carriers should be generated deep within a silicon solar cell. Absorption coefficient data are from [14]

layer suffers from the reduced carrier collection efficiency. Such a trade-off is related to the coupling of photon absorption and carrier collection in a planar solar cell, as shown in Fig. 7.25. Figure 7.25b shows various events that can occur in the generated carriers. The carriers generated within the depletion region are efficiently collected. The minority carriers generated outside of the depletion region should diffuse to the depletion region to be collected. During diffusion, however, the minority carriers can recombine with the majority carriers, especially in a thick solar cell. In this case, the minority carrier cannot be collected efficiently to generate electric power. Since the recombination process is mediated by traps, high-quality (low trap density) material should be used in a thick solar cell. A thicker and higher-quality material unavoidably results in higher cost.

The efficiency of a solar cell can be improved significantly by using a multi-junction cell where a few p–n junctions made of materials with different bandgaps are stacked. In a multi-junction cell shown in Fig. 7.26, three materials (InGaP,

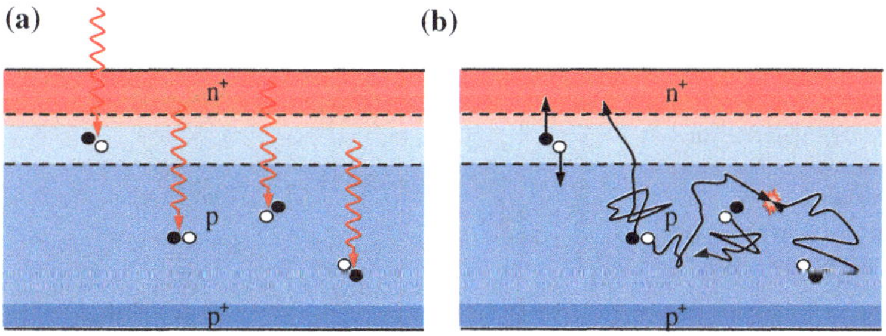

Fig. 7.25 Carrier generation and collection in a solar cell: **a** carrier generation by absorption of light and **b** carrier collection. The carriers generated within the depletion region are efficiently collected. The minority carriers generated outside of the depletion region should diffuse to the depletion region to be collected

GaAs, Ge) are stacked. The top layer made of InGaP absorbs photons with an energy of 1.8 eV or higher, the middle layer made of GaAs absorbs photons with an energy of 1.4 eV or higher, and the bottom layer made of Ge absorbs photons with an energy of 0.67 eV or higher. Since all three junctions are connected in series, the open-circuit voltage of the whole cell is the sum of the open-circuit voltage of each cell. At the same time, photons with a wide range of energy can be absorbed in a relatively thin cell by using a stack of materials with various bandgaps.

In Fig. 7.26, the three materials used in the stack are chosen to have almost the same lattice constant, so that the crystal growth of the heterostructure would not suffer from lattice mismatch. Figure 7.27 shows the bandgap versus lattice constant for various photovoltaic materials. We can see that the indium (In) composition, x, of $In_xGa_{1-x}P$. As should be chosen to match the lattice constant of GaAs. In addition, the GaAs/Ge stack is used, because their lattice constants match almost perfectly. Thus, in the planar multi-junction solar cells, there are severe limitations in the selection of stack materials. As mentioned in Sect. 7.2.4, nanowires can help overcome these limitations, since the small cross-section of nanowires can accommodate the lattice mismatch to a certain extent.

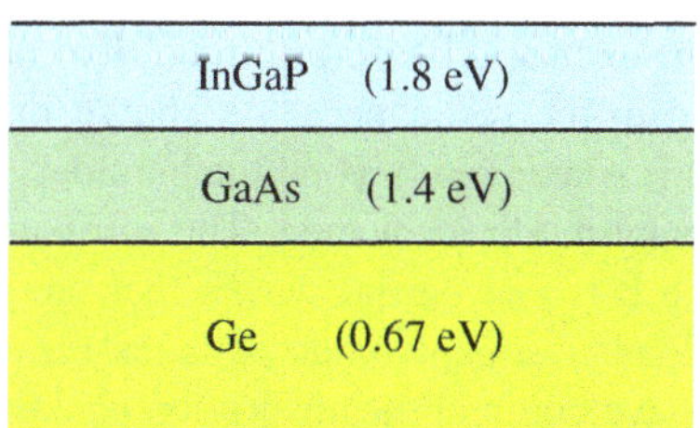

Fig. 7.26 Multi-junction solar cell. The top layer made of InGaP absorbs photons with an energy of 1.8 eV or higher, the middle layer made of GaAs absorbs photons with an energy of 1.4 eV or higher, and the bottom layer made of Ge absorbs photons with an energy of 0.67 eV or higher

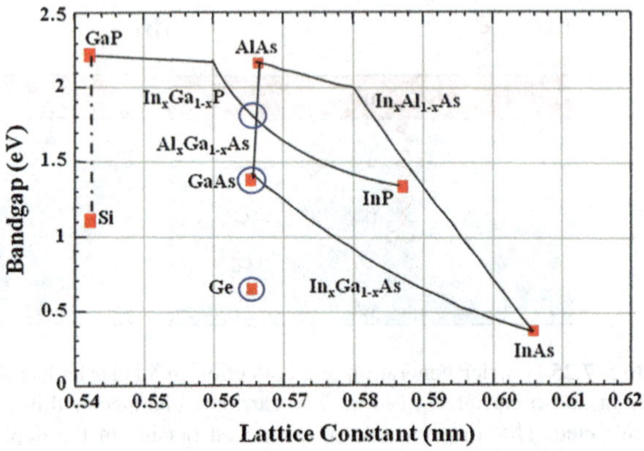

Fig. 7.27 Bandgap versus lattice constant for photovoltaic materials. The characteristics of alloys between two materials are marked by *lines*. The *three circles* indicate *three materials* used in the solar cell shown in Fig. 7.26

7.3.2 Advantages of Nanowire Solar Cells

Nanowire solar cells have several advantages over traditional wafer-based or thin-film planar devices in terms of optical, electrical, and material properties, and cost. Optically, nanowire solar cells can have a decreased reflectance by light trapping, and the resonance effects in the nanowire structure can enhance the light absorption. Electrically, core–shell nanowire structures can increase the junction area and decrease the distance between carrier generation and collection. Due to the strain relaxation effect of nanowires, wider materials, and heterostructures can be used. There are also cost benefits in using nanowires, since less material can be used and material quality standard can be relaxed.

Figure 7.28 shows one advantage of nanowires in optical characteristics. Multiple reflections of photons in an array of nanowires can reduce reflectance and, at the same time, enhance light trapping. This effect, however, can be quite sensitive to the angle of the incident photon in a vertical nanowire array (Fig. 7.28a). For example, the photons incident parallel to nanowires would suffer from low absorption. If we introduce scattering centers between nanowires (Fig. 7.28b) [15] or use randomly aligned nanowires (Fig. 7.28c) [16], the sensitivity to the angle of incidence can be reduced significantly.

Optical resonance effects help enhancing absorption in nanowires. Figure 7.29 shows the leaky-mode resonances (LMRs) in individual nanowire devices and the absorption peaks related to the resonances. As shown in Fig. 7.29a, the electromagnetic wave function is spread out beyond the boundary of nanowire, so that the resonance is called a leaky-mode resonance. The resonance occurs as a result of constructive interference between partial waves that are reflected many times at the nanowire surface. Each resonant mode satisfies the condition of a standing wave and the resonant wavelength at the absorption peak increases as the diameter of the nanowire increases. For a nanowire with a large diameter, multiple absorption peaks can be obtained due to multiple resonance modes.

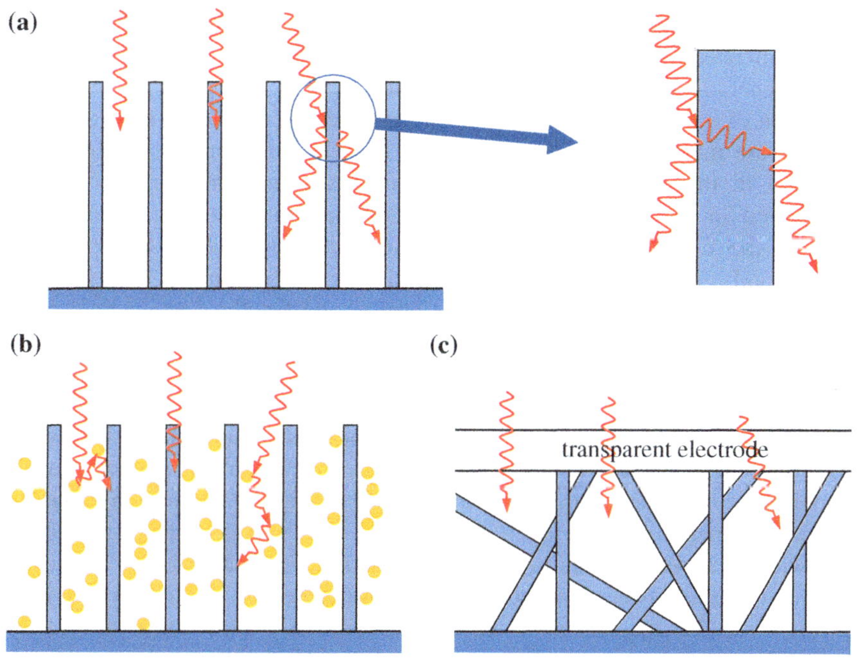

Fig. 7.28 Reduction in reflectance by light trapping: **a** multiple reflections of photons by nanowires, **b** enhancement of light trapping by scattering centers, and **c** enhancement of light trapping by random alignment of nanowires

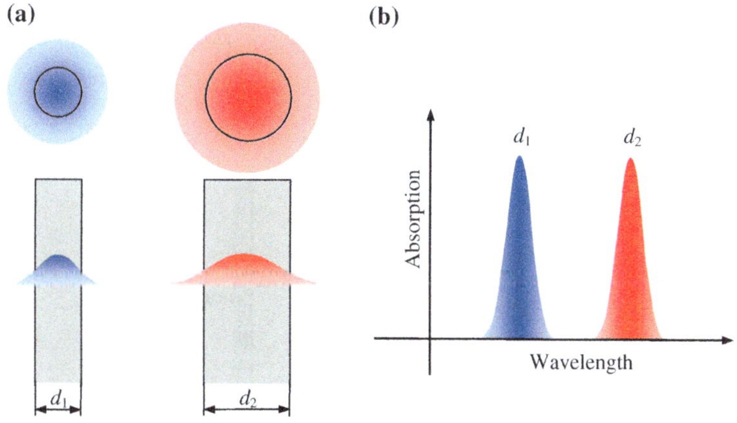

Fig. 7.29 Enhancement of light absorption by optical resonance effects: **a** leaky-mode resonances in nanowires and **b** the related absorption peaks

Vertical core–shell nanowire structures can increase the p–n junction area without increasing the footprint and decrease the distance between the locations of carrier generation and collection. As shown in Fig. 7.30, absorption of photons and collection of carriers are decoupled in the vertical core–shell nanowire structure. Photons are moving mainly in the vertical direction and generate carriers when absorbed in the nanowire. The electron–hole pairs generated by photons are accelerated in the horizontal direction by the horizontal electric field formed between the p-type and n-type regions. Since the directions of movement for photons and carriers are perpendicular to each other, the distance of absorption and collection can be optimized separately. That is, the length of nanowires can be maintained long to provide photons with a sufficient distance for absorption, while the radius of nanowires is maintained small to facilitate the collection of carriers in the p-i-n junction.

As discussed in Sect. 7.3.1, multi-junction heterostructures are very useful in increasing the efficiency of solar cells. Since it is not easy to find a lattice-matched combination of materials that have appropriate bandgaps, the growth of lattice-mismatched heterostructure is often necessary. Due to their small cross-section, nanowires are very useful for growing the lattice-mismatched heterostructure. Heterojunctions between a nanowire and a substrate are possible. Many high-quality direct bandgap materials have been epitaxially grown on a silicon substrate. The use of less expensive substrate is quite effective for the cost reduction. Heterojunctions within a nanowire can also be grown. Both axial and radial (core–shell) heterostructures have been successfully grown in spite of lattice mismatch.

The use of nanowires as an active layer can also contribute to cost reduction. By improving the efficiency of the solar cell and reducing the density of active material, less material with somewhat lower quality can be used. The reuse of substrate is also useful in cost reduction. For example, nanowires are grown on a silicon substrate and a polymer material is filled between nanowires. Then, the layer of nanowire and polymer is detached from the substrate, so that the substrate can be reused for the next nanowire growth.

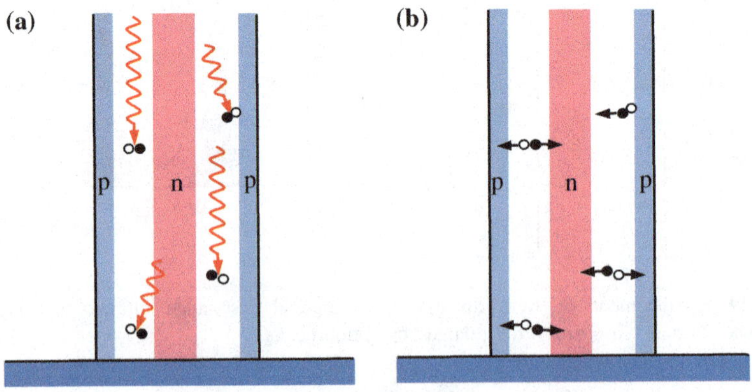

Fig. 7.30 Decoupling of absorption and collection: **a** increased p–n junction area and **b** decreased distance between the locations of carrier generation and collection

7.3.3 Types of Junctions

Figure 7.31 shows the two types of solar cell junctions. In the axial junction structure (Fig. 7.31a), the p–n junction and/or heterojunction are formed in the direction of the nanowire axis. In an array of vertical axial junction nanowires, the absorption length of light can be increased significantly by growing long nanowires and the reflectance can be reduced by light trapping. Resonance effect can further enhance the absorption of light in this array. In the radial (core–shell) junction structure (Fig. 7.31b), the p–n junction and/or heterojunction are formed in the radial direction. In this case, in addition to the enhancement of absorption in an array of vertical nanowires, the radial junction structure enables the decoupling of carrier generation and collection.

For nanowire arrays, a large number of growth or formation techniques have been developed. We will focus on two techniques commonly used in nanowire solar cells: (1) chemical vapor deposition and (2) patterned etch.

In chemical vapor deposition, nanowires are formed by flowing precursor gases into a substrate, often with the assistance of a metal catalyst nanoparticle. There are a number of mechanisms that promote nanowire growth instead of uniform thin-film deposition. The most commonly used is the vapor–liquid–solid (VLS) mechanism, which utilizes a metal catalyst that forms a eutectic liquid with the desired nanowire material (Fig. 7.32) [17, 18]. After chemical decomposition and dissolution into the eutectic liquid, the solution becomes supersaturated and overcomes the nucleation barrier to begin precipitation. Additional flux of dissolved species leads to nanowire growth. With the proper conditions, vertical nanowire growth is obtained, which is advantageous for solar cells. The ordered nanowire array fabrication can be realized by using the catalyst patterning technique. Dopants can be introduced during growth (in situ) or in a separate diffusion step (ex situ).

Patterned etch involves a patterning step followed by an etch step. Figure 7.32 shows the fabrication procedure. The mask patterning step is based either on a top-down approach such as e-beam and nanoimprint or a bottom-up approach such as

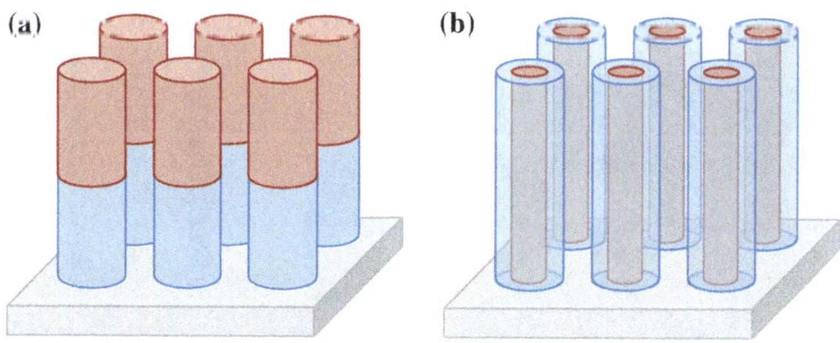

Fig. 7.31 Two types of junctions: **a** axial junction and **b** radial (core–shell) junction

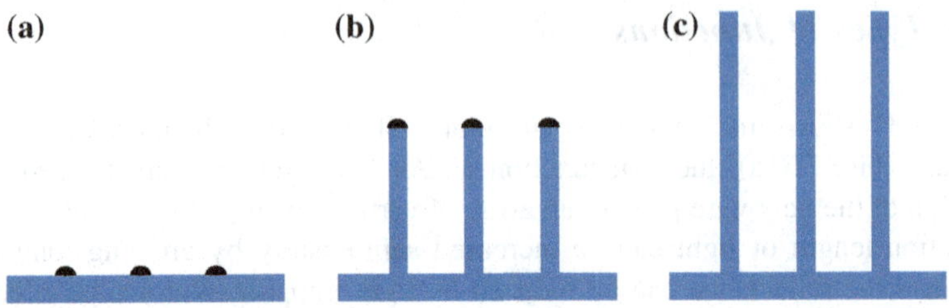

Fig. 7.32 Growth of nanowire array by vapor–liquid–solid (VLS) method: **a** formation of metal catalyst nanodots, **b** vertical nanowire growth by VLS method, and **c** grown nanowires after metal catalyst removal

anodic aluminum oxide (AAO) and block copolymer formation. The unmasked regions of the substrate are removed by an anisotropic etch, so that nanowires would stand out. The patterned etch technique has an advantage in that the starting wafer sets the doping level and material composition, which allows precise control over the material parameters (Fig. 7.33).

Once nanowires are formed, a p–n junction must be introduced to enable carrier separation and collection. As shown in Fig. 7.31, there are two types of junctions in nanowires: axial junction and radial junction. Let us consider the case of axial junction first. If the VLS method is used for the nanowire growth, the p–n junction or heterojunction can be formed by switching the gas during the growth. For the patterned etch method, the junction should be formed before the patterned etch process (Fig. 7.34).

A radial junction should be introduced after nanowires are completely formed. Nanowires can be formed by a variety of methods. Usually, an epitaxial shell is grown conformally on the nanowire core. The isotropic nature of the shell growth is important in this case. If the shell growth has directionality, the uniformity of shell thickness cannot be guaranteed. Sometimes, amorphous material is deposited as a shell and crystallized by annealing (Fig. 7.35).

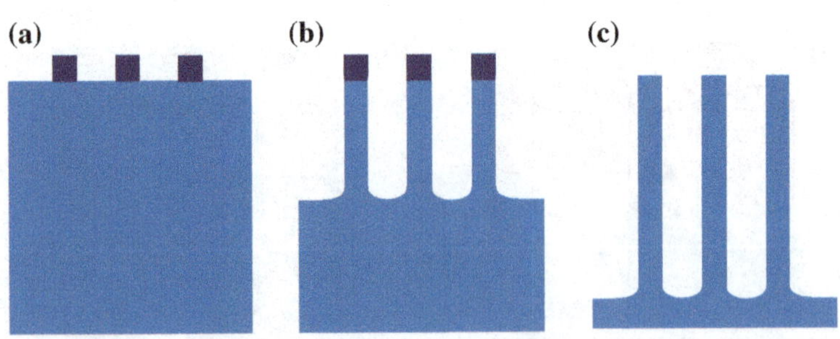

Fig. 7.33 Formation of nanowire array by patterned etch method: **a** nanodot mask patterning on a substrate, **b** anisotropic etch, and **c** nanowires after nanodot mask removal

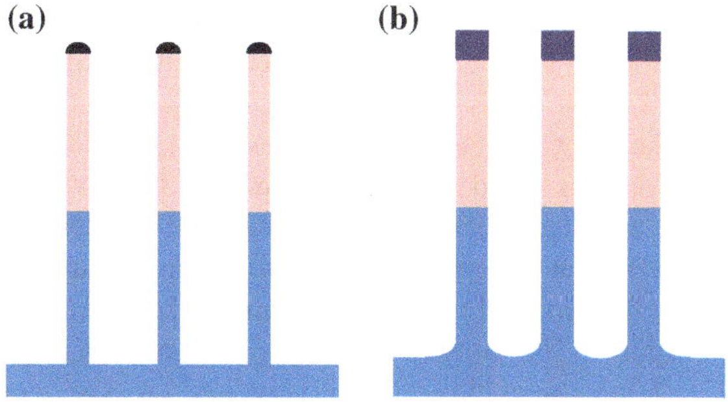

Fig. 7.34 Formation of axial junction: **a** junction formation by switching the gas during the VLS growth and **b** junction formation before the patterned etch process

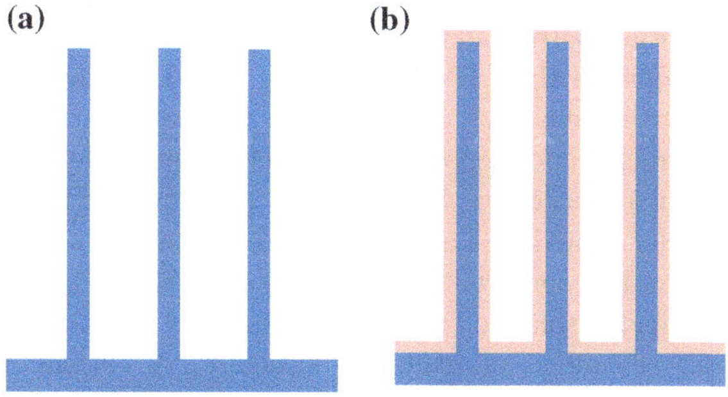

Fig. 7.35 Formation of radial junction: **a** preparation of nanowire array and **b** shell formation by epitaxy or chemical vapor deposition

Sometimes, a heterojunction is used to separate charge carriers. For this case, a staggered bandgap heterojunction in which one material has higher conduction and valence bands than the other is used to separate the photogenerated electron–hole pairs. The staggered bandgap heterojunction is preferred over a straddled bandgap heterojunction in which one material straddles both bands of the other material, because the former helps ensure that electron–hole transfer occurs in the desired direction. Many solar cell materials, including cadmium telluride, copper indium gallium selenide (CIGS), and organic materials, use heterojunctions to separate carriers.

7.3.4 Fabricated Nanowire Solar Cells

For the growth of silicon nanowires, silane (SiH_4), disilane (Si_2H_6), dichlorosilane (SiH_2Cl_2), and silicon tetrachloride ($SiCl_4$) can be used as precursors. The growth temperature ranges from 400 to 1,000 °C. For vertical alignment of nanowires, (111) silicon substrates are used. We can detach nanowires from the substrate and place them on another substrate. The benefits of using nanowire transfer technique include decoupling of high-temperature growth from substrate. It allows the use of low-temperature substrates such as glasses.

The most popular growth method of Si nanowires is metal-catalyzed VLS or vapor–solid–solid (VSS) technique. Gold (Au) and titanium (Ti) are used as typical metal catalysts. Gold makes a liquid eutectic with silicon at about 360 °C and the growth temperature ranges from 400 to 1,000 °C, so that gold nanoparticles stay in liquid state during growth. Thus, with gold as a catalyst, the growth relies on VLS mechanism. The eutectic temperature of titanium and silicon is 1,330 °C, but the growth temperature remains around 600 °C. Thus, VSS must be the growth mechanism for the titanium-catalyzed silicon nanowires.

Figure 7.36a shows a schematic diagram of a Si nanowire solar cell design, and Fig. 7.36b shows typical plan-view and cross-sectional scanning electron micrographs of a Si nanowire solar cell fabricated on a stainless steel foil [16]. The Ta_2N thin film serves as an electrical back contact for the nanowire arrays as well as a diffusion barrier during nanowire growth. Following deposition of a 5-nm-thick Au film, catalytic CVD employing the VLS growth mechanism is used to grow p-type Si nanowires. The array is subsequently coated with a plasma-enhanced chemical vapor deposited (PECVD), conformal n-type amorphous silicon (a-Si:H) layer to create the photoactive p–n junction. After a-Si:H deposition, the array is sputter-coated with a 200-nm-thick transparent conducting indium tin oxide (ITO) layer. Top contacts are created by evaporation of Ti (50 nm)/Al (2,000 nm). The fabricated solar cell shows a significantly reduced reflectance (<1 %), confirming the advantage of nanowire structure. A current density of ~1.6 mA/cm^2 is obtained, and the maximum external quantum efficiency (EQE) of ~12 % is measured at the wavelength of 690 nm. The power conversion efficiency was low (~0.1 %), which may be attributed to the high series and low shunt resistances.

The ordered arrays of silicon nanowires can be formed with the use of patterned etch. It is shown that the ordered arrays of silicon nanowires can increase the path length of incident solar radiation by up to a factor of 73 [19]. This extraordinary

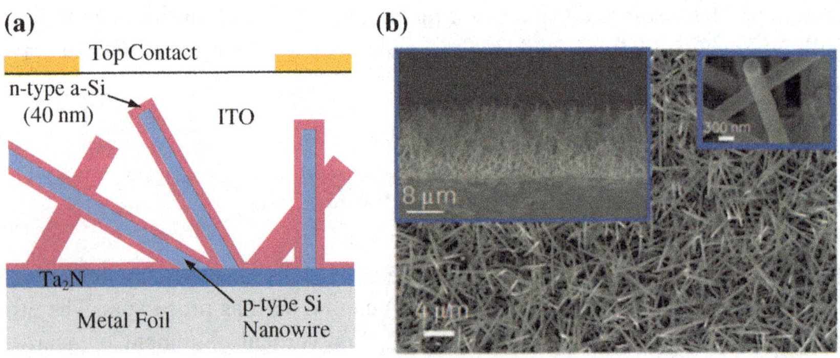

Fig. 7.36 Si nanowire solar cell: **a** schematic cross-section of the Si nanowire solar cell. The nanowire array is coated with a conformal a-Si:H thin-film layer. **b** Scanning electron micrograph (*plan-view*) of a typical Si nanowire solar cell on a stainless steel foil, including a-Si and ITO layers with insets showing a cross-sectional view of the device and a higher magnification of Si nanowires coated with a-Si and ITO. Reproduced with permission from [16]. ©2007 AIP

light-trapping path length enhancement factor is above the randomized scattering limit, and the enhancement scheme is superior to other light-trapping methods. It should be pointed out that, in nanowire solar cells, there is a competition between improved absorption and increased surface recombination. For nanowire arrays fabricated from 8-μm-thick silicon films, the enhanced absorption can dominate over surface recombination. These nanowire devices are capable of attaining the power conversion efficiencies above 5 %, with short-circuit photocurrents higher than planar control samples.

Compound semiconductor nanowire solar cells have been comprehensively reviewed by Sun et al. [20]. Various III–V compound materials, such as GaN, GaAs, and InP, have been used for nanowire growth. The commonly used techniques for III–V nanowire growth include metal–organic chemical vapor deposition (MOCVD), molecular beam epitaxy (MBE), chemical beam epitaxy (CBE), and chemical vapor deposition (CVD). Various metals, such as Au, Pt, Ni, Ag, and Al, have been used as catalysts for VLS nanowire growth of III/V compounds. Tang et al. have reported the fabrication of p-type GaN nanowire arrays on n-type Si substrates, and a maximum conversion efficiency of 2.73 % is achieved [21]. Colombo et al. have demonstrated the fabrication of GaAs core–shell nanowire p-i-n junction solar cells that show a conversion efficiency of 4.5 % with a single nanowire structure [22]. Goto et al. have fabricated high-quality core–shell InP nanowire arrays with an overall conversion efficiency of 3.37 % [23].

The semiconductor nanowires of II–VI compounds, such as CdS, CdTe, and ZnO, can be synthesized by low-cost solution-based methods. These materials maintain the benefits of inorganic materials, such as high carrier mobility, robust material stability, and large interfacial area. Unfortunately, however, II–VI materials usually suffer from the high defect density, and controlled doping is often challenging in II–VI semiconductors. Tak et al. have synthesized a ZnO/CdS core–shell nanowire array and tuned the light absorption spectrum by altering the thickness of the CdS shell, achieving a maximum power conversion efficiency of 3.53 % [24]. Fan et al. have fabricated CdS/CdTe core–shell nanowire solar cells in AAO templates, demonstrating a power conversion efficiency of 6 % [25].

Finally, the nanowire–polymer hybrid solar cells employ the high electron affinity of inorganic semiconductors and the low ionization energy of organic polymers to promote efficient carrier collection. They also utilize large absorption coefficients of polymers and high electron mobilities in inorganic II–VI semiconductors. Kwak et al. have fabricated CdS/polymer hybrid solar cells operated with 1.73 % power conversion efficiency [26]. Hybrid systems based on metal oxide, such as ZnO and TiO_2 nanostructures, are under active investigation. Zhang et al. have reported TiO_2 nanotube-buffered solar cells covered by a (6,6)-phenyl-C_{61}-butyric acid methyl ester (PCBM) layer, which achieve a power conversion efficiency of 3.32 % [27].

7.4 Problems

7.1. Answer the following questions related to the calculation of power consumption in a CMOS inverter circuit.

(a) Derive Eq. (7.2) by integrating Eq. (7.1).

(b) Before the high–low transition, how much energy was stored in the capacitor C?

(c) Now, let us calculate the power consumption from the viewpoint of the power supply. What is the energy that the power supply provides during the low–high transition. What energy should the power supply provide during the high–low transition? Calculate the total energy that the power supply provides during one clock cycle and show that this energy is the same as the energy consumed by the MOSFETs during the cycle.

7.2. Consider an n-channel MOSFET with 2-nm gate oxide and 10^{18} cm^{-3} channel doping. Assuming that the MOSFET has a planar structure, let us calculate its subthreshold swing as follows.

(a) Calculate the oxide capacitance per unit area, C_{ox}. The permittivity of silicon dioxide is 3.45×10^{-13} F/cm.

(b) Calculate the channel capacitance per unit area, C_{ch}. We assume that C_{ch} is the depletion capacitance per unit area for 1 V voltage drop across the depletion region. The permittivity of silicon is 1.04×10^{-12} F/cm.

(c) Now calculate the subthreshold swing, SS.

7.3. Tunneling probability determines the exponential factor of tunneling current given in Eq. 7.11. Let us calculate the tunneling probability and consider the effect of bandgap on the current in a TFET.

(a) For the triangular potential given in Fig. P7.3, calculate the tunneling probability as a function of the electric field, E, using Wentzel–Kramers–Brillouin (WKB) approximation. According to the WKB approximation, the wave function in a potential barrier can be approximated as $\psi(x) = \psi(0)\exp\left[-\int_0^x \kappa(x)dx\right]$, where $\kappa(x) = \sqrt{2m^*[U(x) - E]}$ ($U(x)$ is the

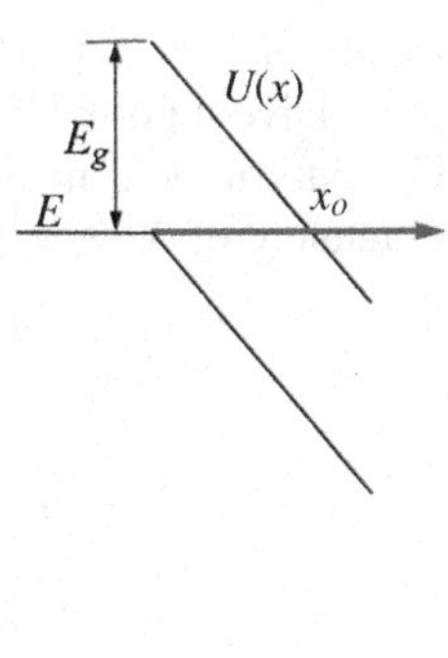

Fig. P7.3 Triangular potential approximation for tunneling probability calculation. E_g is the bandgap, E is the electron energy, $U(x)$ is the potential energy, and x_o is the exit point of electron

potential energy and E is the energy of the tunneling electron). The tunneling probability is given as $T = |\Psi(x_o)/\Psi(0)|^2$.

(b) Compare the tunneling probability obtained in (a) with Eq. 7.11 and confirm that the tunneling probability is the exponential factor of the tunneling current. Explain why the tunneling probability is a dominant factor in determining the magnitude of the tunneling current.

(c) Now, plot the tunneling probability as a function of electric field, E (1 kV/cm < E < 4 MV/cm), for silicon and germanium, on the same semilogarithmic graph. The abscissa of the graph should be log T. The bandgap of silicon is 1.1 eV, and the bandgap of germanium is 0.67 eV. Assume that the effective mass is the same as the free electron mass (9.1×10^{-31} kg), the electronic charge is 1.6×10^{-19} C, and the reduced Planck's constant is 1.06×10^{-34} Js.

7.4. Let us derive Eq. (7.13) as follows.

(a) Using Gauss's law, find the electric field in the oxide as a function of radius, x. Assume that the charge per unit length in the channel is Q.

(b) By integrating the electric field from R_{ch} to $R_{ch} + t_{ox}$, calculate the voltage across the oxide.

(c) Calculate the oxide capacitance per unit length. Dividing this value by the perimeter of the channel cross-section, we can obtain Eq. (7.13).

7.5. The typical I–V characteristic of a solar cell is shown in Fig. P7.5. Solar cells operate in the 4th quadrant of the I–V characteristic (i.e., forward voltage and reverse current) such that power is extracted rather than consumed in the device. Answer the following questions.

(a) Open-circuit voltage is defined as the voltage that appears at the output terminals of the solar cell when no current flows through the external circuit. Mark the open-circuit voltage as V_{oc} on the graph.

(b) Short-circuit current is defined as the current that flows when the output terminals of the solar cell are shorted. Mark the short-circuit current as $-I_{sc}$ on the graph.

(c) In general, the load connected to the solar cell has a finite value, so that the output voltage stays between 0 and V_{oc}. If the voltage is V and the current is I, the power extracted from the solar cell is $P = -VI$. By

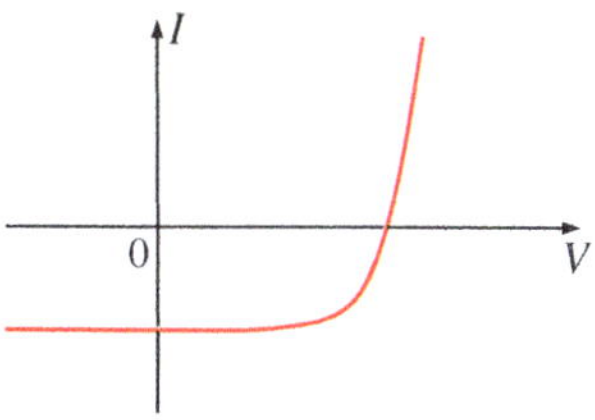

Fig. P7.5 Typical I–V characteristic of a solar cell. Solar cells operate in the 4th quadrant of the I–V characteristic

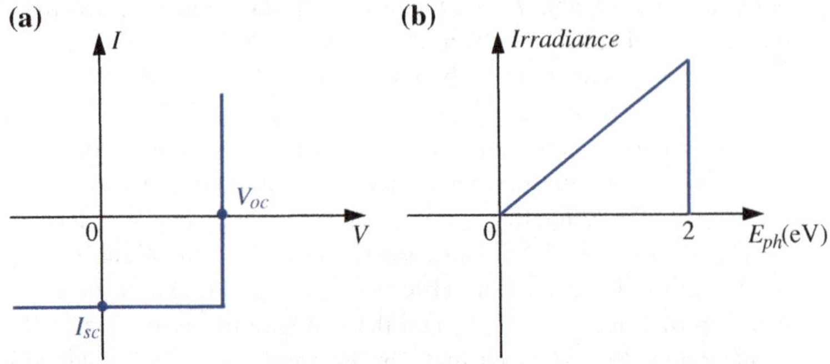

Fig. P7.6 **a** I–V characteristic of a solar cell and **b** irradiance of sun

adjusting the load, we can extract the maximum power, P_{max}, from the solar cell. Now, we define the fill factor as $FF = P_{max}/(V_{oc}I_{sc})$. Explain what the fill factor means graphically on the graph in Fig P7.5.

(d) Let us approximate the I–V characteristic of a solar cell with a piecewise linear model, $I = -I_{sc} + (V - V_{on})s(V - V_{on})$, where $V_{on} < V_{oc}$ and $s(x)$ is a step function defined as

$$s(x) = \begin{cases} 1, & x \geq 0 \\ 0, & x < 0 \end{cases}.$$

Calculate the fill factor for this I–V characteristic.

7.6. Consider the methods of improving the conversion efficiency of a solar cell. We will assume a simplified current–voltage characteristic and solar spectrum as shown below. The open-circuit voltage of the solar cell is expressed as

$$V_{oc} = \frac{kT}{q} \ln\left(\frac{I_{op}}{I_o}\right).$$

The unit of irradiance is [$cm^{-2}s^{-1}$], and E_{ph} is the photon energy.

(a) If we assume the current–voltage characteristic shown above, what would be the value of the fill factor, FF? In addition, calculate the maximum output power, $P_{\max}$, that can be delivered by the solar cell.

(b) Let us first calculate the efficiency of the solar cell with one junction. If we assume 100 % absorption of photons with energy above the bandgap, what is the efficiency of the solar cell with a bandgap of 0.5 eV? For simplicity, we assume that the solar cell is designed to have the relationship $I_o = 10^7 I_{op} \exp[-E_g/(k_B T)]$.

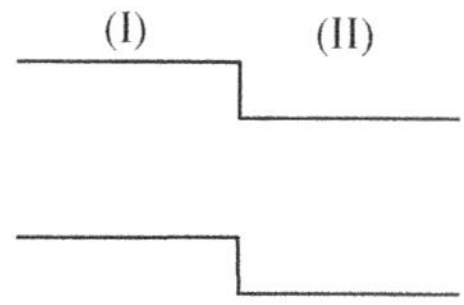

Fig. P7.7 Staggered bandgap heterojunction

(c) In (b), what is the value of the bandgap that maximizes the efficiency? Calculate the maximum efficiency.

(d) Now let us consider a multi-junction case where we have *m* junctions. Since all the junctions are connected in series, there should be a constraint on the value of the bandgaps. What should be the bandgaps for each junction, if the smallest bandgap is fixed at 0.5 eV? Each junction is designed in the same way as in (b). We assume that we can always find an appropriate material (or alloy) as long as the bandgap is between 0.5 and 2 eV.

(e) Calculate the efficiency for *m*-junction solar cell, assuming that the smallest usable bandgap is 0.5 eV.

(f) What is the maximum efficiency we can achieve, if there is no limit in the number of junctions? What do you think is the major reason why there is a limit in the number of junctions?

7.7. Consider a staggered bandgap heterojunction shown in Fig. P7.7. Answer the following questions regarding solar cells with a staggered bandgap heterojunction.

(a) How a staggered bandgap heterojunction can be used to separate charges in a solar cell.

(b) If we want to use a p–n junction in addition to the band offset for charge separation, which of the following combinations is preferable? Explain why that is the case.

(I) n-type and (II) p-type
(I) p-type and (II) p-type

References

1. Wikipedia Global warming. http://en.wikipedia.org/wiki/Global_warming.
2. Gopalakrishnan, K., Griffin, P. B., & Plummer, J. D. (2002). I-MOS: A novel semiconductor device with a subthreshold slope lower than kT/q. IEDM Tech Dig 289.
3. Choi, W. Y., Park, B.-G., Lee, J. D., & King Liu, T.-J. (2007). Tunneling field-effect transistors (TFETs) with subthreshold swing (SS) less than 60 mV/dec. *IEEE Electron Device Letters, 28*(8), 743.
4. Fair, R. B., & Wivell, H. W. (1976). Zener and avalanche breakdown in as-implanted low voltage Si n-p junctions. *IEEE Transactions on Electron Devices, 23*, 512.

5. Vallett, A. L., Minassian, S., Kaszuba, P., Datta, S., Redwing, J. M., & Mayer, H. S. (2010). Fabrication and characterization of axially doped silicon nanowire tunnel field-effect transistors. *Nano Letters, 10*, 4813.
6. Chen, Z. X., Yu, H. Y., Singh, N., Shen, N. S., Sayanthan, R. D., Lo, G. Q., et al. (2009). Demonstration of tunneling FETs based on highly scalable vertical silicon nanowires. *IEEE Electron Device Letters, 30*(7), 754.
7. Kwong, D.-L., Li, X., Sun, Y., Ramanathan, G., Chen, Z. X., Wong, S. M., et al. (2012). Vertical silicon nanowire platform for low power electronics and clean energy applications. *Journal of Nanotechnology, 2012*, 492121.
8. Yang, B., Buddharaju, K. D., Teo, S. H. G., Singh, N., Lo, G. Q., & Kwong, D. L. (2008). Vertical silicon-nanowire formation and gate-all-around MOSFET. *IEEE Electron Device Letters, 29*(7), 791.
9. Lee, M., Jeon, Y., Jung, J.-C., Koo, S.-M., & Kim, S. (2012). Multiple silicon nanowire complementary tunnel transistors for ultralow-power flexible logic applications. *Applied Physics Letters, 100*, 253506.
10. Verhulst, A. S., Vandenberghe, W. G., Maex, K., & Groeseneken, G. (2008). Boosting the on-current of a n-channel nanowire tunnel field-effect transistor by source material optimization. *Journal of Applied Physics, 104*, 064514.
11. Ionescu, A. M., & Riel, H. (2011). Tunnel field-effect transistors as energy-efficient electronic switches. *Nature, 479*, 329.
12. Tomioka, K., & Fukui, T. (2011). Tunnel field-effect transistor using InAs nanowire/Si heterojunction. *Applied Physics Letters, 98*, 083114.
13. Sunlight Wikipedia, http://en.wikipedia.org/wiki/Sunlight.
14. Green, M. A., & Keevers, M. J. (1995). Optical properties of intrinsic silicon at 300 K. *Progress in Photovoltaics: Research and Applications, 3*, 189.
15. Kelzenberg, M. D., Boettcher, S. W., Petykiewicz, J. A., Turner-Evans, D. B., Putnam, M. C., Warren, E. L., et al. (2010). Enhanced absorption and carrier collection in Si wire arrays for photovoltaic applications. *Nature Materials, 9*, 239.
16. Tsakalakos, L., Balch, J., Fronheiser, J., Korevaar, B. A., Sulima, O., & Rand, J. (2007). Silicon nanowire solar cells. *Applied Physics Letters, 91*, 233117.
17. Schmidt, V., Wittemann, J. V., & Gösele, U. (2010). Growth, thermodynamics, and electrical properties of silicon nanowires. *Chemical Review, 110*(1), 361.
18. Wacaser, B. A., Dick, K. A., Johansson, J., Borgström, M. T., Deppert, K., & Samuelson, L. (2009). Preferential interface nucleation: an expansion of the VLS growth mechanism for nanowires. *Advanced Materials, 21*(2), 153.
19. Garnett, E., & Yang, P. (2010). Light trapping in silicon nanowire solar cells. *Nano Letters, 10*, 1082.
20. Sun, K., Kargar, A., Park, N., Madsen, K. N., Naughton, P. W., Bright, T., et al. (2011). Compound semiconductor nanowire solar cells. *IEEE Journal of Selected Topics in Quantum Electronics, 17*(4), 1033.
21. Tang, Y. B., Chen, Z. H., Song, H. S., Lee, C. S., Cong, H. T., Cheng, H. M., et al. (2008). Vertically aligned p-type single crystalline GaN nanorod arrays on n-type Si for heterojunction photovoltaic cells. *Nano Letters, 8*, 4191.
22. Colombo, C., Heibeta, M., Gratzel, M., & Fontcuberta i Morral, A. (2009). Gallium arsenide p-i-n radial structures for photovoltaic applications. *Applied Physics Letters, 94*, 173108.
23. Goto, H., Nosaki, K., Tomioka, K., Hara, S., Hiruma, K., Motohisa, J., et al. (2009). Growth of core–shell InP nanowires for photovoltaic application by selective-area metal organic vapor phase epitaxy. *Applied Physics Express, 2*, 035004.
24. Tak, Y., Hong, S. J., Lee, J. S., & Yong, K. (2009). Fabrication of ZnO/CdS core/shell nanowire arrays for efficient solar energy conversion. *Journal of Materials Chemistry, 19*, 5945.
25. Fan, Z., Razavi, H., Do, J.-W., Moriwaki, A., Ergen, O., Chueh, Y.-L., et al. (2009). Three-dimensional nanopillar-array photovoltaics on low-cost and flexible substrates. *Nature Materials, 8*, 648.

26. Kwak, W.-C., Kim, T. G., Lee, W., Han, S.-H., & Sung, Y.-M. (2009). Template-free liquid-phase synthesis of high-density CdS nanowire arrays on conductive glass. *Journal of Physical Chemistry C, 113*, 1615.
27. Zhang, Q., Yodyingyong, S., Xi, J., Myers, D., & Cao, G. (2012). Oxide nanowires for solar cell applications. *Nanoscale, 4*, 1436.

Chapter 8
Nanowire Field Effect Transistors in Optoelectronics

Mehrdad Shaygan, M. Meyyappan and Jeong-Soo Lee

Abstract The nanowire FETs are discussed in this chapter as key components for optoelectronic circuits. The features of photonic devices are elaborated such as light absorption, generation of e–h pairs and the carrier transport, and the parameters dictating sensitivities, e.g., the wavelengths of light and the materials used are examined. Moreover, the 1-D nanostructures as a platform for photodetectors are highlighted, in conjunction with carbon nanotubes, nanobelts, nanoribbons, nanorods and nanowires. Also, the issues are addressed to regarding the process complexities, light absorption and adsorption, nanowire cross-section, quantum efficiency, threshold voltage shifts, the surface to volume ratio, carrier collection, etc. Moreover, the materials used for 1-D photodetector are examined, including compound semiconductors, Sulfides, Selenides, Tellurides, Metal Oxides, Zinc Oxide, Tin Oxide, Copper Oxide, Gallium Oxides and Indium Oxides, etc. These materials are characterized in terms of Schottky and Ohmic contacts, photoluminescence, photoconductivity, efficiency and speed of detection, etc.

Abbreviation

FET	Field effect transistor
CVD	Chemical vapor deposition
VLS	Vapor–liquid–solid

M. Shaygan (✉)
Nano Sensors and Systems Lab, Division of IT Convergence Engineering, Pohang University of Science and Technology, Pohang, South Korea
e-mail: mshaygan@postech.ac.kr; m.shaygan@gmail.com

M. Meyyappan
Moffett Field, NASA Ames Research Center, California, USA
e-mail: m.meyyappan@nasa.gov

J.-S. Lee
Department of Electronics and Electrical Engineering, Pohang University of Science and Technology, Pohang, Republic of Korea
e-mail: ljs6951@postech.ac.kr

D. M. Kim and Y.-H. Jeong (eds.), *Nanowire Field Effect Transistors: Principles and Applications*, DOI: 10.1007/978-1-4614-8124-9_8,

8.1 Introduction

Photonic devices function based on phenomena such as generation, emission, transmission, modulation, signal processing, switching, amplification, and detection of light. They can be divided into three groups: light emitting diodes (LEDs) and laser diode, photodetectors, and photovoltaic devices or solar cells. In LEDs, electrical energy is converted into optical radiation, while in photodetectors light is detected. The photovoltaic devices or solar cells generate electrical energy from the incident optical radiation. Optoelectronics is a subfield of photonics that deal with the application of electronic devices that generate, detect, and control light. Their operation can cover invisible radiation such as gamma rays, X-rays, ultraviolet and infrared, in addition to visible light. The devices include electrical-to-optical or optical-to-electrical transducers, or instruments incorporating such devices in their operation.

Photoconductivity of semiconductors is defined by the increase in conductivity resulting from the incident light. A sequence of physical events leading to conductivity change includes absorption of incident light, generation of electron–hole pairs, and carrier transport. Hence, there are different physical parameters that determine the magnitude of photoconductivity. The carrier generation quantum yield is defined by the number of carriers generated per absorbed photon, and the mobility of photogenerated carriers [1]. The mobility of the charge carriers determines the response speed and the low dark current determines the sensitivity. The following processes are involved in photodetection:

- Absorption of photons and generation of electron–hole pairs
- Drift of the charge carriers under the electric field
- Collection of charge carriers at the respective contacts providing an output current

In the photoemission, the energy of light must satisfy the photoelectric threshold condition ($h\nu > E$), while the threshold wavelength is defined as:

$$\lambda_s[\mu\mathrm{m}] = hc/E = 1.24/E[\mathrm{eV}] \tag{8.1}$$

where λ is the wavelength, c is the speed of light, and E is the energy gap. For *intrinsic* detectors, $E = E_g$ in Eq. 8.1, where E_g is defined by the positions of the conduction and valence bands of the material; for *extrinsic* detectors useful in far-infrared applications, the energy level of the donor (or acceptor) E_l and conduction band E_c (or valence band E_v) determine the energy gap as $E = E_c - E_l$ (or $E = E_l - E_v$) in Eq. 8.1.

The material choices for semiconductor detectors include elemental (Si, Ge, Se, Te, etc.) and compound semiconductors such as binary (GaAs, CdSe, ZnTe, etc.), ternary (GaAlAs, InGaP, HgCdTe, etc.), and quaternary (InGaAsP) materials. The wide range of available energy gaps allows detectable wavelengths from ultraviolet (UV) to infrared (IR). The solid-state photodetectors exhibit advantages such as:

- suitable size, higher sensitive area due to reasonable aspect ratio
- low bias voltage
- wide range of wavelengths from UV to IR
- higher quantum efficiency
- wide range of operating temperatures
- significant mean time to failure (MTTF)
- low cost.

The photoconductivity effect was first observed in selenium by Smith [2] in 1873. In 1905, Einstein observed photoelectric effect in metals and Plank succeeded in finding a solution for the blackbody emission problem by quanta hypothesis. The type of conversion effect (either optical or electrical) is one of the criteria for the classification of photodetectors. In external photoelectric devices including vacuum diodes, photomultipliers, and image-intensifier tubes, absorption of a photon and generation of electron–hole pairs are followed by emission of further photogenerated electrons, thus leading to multiplication. In internal photoelectric devices such as semiconductor photodetectors, junction and avalanche diodes, phototransistors etc., the generated electron–hole pair, without any emission, is available for current in the external circuit.

Thermal detectors—which constitute another class of photodetectors—operate based on two main processes: dissipation of radiation in an absorber and measurement of the increase in temperature. The sensitivity of thermal detectors is much lower than the other two groups, but they can cover a wide range of spectra from UV to far IR in a single unit. Their current applications include systems for calibrating other detectors and hand-held infrared cameras [3].

8.2 Basic Concepts of Photodetection

Absorption of a photon with energy higher than the bandgap of the semiconductor leads to the generation of electron and hole pairs, which results in an increase in the conductivity of the material. The absorption coefficient determines the possibility of photoexcitation as well as the position of absorption. Light will be absorbed near the surface of its entrance if the absorption coefficient is high. A low value of absorption coefficient implies that the light can penetrate deeper into the semiconductor. For much lower values, photoexcitation does not happen, and thus, light can be transparent for long wavelengths. Furthermore, the quantum efficiency of the semiconductor is determined by the absorption coefficient. The sensitivity of the photodetector can be defined by its quantum efficiency, which shows the number of generated carriers per photon.

$$\eta = I_{ph}/q\phi = (I_{ph}/q) \times (hv/P_{opt}) \tag{8.2}$$

where I_{ph} is the photocurrent, ϕ is the photon flux, and P_{opt} is the power, which determines the photon flux ($\phi = P_{opt}/hv$). Some phenomena including recombination, incomplete absorption, reflection and etc., might result in a reduction in efficiency by causing some current loss. The responsivity R (A/W) is defined as follows:

$$R = \frac{I_{\mathrm{ph}}}{P_{\mathrm{opt}}} = \frac{\eta q}{h\nu} = \frac{\eta\lambda(\mu\mathrm{m})}{1.24} \tag{8.3}$$

The key parameters defining the performance of photodetectors and photoconductors are the quantum efficiency and gain, response time, and sensitivity or detectivity. A simple configuration of the photoconductor shown in Fig. 8.1 consists of a slab of semiconductor, either bulk or thin film, with ohmic contacts on the two ends. Internal gain is the mechanism by which some photoconductors increase the photocurrent signal.

In order to define photocurrent gain as a critical parameter, carrier generation rate G_e at steady-state must be equal to the recombination rate, which can be given by:

$$G_e = \frac{n}{\tau} = \frac{\eta(P_{\mathrm{opt}}/h\nu)}{WLD} \tag{8.4}$$

where n is the excess carriers density, τ is the carrier lifetime, η is the quantum efficiency, P_{opt} is the incident optical power, $h\nu$ is the energy of photon. Incident light on the surface of the photoconductors generates carriers thus causing an increase in conductivity. As mentioned before, depending on the type of semiconductor (either intrinsic or extrinsic), carriers are generated by band-to-band transition or photoexcitation between impurity level and band, respectively (Fig. 8.2).

For an intrinsic photoconductor, the photocurrent between the two contacts can be defined by:

$$I_p = \sigma\xi WD = (\mu_n + \mu_p)nq\xi WD \tag{8.5}$$

where ξ is the applied electric field. By substituting n from Eq. 8.4, the photocurrent is given by:

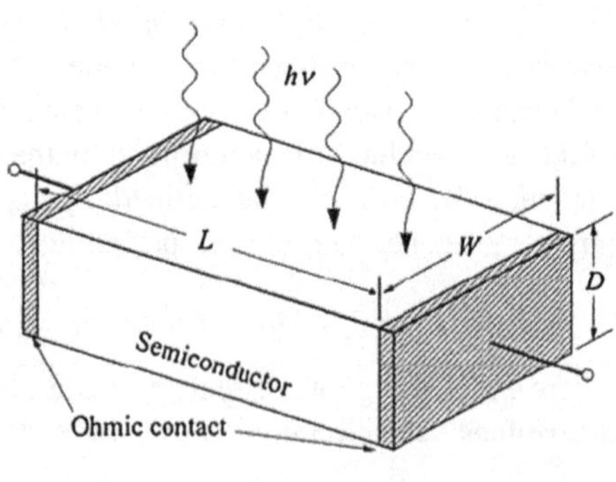

Fig. 8.1 Schematic of a photoconductor consisting of a slab of semiconductor [4]

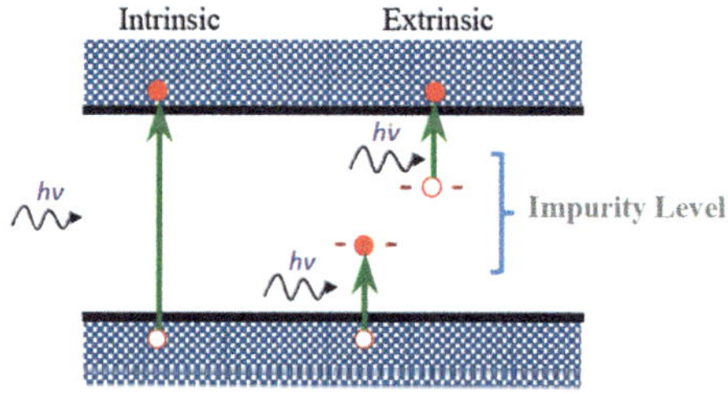

Fig. 8.2 Different processes for carrier generation in intrinsic and extrinsic semiconductors [4]

$$I_p = q\left(\mu\frac{P_{\text{opt}}}{h\nu}\right)\frac{(\mu_n + \mu_p)\tau\xi}{L} \tag{8.6}$$

By defining the new term as primary photocurrent $\left(I_{\text{ph}} \equiv q\left(\mu\frac{P_{\text{opt}}}{h\nu}\right)\right)$, the photocurrent gain G_a is:

$$G_a = \frac{I_p}{I_{\text{ph}}} = \frac{(\mu_n + \mu_p)\tau\xi}{L} = \tau\left(\frac{1}{t_{\text{rn}}} + \frac{1}{t_{\text{rp}}}\right) \tag{8.7}$$

where the electron and hole transit times across the electrodes are given by $t_{\text{rn}}(= L/\mu_n\xi)$ and $t_{\text{rp}}(=L/\mu_p\xi)$, respectively. The typical value of gain is 1,000 but higher values up to 10^6 can be seen in Table 8.1. Long lifetime, short distance between electrodes and high mobilities can help to obtain a high gain. Interestingly, lifetime not only determines the gain value but also determines the response time of a photoconductor. Generally, the response time of a photodiode is much shorter than that of a photoconductor, which can reach high gain values but the accompanying high noise may be a drawback.

As for signal-to-noise ratio, low noise has the same importance as large noise while it could ascertain the minimum detectable signal. Parameters determining the noise value are dark current (the leakage current of the photodetector under bias without illumination), background radiation such as black-body radiation, and

Table 8.1 Values of gain and response time for typical photodetectors [1, 4]

Photodetector		Gain	Response time (s)
Photoconductor	Bulk	1–10^6	10^{-8}–10^{-3}
	Nanowires	1–10^{10}	10^{-11}–10^2
Photodiodes	*p–n* junction	1	10^{-11}
	p–i–n junction	1	10^{-10}–10^{-8}
	Metal–semiconductor diode	1	10^{-11}
CCD		1	10^{-11}–10^{-4}
Avalanche photodiode		10^2–10^4	10^{-10}
Phototransistor		$\sim 10^2$	10^{-6}
Photoemissive	Diode	1	10^{-11}
	Photomultiplier	10^6	10^{-10}–10^{-8}

internal device noise such as thermal noise (Johnson noise), shot noise (due to some distinct phenomena of the photoelectric effect), flicker noise (because of random effects related to surface traps), and the generation-recombination noise, which can come from optical and thermal processes. The sum of all these noises is given as a new term, namely the noise-equivalent power (NEP). It represents the required optical power to generate one signal-to-noise ratio in a 1 Hz bandwidth, which is the minimum optical power that can be detected.

The detectivity $D*$ is defined as follows:

$$D^{*} = \frac{\sqrt{AB}}{\mathrm{NEP}} \tag{8.8}$$

where A and B are the area and bandwidth, respectively. The unit of detectivity is cm–Hz$^{1/2}$/W. Several factors influence detectivity, namely the detector's sensitivity, spectral response, noise, wavelength, modulation frequency and bandwidth [4].

8.3 Photodetection

Figure 8.3 depicts the electromagnetic spectrum covering the ultraviolet (UV), visible and infrared (IR) regimes [4]. The visible portion is expanded at the top exhibiting major color bands, and the human eye can detect only from about 0.4 to 0.7 μm. The UV region is divided into near UV (400–300 nm), mid UV (300–200 nm), far UV (200–100 nm), and extreme UV (100–10 nm). UV

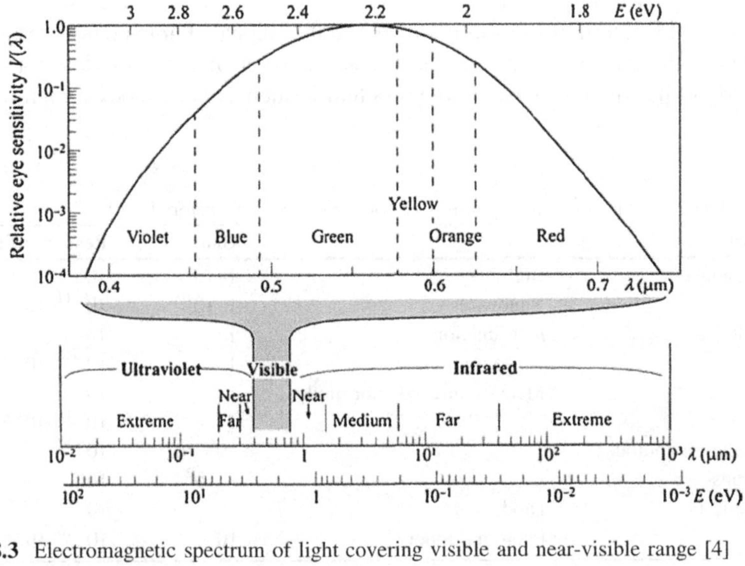

Fig. 8.3 Electromagnetic spectrum of light covering visible and near-visible range [4]

Table 8.2 The bandgap energies and cutoff wavelengths of promising photodetector materials [9]

Type	Bandgap energy (eV)	Cutoff wavelength (nm)	Band
Si	1.12	1,100	Visible
GaAs	1.42	875	Visible
Ge	0.66	1,800	Near-infrared
InGaAs	0.73–0.47	1,700–2,600	Near-infrared
InAs	0.36	3,400	Near-infrared
InSb	0.17	5,700	Medium-infrared
HgCd	0.7–0.1	1,700–12,500	Near-to-far-infrared

photodetectors are widely used in medicine, space communication, high-temperature plasma research, chemical and biological sensing, and in military as flame and missile launching detection. They can also be utilized in monitoring the solar UV radiation [5, 6]. A variety of materials have been employed in the fabrication of UV photodetectors including GaN and ZnO. The visible light detectors attracted much attention after the invention of the image sensor in 1964. The two major image sensors are: area image sensors that include cameras, camcorders, machine vision; and linear sensors in scanning applications [7].

The infrared sensors have been around since the 1940s to detect and image the thermal radiation emitted from the target objects. The infrared region can be divided into short wavelength infrared (1–3 μm), medium wavelength infrared (~3–5 μm), and long wavelength infrared (8–14 μm). These detectors are widely used in military applications; for example, HgCdTe photodetector arrays were first produced in the 1970s [8]. Table 8.2 lists some important infrared materials, their bandgaps, and cutoff wavelengths for photodetection at room temperature.

8.4 One-Dimensional Nanostructures for Photodetectors

One-dimensional (1-D) nanostructures have received much interest due to their significant properties. They can be synthesized with controlled chemical composition, shape, diameter, and length. Reducing the size results in the improvement of electrical properties such as carrier transport, thus making them ideal candidates for nanoscale optoelectronics [10]. Various 1-D nanostructures have been investigated in optoelectronics including nanotubes, nanobelts, nanoribbons, nanorods, and nanowires, which are discussed in detail below.

8.4.1 Carbon Nanotubes

Single-walled carbon nanotube (SWCNT) has been used as a conducting channel for field effect transistors (FETs) [11, 12]. Freitag et al. [13] synthesized a 1.3 nm-diameter nanotube exhibiting 0.7 eV bandgap by laser ablation and fabricated CNT-FETs on Si with a thermal oxide (150 nm), titanium contacts and a thin layer of carbon serving as a barrier to prevent the diffusion of titanium atoms as seen in Fig. 8.4a. The photocurrent depends on the polarization of incident infrared radiation, with the maximum current occurring when the polarization of the light is in the direction of the nanotube axis (Fig. 8.4b). The current due to polarization parallel to the axis of the nanotube is twice that due to perpendicular polarization. The photocurrent also depends on the drain voltage and the power of IR light (Fig. 8.4c and d). The estimated quantum efficiency is 0.1, while polarization angle of infrared light has a significant effect on the measured photocurrent.

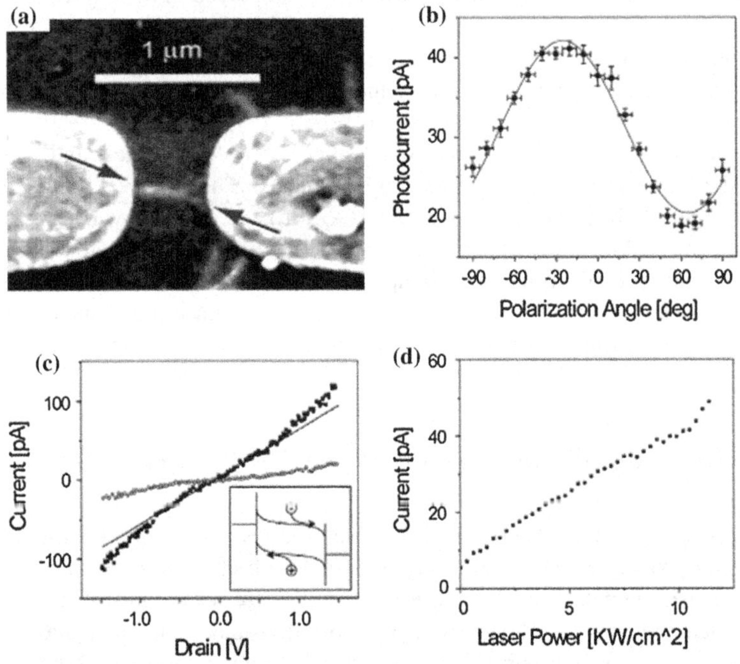

Fig. 8.4 **a** SEM image of a nanotube-based field effect transistor, **b** measured photocurrent versus polarization angle of incident infrared light (excitation energy of 1.28 eV, $V_d = +1$ V, $V_g = -1.5$ V), **c** Drain-voltage dependence of the photocurrent (O) without light and *filled square* (■) under illumination ($V_g = -2$ V, IR intensity 5.6 KW/cm^2). *Inset*: The band configuration in carbon nanotube and the contacts, **d** Photocurrent as a function of the laser power density ($V_g = -2$ V, $V_d = -0.6$ V) [13]

8.4.2 Nanobelts

Nanobelt structures with their rectangular cross-sections exhibit distinct geometries and crystallinities, and their properties are different from nanotubes and nanowires. Among semiconductor nanobelts, various materials have been investigated for photodetectors such as CdS [14, 15], Nb_2O_5 [16], ZnO [17, 18], ZnS [19], ZnSe [20], and ZrS_2 [21]. Catalyzed chemical vapor deposition (CVD) has been used for growing nanobelts. For example, CdS nanobelts were grown with gold nanoparticles as catalyst (see Fig. 8.5). The growth temperature can change the type of nanostructures from nanoribbons to nanowires at lower temperatures. The TEM image of the single CdS nanobelt and the corresponding SAED in Fig. 8.5b reveal that the synthesized nanobelts show single-crystalline structure and ⟨101⟩ growth direction. The existence of gold nanoparticles at the end of the nanobelt, appearing darker with higher contrast compared to the stem of the nanobelt in Fig. 8.5c, confirms the vapor–liquid–solid growth mechanism. The nanobelt under illumination showed a significant enhancement in conductance about four times larger

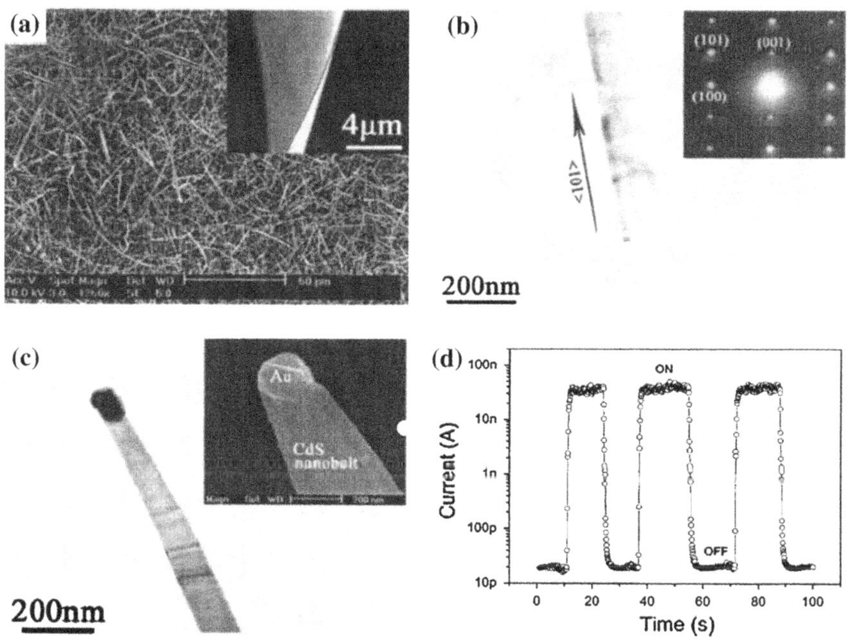

Fig. 8.5 **a** SEM image of the CdS nanobelts synthesized by VLS technique. Inset: higher magnification SEM image of the curved nanobelt. **b** TEM image of an individual CdS nanobelt with a width of 200 nm. *Inset*: SAED pattern depicts a single-crystalline structure of the CdS nanobelt. **c** TEM image of the CdS nanobelt containing a gold nanoparticle on top. **d** The reversible switching behavior of nanobelt while the light is on and off [14]

than when the light was turned off. The rise and decay times of the detector were 1 s and 3 s, respectively (Fig. 8.5d) [14].

Figure 8.6 shows an FET based on CdS nanobelts fabricated by Ye et al. [15]. They synthesized the nanobelts by CVD and deposited two In/Au (10/100 nm) ohmic contacts and a Au Schottky contact using a sequence of steps including photolithography, thermal evaporation and lift-off processes. The dark current and photocurrent were about 26 fA and 70 nA, respectively, yielding a ratio of 2.7×10^6, one of the highest values for photodetectors (Fig. 8.6c and d). On the other hand, they observed a significant shift in V_{th} after illumination due to either the increase in channel conductance or the effect of local electric field. Considering the high responsivity (2.0×10^2 A/W), external quantum efficiency (5.2×10^2 %) and other noticeable properties of the fabricated device could be attributed to the structure of the device. As an improvement of this type of MESFET under illumination, applying a voltage close to the V_{th} as the gate voltage could enable getting lower dark current and faster response time.

In evaluating the performance of photodetectors, selectivity and sensitivity must be considered. ZrS_2 nanobelt photodetectors showed a high selectivity for a wide range of wavelengths from 450 to 700 nm while the photocurrent decreased.

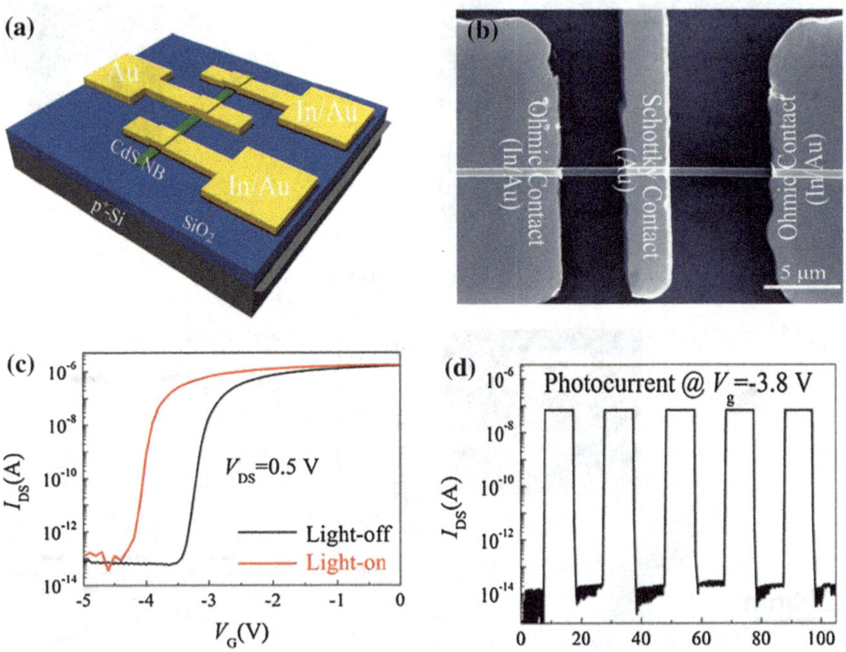

Fig. 8.6 **a** Schematic of the CdS nanobelt-based photodetector. **b** SEM image of the fabricated device. **c** Photoresponse of CdS nanobelt FET measured in dark (*black line*) and under illumination (*red line*). **d** On/off response of the fabricated photodetector with $V_G = -3.8$ V as a function of time [15]

A significant increase in photocurrent (10 times) upon illumination with 450 nm light confirmed its high sensitivity. The responsivity and quantum efficiency of nanobelt photodetectors can be calculated by the following equations:

$$R_\lambda = \Delta I / PS \quad (8.9)$$

$$QE = hcR_\lambda / e\lambda \quad (8.10)$$

where ΔI is the difference between photocurrent and dark current, P is the power of illuminated light, S is the illuminated area of the nanobelt, and λ is the wavelength. The corresponding responsivity of the photocurrent for ZrS_2 was measured as 7.1×10^5 A/W, resulting in a quantum efficiency of 1.8×10^8 %. These high values can be attributed to the presence of oxygen, large surface-to-volume ratio of the nanobelt, high crystallinity and short distance between the electrodes.

Free electrons existing on the surface of the nanobelts, when captured by oxygen molecules, result in not only forming a surface depletion layer but also a decrease in the mobility and carrier density. The incident light on an *n-type* material such as ZrS_2 generates electron–hole pairs, and then, the holes move to the surface of the nanobelt and discharge the negative oxygen ions. Therefore, the number of free-electrons increases and the responsivity and external quantum efficiency are enhanced. Li et al. [21] investigated the effect of oxygen on photocurrent variation of ZrS_2 nanobelts and found an increase in measured photocurrent by decreasing the pressure from 93 kPa to 1 Pa.

The large surface-to-volume ratio of nanobelts can enhance the number of surface trap states, which affects the rate of adsorption or desorption of oxygen. As quantum efficiency depends on the balance between lifetime of carriers and their transit time, reducing the distance between the electrodes and ensuring single-crystalline properties could help to decrease the transit time and increase the quantum efficiency [21]. Interestingly, single crystallinity has a significant effect on photo response of nanobelts as reported by Fan et al. [20]. They found a significant decrease in dark current (1,000 times less) of the ZnSe-nanowire photodetector when twinning type defects were absent, and those with the defects did not show a stable photoresponse behavior.

Nanobelt-based photodetectors have been reported using various materials such as ZnO, ZnS, ZnSe, and Nb_2O_5 showing good performance including high photocurrent, high responsivity, and fast response time. ZnSe nanobelts exhibit a significant selectivity for various wavelengths from 250 to 630 nm, while its response for deep-UV light (250 nm) is three orders of magnitude higher compared with visible light (600 nm). Two orders of magnitude higher photoresponse at 400 nm over that at 600 nm make them a good candidate for detection of UV and short-wavelength blue light. Fang et al. [22] reported a lower responsivity (0.12 A/W at 30 V) for nanobelts compared with the ZnSe nanowire synthesized by vapor-phase growth technique (22 A/W at 2 V). They noticed a fast response time for ZnSe-nanobelt photodetector (<0.3 s), while nanowire photodetectors fabricated with the same material by MBE showed a longer response time (<1 s).

8.4.3 Nanoribbons

Nanoribbons with a rectangular cross-section and well-defined geometry, high crystallinity and significant morphology could be a candidate for photodetectors to obtain high sensitivity and fast response time [23]. Among compound semiconductor nanoribbons, CdS with a bandgap of 2.4 eV is the most promising material for visible light detection. Jie et al. [24] grew CdS nanoribbons on a silicon wafer by CVD using gold as the catalyst. The source temperature was set at 860 °C with a carrier gas mixture of high-purity argon and hydrogen (20 sccm) at a pressure of 150 Torr. The SEM image in Fig. 8.7 shows uniform morphology of CdS nanoribbons with smooth and clean surfaces. The synthesized nanoribbons are 2–40 μm long and 10–60 nm thick. The [112] orientation as growth direction is evident from the HRTEM and selected-area-electron diffraction (SAED) in Fig. 8.7b.

As seen in Fig. 8.7c, the nanoribbons-based photodetector was fabricated by dispersing CdS nanoribbons on Si/SiO_2 wafer and patterning the electrodes using

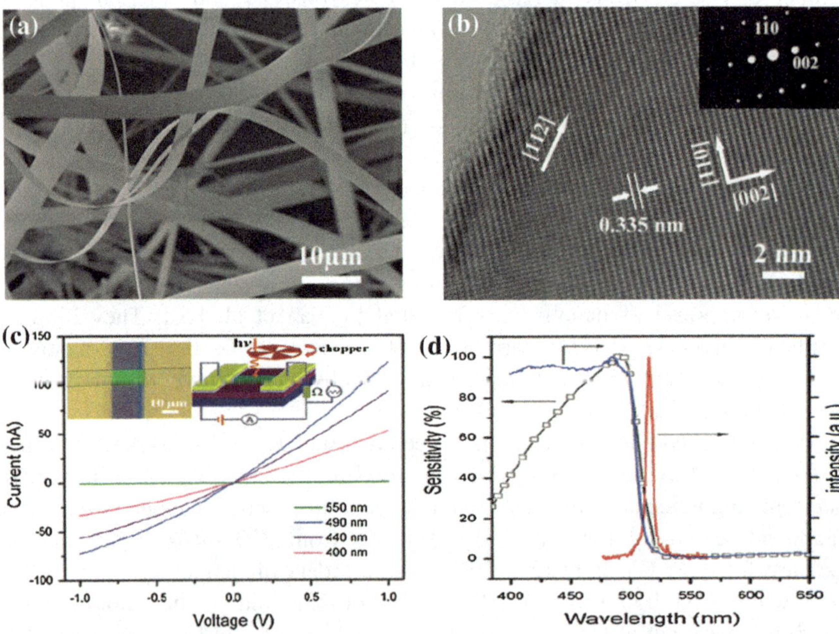

Fig. 8.7 **a** SEM image of the as-synthesized CdS nanoribbons. **b** HRTEM of a CdS nanoribbon and SAED pattern (inset image) reveal the hexagonal structure of the nanoribbons grown along [112] direction. **c** Current–voltage characteristics of a nanoribbon under illumination with different wavelengths with intensity of 1.8 mW/cm^2. *Inset*: schematic of the measurement system (*right*) and the optical microscope image of the single nanoribbon device (*left*) **d** Measured sensitivity of the CdS nanoribbon versus wavelength, PL (*red line*) measurement of the nanoribbon and absorption spectra (*blue line*) [24]

Ti (100 nm) and Au (25 nm). The photodetector showed a high sensitivity to various wavelengths with a conductance of 0.793 nS, 128 nS, 96 nS, and 54 nS for 500, 490, 440, 400 nm, respectively, and 0.021 nS under dark conditions. This indicates a maximum of sensitivity near 490 nm and minimum sensitivity for wavelengths longer than 520 nm. Regarding electron–hole pair generation as the dominant mechanism, longer wavelengths do not have the capability to excite electrons from the valence band to the conduction band, which results in lower photocurrent. On the other hand, the sensitivity of the CdS nanoribbon photodetector decreased for wavelengths below 490 nm, which could be due to the absorption of photons with high energy near the surface area of the semiconductor. Photoluminescence analysis and absorption spectrum of CdS nanoribbons in Fig. 8.7d emphasize the effect of localized states such as surface states and trap states to originate a Stock shift in CdS nanoribbons [25]. The response time of CdS nanoribbons was found to be greater than that of CdS films and bulk forms. The high crystallinity nanoribbons with lower number of defects leads in a steady-state in rise and decay time. On the other hand, defects can play a role in recombination rate of nanostructures. The high surface-to-volume ratio is another important factor in determining the performance of photodetectors. The band bending due to Fermi-level pinning on the surface of the semiconductor results in an energy barrier for recombination of electron–hole pairs.

The dependency of photocurrent on the power of incident light for nanostructures was given by Kind et al. [26]:

$$I = AP^{\theta} \tag{8.11}$$

where I is the photocurrent, A is a constant, P is the power of illumination. In Eq. 8.11, the dependency of photocurrent on light intensity is determined by the θ value, which is wavelength-independent; for CdS nanoribbons, it was calculated as 0.76, 0.74, and 0.77 for 490, 450, and 400 nm, respectively. This power-law dependency of photocurrent on light intensity has been observed for CdSe nanoribbons as well. Jiang et al. [23] obtained the exponent θ as 0.770, 0.754, and 0.762 for the wavelengths of 700, 650 and 600 nm. They noted that the non-integer dependence of photocurrent to power of light ($0.5 < \theta < 1$) was due to a complex electron–hole recombination process.

CdSe nanoribbons photodetectors by Jiang et al. [23] exhibited a high photo-to-dark current ratio (five orders of magnitude at 650 nm) with great stability, reproducibility and fast response speed (<1 ms). Enhancing the light intensity increased the photoconductance but decreased the rise and fall times. Increasing the light intensity causes the quasi-Fermi levels for electron shifting toward conduction band and that for holes moving toward the valence band. Consequently, the recombination centers created by the traps lead to a reduction in rise and fall times. Furthermore, they investigated the effect of gas adsorption and surface defects on photoconductance of CdSe nanoribbons by coating the surface of the nanoribbons with 300 nm SiO_2 (as a transparent coating), which not only isolated the photodetector from the air but also passivated the surface defects. The conductance of the fabricated device was enhanced both under dark and illuminated conditions but the response time decreased since the coating layer reduces the surface adsorption and the number of recombination centers.

Kum et al. [27] noticed that the morphology of CdTe nanoribbons had a significant effect on photoconductive gain, which was measured as 1.13 and 12.2 for a FET device at $V_{sd} = 1$ V. The polycrystalline morphology of CdTe nanoribbons, with many grain boundaries is believed to be the main reason for the lower photoconductive gain of CdTe nanoribbons compared with that for single-crystal nanostructures. Grain boundaries serve as recombination centers for electron–hole pairs, which can decrease the value of photocurrent. On the other hand, the FET mobility has a significant effect on the change of photo conductance given by $\Delta G = \left(I_{light} - I_{dark}\right)/V_{sd}$. The dependency of photoconductive gain on mobility is expressed by the following relation:

$$G = \left(\frac{I_{ph}}{P}\right)\left(\frac{hv}{q}\right) = R\left(\frac{hv}{q}\right) = \frac{\mu\tau\xi}{L^2} \tag{8.12}$$

where I_{ph} is the photocurrent, P is the power of light absorbed by the nanostructure, hv is the photon energy, q is the elementary charge, R is the responsivity, τ is the lifetime of carrier, μ is the carrier mobility, ξ is the electric field, and L is the length of the nanostructure. According to Eq. 8.12, increasing the mobility will enhance the photocurrent. The change in photocurrent of CdTe nanoribbons with mobility indicated that higher photocurrents could be achieved by increasing the mobility of the device. The mobility-lifetime ($\mu\tau$) product plays a key role in devices that work on carrier collecting mechanism. It means that higher mobility and longer lifetime can lead to collecting the photogenerated carriers efficiently. Kum et al. showed that the photocurrent obtained from higher mobility devices is more than three orders of magnitude compared to the devices with lower mobility. On the other hand, the response and decay times for the devices with high mobilities (0.32 cm^2/V s) were estimated as 11 and 21 s, respectively, while those of for lower mobilities (<0.01 cm^2/V s) were 3.5 and 2.5 s, which means the device with a lower mobility responds faster.

8.4.4 Nanorods

Nanorods employed to fabricate photodetectors have been grown by various methods including metal–organic chemical vapor deposition [28, 29], chemical vapor deposition [30], laser ablation [31] and hydrothermal synthesis [32–34]. Most of the reported nanorod-based photodetectors have used ZnO [30–34] but other materials such as TiO_2 [29] and InN [28] have also been employed to construct photodetectors. Chaia et al. [32] synthesized ZnO nanorods of hexagonal cross-section and single-crystalline structure without any dislocations or stacking faults using an aqueous solution deposition in a hydrothermal reactor. The photoresponse of the fabricated device revealed a significant sensing behavior for UV radiation from sunlight. The dominant mechanism is considered to be oxygen chemisorption on ZnO surface [35]. Under dark conditions, adsorption of oxygen molecules on the surface results in taking the free electrons from ZnO by:

$$O_2(g) + e^- \rightarrow O_2^-(ad) \tag{8.13}$$

$$O_2(g) + 2e^- \rightarrow O_2^{2-}(ad) \tag{8.14}$$

This forms a depletion layer near the surface of the ZnO nanorod which has a low conductivity. The incident UV light with higher energy than the bandgap of ZnO creates electron–hole pairs $[hv \rightarrow e^- + h^+]$; due to the existence of a potential slope made by band bending, holes migrate to the surface according to Eqs. (8.15) and (8.16) by surface recombination, and discharge the adsorbed oxygen ions, which leads to photodesorption of oxygen on the surface.

$$O_2^-(ad) + h^+ \rightarrow O_2(g) \tag{8.15}$$

$$O_2^{2-}(ad) + 2h^+ \rightarrow O_2(g) \tag{8.16}$$

For the remaining electrons, they are either collected by the anode or recombined with holes generated when oxygen molecules are reabsorbed or ionized on the surface. As discussed before, the number of trap states is a crucial issue in the hole-trapping mechanism via adsorption and desorption of oxygen, and nanowires due to the dangling bonds at the surface have a high density of traps [32, 35]. Consequently, the photoconductivity of ZnO nanorod increases due to the

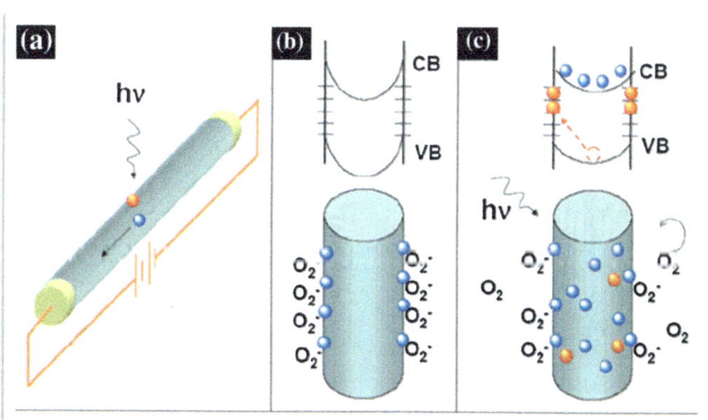

Fig. 8.8 **a** Schematic of a nanowire photoconductor. **b** Trapping mechanism of the ZnO nanowire by adsorption of oxygen molecules at the surface of nanowire and electron capture which leads to a depletion layer near the surface. *Inset*: schematic of conduction band (CB) and valence band (VB) of a nanowire in dark condition. **c** By exposure of UV light, photogenerated holes are trapped and unpaired electrons resulting in photocurrent [35]

remaining electrons in the conduction band. The same mechanism has been reported for ZnO nanowires as shown in Fig. 8.8.

Besides single ZnO nanorods, some reports have utilized an array of ZnO nanorods to fabricate the device [31, 33, 34]. Li et al. [34] fabricated a device with densely packed vertical nanorods. They synthesized the ZnO nanorods with a simple hydrothermal method by using a F-doped SnO_2 substrate. By dispersing uniform ZnO seeds on the substrate, ZnO nanorods 60–100 nm in diameter and 1.5 μm long were grown (Fig. 8.9a and b). The hexagonal morphology of vertical ZnO nanorods, grown along the [001] direction, is evident in Fig. 8.9b. A thin film of gold as electrodes ($1.0 \times 1.0\ cm^2$) was deposited on top of the nanorod array (Fig. 8.9c). Figure 8.9d shows the current–voltage characteristics of the ZnO nanorod-based photodetector under illumination of 365 nm UV light; the bandgap of ZnO is about 3.2 eV according to the PL spectrum shown in the inset of Fig. 8.9d. By comparing with the dark current and applying the same bias voltage (0.4 V), the photocurrent under illumination is about 4.7 mA larger. The non-linear characteristics of the measured *I–V* curve in dark was attributed to Schottky barrier between gold electrodes and nanorods; when illuminated by light, the height of the Schottky barrier decreases.

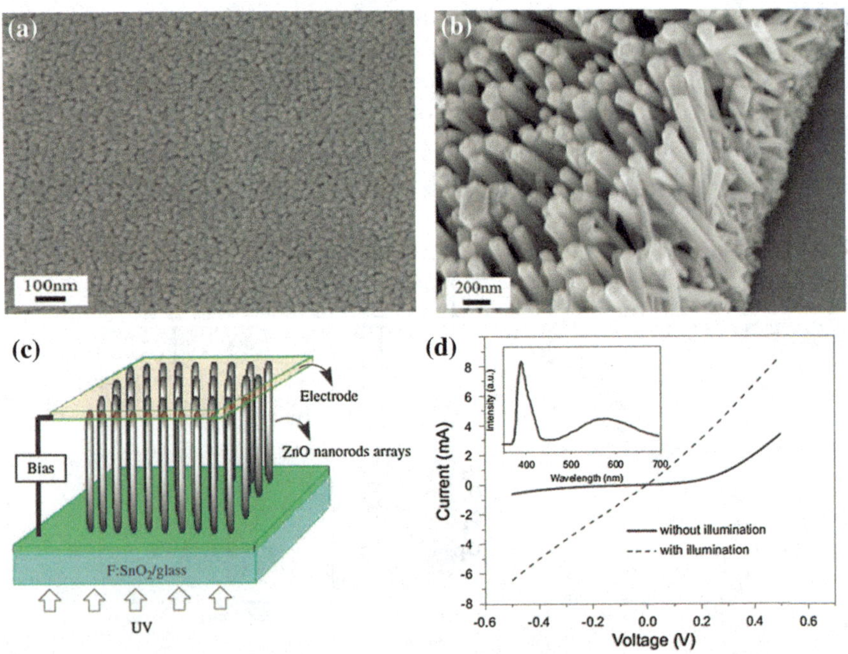

Fig. 8.9 **a** SEM image of ZnO seeds. **b** A cross-section view of the ZnO nanorod arrays. **c** Schematic of the fabricated device based on ZnO nanorod array. **d** Current–voltage curves of the ZnO nanorod arrays in dark and under illumination ($\lambda = 365$ nm). *Inset*: PL spectrum at room temperature [34]

8.4.5 Nanowires

Recently, nanowires have attracted much attention as potential building blocks for nanoelectronics. The sensitivity of nanowires to light makes them attractive for photodetectors, photovoltaics, and optical switches [1]. The contributing factors to the high sensitivity of nanowires are their large surface-to-volume ratio and reduced dimension of the active area. The photocarrier lifetime would be prolonged due to the presence of deep surface trap states as a result of large surface-to-volume area, and short transit time of carriers could be achieved by reduced active area [35].

A schematic of the photoconducting mechanism in a single nanowire is shown in Fig. 8.10. The intrinsic conductivity of the nanowire in dark is given by [1]:

$$\sigma = en\mu \tag{8.17}$$

where e is the electronic charge, n is the charge carrier density, and μ is the carrier mobility.

The conductivity of nanowire will change upon illumination due to the change in either carrier concentration or mobility:

$$\Delta\sigma = \sigma_{\text{light}} - \sigma_{\text{dark}} = e(\mu\Delta n + n\Delta\mu) \tag{8.18}$$

As in many semiconductors, the change in mobility compared with the carrier concentration ($\Delta n \gg \Delta\mu$) is negligible as well as its time dependency; the photocurrent density of nanowire can be deduced from the following formula:

$$J_{\text{PC}}(t) = \Delta\sigma\xi = e\mu\Delta n(t)\xi \tag{8.19}$$

where ξ is the applied electric field ($\xi = V/l$, where V is the voltage applied across the nanowire and l is the length of nanowire). Considering the geometrical properties of a cylindrical nanowire photodetector including exposed surface area $\left(A = \frac{1}{2}2\pi rl = \pi dl/2\right)$, cross-sectional area ($A_{\Phi} = \pi r^2 = \pi d^2/4$), and volume $\left(\text{Vol} = \pi r^2 l = \pi d^2 l/4 = A_{\Phi} l = Ad/2\right)$, the average optical generation rate over the thickness of the nanowire could be derived by:

$$g_d = \eta^*(P_{\text{opt}}/\hbar\omega)/Vol = 2\eta^*(P_{\text{opt}}/\hbar\omega)/Ad \tag{8.20}$$

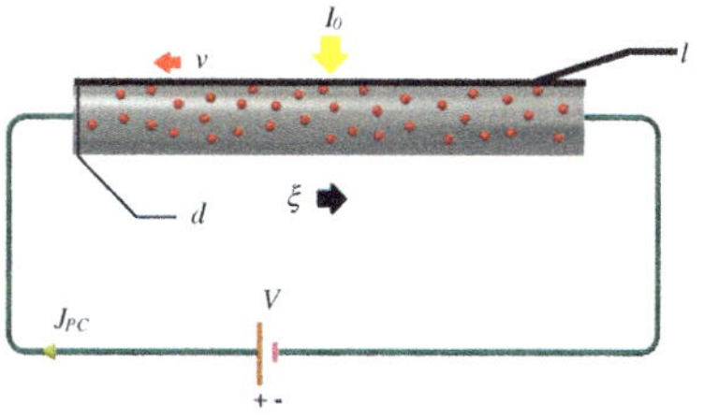

Fig. 8.10 Schematic describing photoconductivity in a nanowire, I_o is the light intensity, J_{pc} is the current density, ξ is the applied electric field, and v is the charge carrier drift velocity [1]

where P_{opt} is the optical power of incident light, and $\eta * = \eta\eta^{+}$ is the quantum efficiency of the effective carrier photogeneration. With constant density of photogenerated carriers ($\Delta n(t) = g_d\tau = $ const.) due to the steady-state condition of illumination, and the photoexcited carriers being relaxed to the ground state with a specific lifetime (τ), the photoconductivity of nanowire is given by:

$$\Delta\sigma(t) = g_d e(\mu\tau) \tag{8.21}$$

where $\mu\tau$ as the mobility-lifetime product determining the photosensitivity of nanowire can be considered as a figure of merit for photoconductors. By substitution of Eqs. (8.19) and (8.20) in Eq. (8.21), the photocurrent density and photocurrent of nanowire for a specific photon energy is given by:

$$J_{PC}(t) = \Delta\sigma\xi = 2\eta^*(P_{opt}/\hbar\omega)e\xi\mu\tau/Ad \tag{8.22}$$

$$I_{PC} = J_{PC}A_{\Phi} = \eta^*(P_{opt}/\hbar\omega)e\xi(\mu_n\tau_n + \mu_p\tau_p)d \tag{8.23}$$

The photoconductive gain of the nanowire is defined as number of charge carriers, which pass between the electrodes for each photon absorbed per second:

$$G = \frac{I_{PC}/e}{P_{abs}/\hbar\omega} \tag{8.24}$$

where P_{abs} is the power adsorbed in the photoconductor, derived by ($P_{abs} = \eta^* P_{opt}$), causes an electron–hole pair generation in the nanowire; the photoconductive gain can be written by combining Eqs. (8.23) and (8.24):

$$G = \frac{\xi\mu\tau}{l} = \frac{\tau}{\tau_t} \tag{8.25}$$

where τ_t is the carrier transit time derived by $\tau_t = l/v = l/\mu\xi = l^2/\mu V$. The nanowire-based photoconductor exhibits a high photoconductive gain, which is attributed to prolonged carrier lifetime and reduced transit time [1]. It is worth mentioning that the nanowire photoconductors have a higher photoconductive gain, more than three orders of magnitude, compared with their thin film counterparts [36].

The unique properties of nanowires such as high sensitivity, high wavelength selectivity, fast response, light polarization sensitivity, light absorption enhancement, and high photoconductive gain could be exploited to fabricate the nanowire-based photodetectors used in applications such as optical switches, optical interconnects, image sensors and etc. Below we will describe the nanowire photodetector developments based on the material categories.

8.5 Materials for Nanowire-Based Photodetectors

8.5.1 Group IV

Group IV semiconductors, including Si and Ge, have been considered for visible light detection. Si photodetectors can be used in diverse applications such as image processing units and image sensors due to their advantages including simple fabrication method, technology compatibility, integration with other circuits and low cost [1, 37]. Vapor–liquid–solid (VLS) technique [38], template-assisted chemical etching [39], electron beam lithography [37], thermal chemical vapor deposition [40], and plasma-enhanced chemical vapor deposition (PECVD) [41] are the synthesis routes used for growing Si nanowires for these applications.

Besides the influence of light intensity on the photocurrent of Si nanowires [37], Ahn et al. [38] found a significant dependency on the angle of incident light to the axis of the nanowire. The schematic view of their measurement setup is shown in Fig. 8.11a. The photocurrent is lower if the light is polarized perpendicular to the

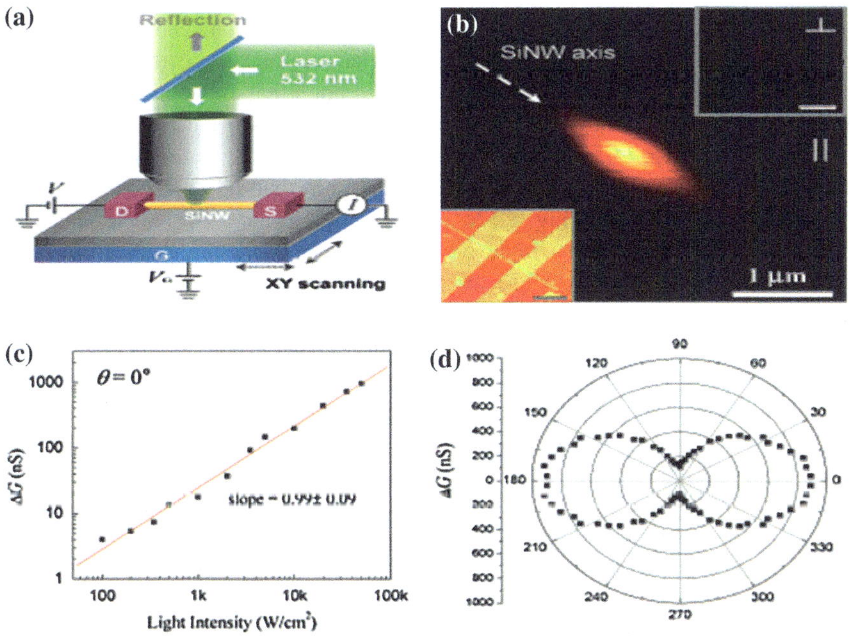

Fig. 8.11 **a** The setup for scanning photocurrent measurement. **b** The scanning photocurrent image of Si nanowires while the light polarization is parallel to the axis of nanowire as the photocurrent enhanced from 0 *nS* (*red color*) to 1,000 *nS* (*yellow color*). *Upper inset*: perpendicular orientation of light, *Bottom inset*: AFM image of the device. **c** The measured photocurrent of Si nanowire as a function of light intensity. **d** The photocurrent of Si nanowire by varying the angle of light from 0 to 180° [38]

axis of the nanowire instead of parallel (Fig. 8.11b and d). The photocurrent anisotropy ratio of nanowire to the polarization of the light is given by:

$$\sigma = (\Delta G_{||} - \Delta G_{\perp})/(\Delta G_{||} + \Delta G_{\perp}) \quad (8.26)$$

where $\Delta G_{||}$ and $\Delta G_{\perp}$ is the photocurrent change when the emitting light is parallel and perpendicular to the axis of nanowire, respectively. This dependency of photocurrent on anisotropy ratio could be attributed to the anisotropy in light absorption, arising from the difference in dielectric constant of Si nanowire and its surroundings (11.8 compared to 1) [38].

Kim et al. [40] investigated the effect of different dopants on the photoresponse of Si nanowires by mixing B_2O_3 and P powder with SiO for *p-type* and *n-type* nanowires, respectively. They placed the above mixture at the center of an alumina boat, followed by heating the furnace system at 1,380 °C for one hour. The carrier gas consisted of Ar and H_2 flowing at a rate of 500 sccm at a pressure of about 250 Torr. The Si nanowires were 50–100 nm in diameter and 10–50 μm long. The photodetectors on Si/SiO_2 substrate with Ti and Au contacts had a separation distance about 3 μm between the electrodes (Fig. 8.12a and b). The current–voltage curves of fabricated undoped Si nanowire-based photodetectors in dark current and under illumination of 325 and 633 nm wavelengths showed an increase in photocurrent with increasing wavelength. The mechanism of photoresponse of Si nanowires is different from that of ZnO since the existence of oxygen ions on the surface of Si nanowire could not affect the dominant process of electron–hole generation by light. In other words, not only electrons can participate in photocurrent of Si nanowire but also holes might have the same function. The doped nanowires are more sensitive to 325 nm than 633 nm compared with the undoped Si nanowires (Fig. 8.12c–e). The *n-type* Si nanowire might be heavily doped and the Fermi levels might be merged deeply into the conduction band; then, the energy of photons at 633 nm might not be enough to pump the electrons from the valance band to the conduction band. For *p-type* Si nanowire, which is doped heavily, the significant difference between Fermi level and conduction band might be the corresponding reason.

Another group IV element, Ge ($E_g = 0.68$ eV) is a promising material for photodetectors in the visible and near-IR regions [42]. Single Ge nanowire-based photodetectors exhibited sensitivity to visible light even in low intensity due to their large absorption of light, which makes them ultrasensitive photodetectors in the normal intensity of sunlight (100 mW/cm^2). Their high gain is attributed to the charge trapping on the surface. Compared to Si, Ge nanowires have larger quantum efficiency due to their short penetration depth in the visible light spectra [43]. The performance of the photodetector system can be improved by employing a nanowire array-based photodetector instead of a single-nanowire detector. Yan et al. [42] compared the two configurations and as seen in Fig. 8.13, the reset time of the photodetector can be improved greatly by using a network of nanowires due to changing the active detection channel. The reset time of single-nanowire-based

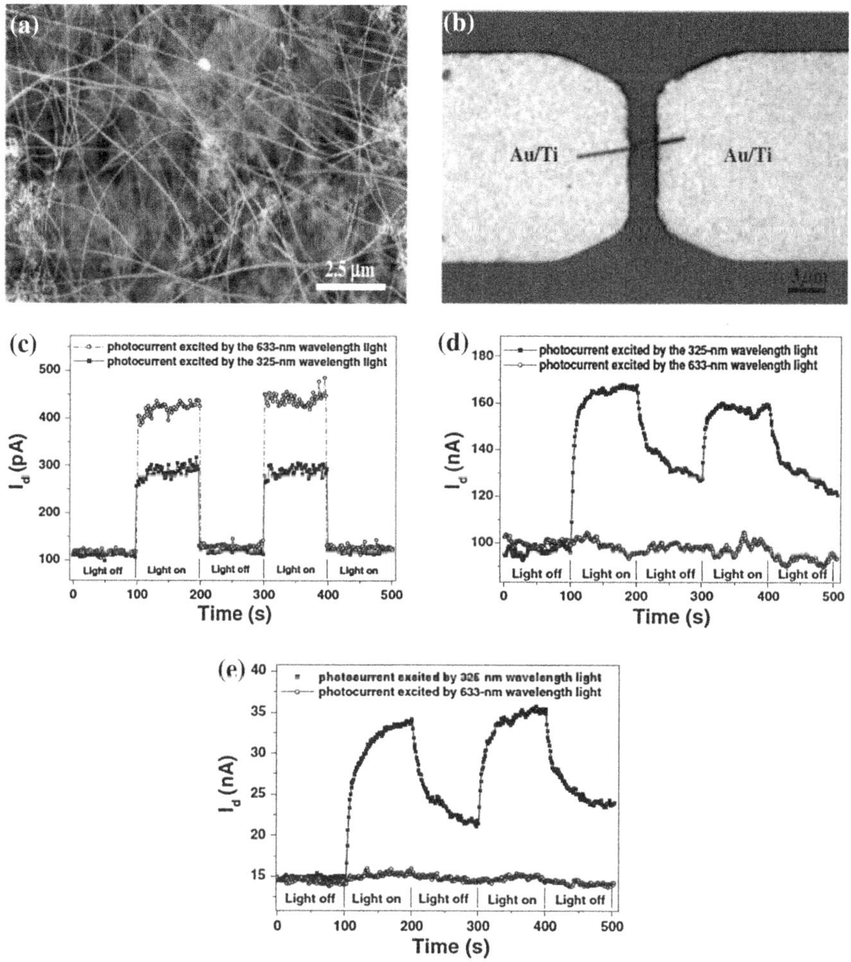

Fig. 8.12 **a** SEM image of Si nanowires synthesized by thermal CVD. **b** Optical microscope image of Si nanowire-based photodetector; Time-resolved photocurrent measured under illumination of 325 nm and 633 nm. **c** Undoped Si nanowire. **d** *N-type* Si nanowire. **e** *P-type* Si nanowire [40]

devices with an ohmic contact is more than 70 s while the network devices with Schottky contact show faster reset time of just 1 s. The network device is characterized by nanowire–nanowire junctions, which dominate the electron conduction. This intrinsic junction barrier depicts Schottky barrier exhibiting fast modulation rates under illumination (Fig. 8.13e). Contacts can affect the photocurrent of Ge nanowires, as reported by Kim et al. [44] who studied both Schottky and ohmic contacts for a single Ge nanowire-based photodetector. In the case of Schottky contact, the local electric field at the contact accelerates the electron–hole

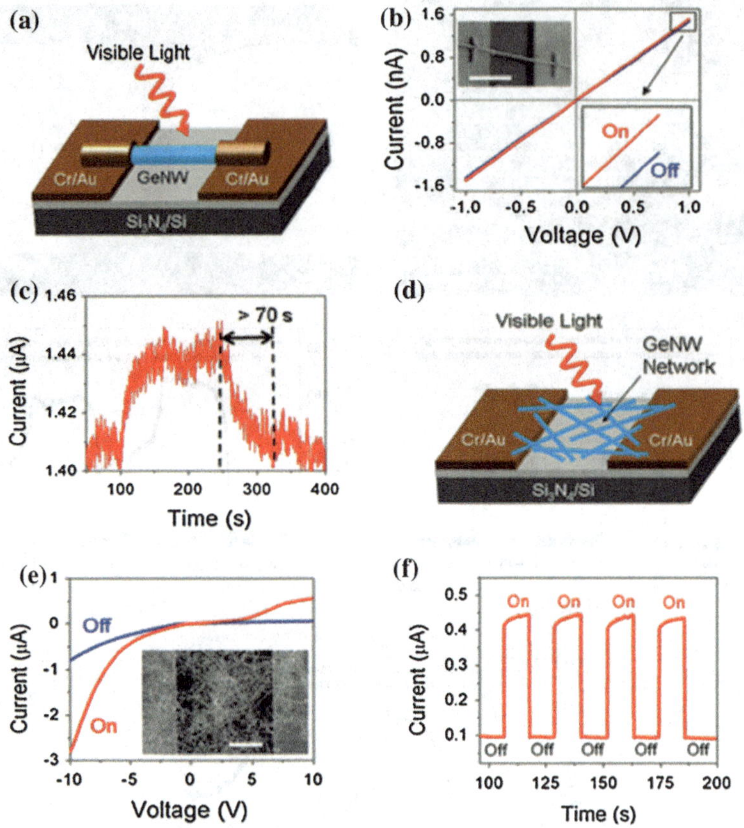

Fig. 8.13 Photocurrent and photoresponse behaviors of Ge nanowire by fabricating single-nanowire based **a–c** and network-based devices **d–f**. The scale bars in the *inset* images of **b** and **e** are 5 µm [42]

pairs generated by the illumination, when the remaining carrier transport is diffusion-limited. In an ohmic contact, the carrier generation rate is uniform due to the uniform applied field along the nanowire channel and the carriers move along the linear potential profile through drift mechanism. This work also showed an inverse relation between the gain and photoconductivity with the nanowire diameter.

8.5.2 Group VI

Selenium exhibits unique properties such as high photoconductivity, large piezoelectric and thermoelectric effects. Solution-based techniques have been used for synthesizing Se nanowires [45, 46]. Gates et al. [45] described a facile method to synthesize Se nanowires 10–800 nm in diameter and up to 100 s of microns long. They observed an enhancement by $\sim$150 times in the measured conductivity between dark and illuminated conditions of 3 μW/μm^2. By increasing the light intensity to 10 μW/μm^2, the photoconductivity increased nearly 70 times compared with dark conditions. The enhancement in conductance of Se nanowire (as a *p-type* semiconductor) after illumination could be attributed to the existence of oxygen. The electronegativity of electron is higher than that of selenium, which means it adsorbs electron more easily. Therefore, oxygen can increase the lifetime of the holes and decrease the electron lifetime by disturbing the initial balance of electrons and holes [46].

Tellurium as a typical *p-type* narrow-bandgap (0.35 eV) semiconductor material shows some unique features including photoconductivity, non-linear optical response, and thermoelectric effects [47]. The photoresponse of Te nanowires, grown by Langmuir–Blodgett technique, showed a significant enhancement in photocurrent by increasing the light intensity [48].

8.5.3 Group III–V

III–V compound semiconductors are one of the most important materials for applications in electronics and optoelectronics. The promise of nanowire-based photodetectors in this material system is due to their very good transport properties and easy processes to make alloy compounds, thus allowing tuning of the optical absorption by varying the bandgap energy [1].

8.5.3.1 III-Arsenides

Thunich et al. [49] investigated the photocurrent and photoconductance of GaAs nanowires, synthesized by MBE technique. The origin of photocurrent generation was the Schottky contact between the nanowire and the source–drain electrodes. They observed photogating phenomenon involved in photoconductance, which means photoexcited electrons get trapped at the nanowire surface, acting as negative voltage on the *p-type* nanowire. On the other hand, free excess holes excited by optical source can play a role in photoconductance of the nanowire by raising the Fermi energy of holes inside the nanowires (photo-doping effect). The GaAs nanowires exhibit a polarization-dependent behavior, while their photocurrent changed by varying the light direction from parallel to perpendicular. GaAs and AlGaAs nanowires synthesized through VLS route showed an ultrafast photoconductive response (<100 ps) to 400 nm laser light [50].

InP nanowire-based photodetectors showed fast photoconductive response of 14 ps at 780 nm by using an array of InP nanowires, which were grown in a coplanar waveguide transmission line [51]. Polarization-sensitive photoresponse of

InP nanowire was investigated by Wang et al. [52], who fabricated photodetectors using a single nanowire grown via a laser-assisted catalytic growth. They observed that polarization ratio of InP nanowires is around 0.96, which is independent of the nanowire diameter between 10 and 50 nm. By increasing the light intensity, the conductance of the individual InP nanowires can be enhanced by two to three orders of magnitude, which is attributed to the direct carrier collection at the nanowire–metal contact. The wavelength does not affect the polarization anisotropy of photocurrent for energies higher than the bandgap of nanowire.

8.5.3.2 III-Nitrides

III-nitride-based nanowires have been employed to fabricate photodetectors for UV detection, which can be extended to visible spectrum by an alloying process [36, 53–55]. GaN with 3.4 eV is an ideal candidate for UV detection due to its attractive properties including relatively high mobility, a sharp cutoff wavelength and high-quantum efficiency [56]. CVD and MBE have been used to synthesize GaN nanowires. Photodetectors fabricated using GaN from these two methods differ significantly in the photocurrent gain, mainly due to the resulting material properties. First, the carrier concentration (n) of the CVD-NWs is in the range of 10^{18}–10^{20} cm^{-3}, while that of the MBE-NWs is one or two orders of magnitude lower ($n = 10^{17}$–10^{18} cm^{-3}). The carrier concentration affects the width of the depletion layer, which is estimated to be 10–30 nm and 50–100 nm for CVD-NWs and MBE-NWs, respectively. Second, the orientation of the nanowire either m- or a-axial for CVD-NWs or c-axial for MBE-NWs also may have an effect on the photocurrent [36, 57].

Han et al. [53] synthesized GaN nanowires by CVD and then fabricated a single-nanowire-based device. The morphology of grown nanowires is shown in Fig. 8.14a. The photoresponse of the GaN nanowire-based device was measured at two wavelengths of 254 and 365 nm with a power density of 3 mW/cm^2. As seen in Fig. 8.14b, the GaN nanowires respond to the UV light at these wavelengths significantly, although the photoconductance decreases from 258.3 to 23.4 nS when the wavelength is increased. The nanowires exhibit a fast recovery because the recombination of electron–hole pairs is fast (Fig. 8.14c and d).

The diameter-dependent photoconductivity of GaN nanowires was investigated by Chen et al. [58]. They observed three orders of magnitude decrease in photocurrent from 10^{-5} to 10^{-8} A when GaN nanowire diameter is reduced from 120 to 20 nm. They reported the critical diameter (d_{crt}) for GaN nanowires grown by CVD to be 30–40 nm, while that for MBE is around 90 nm. Below the d_{crt}, the product of carrier lifetime and mobility ($\mu\tau$), which defines the transport and collection efficiency of photocarriers, decreases significantly. Nanowires, compared with thin films, show three orders higher $\mu\tau$ values, and hence higher photoefficiency. As the nanowire mobility does not change very much by varying the diameter of the nanowire (10–50 cm^2/Vs for 10–200 nm diameter), the main

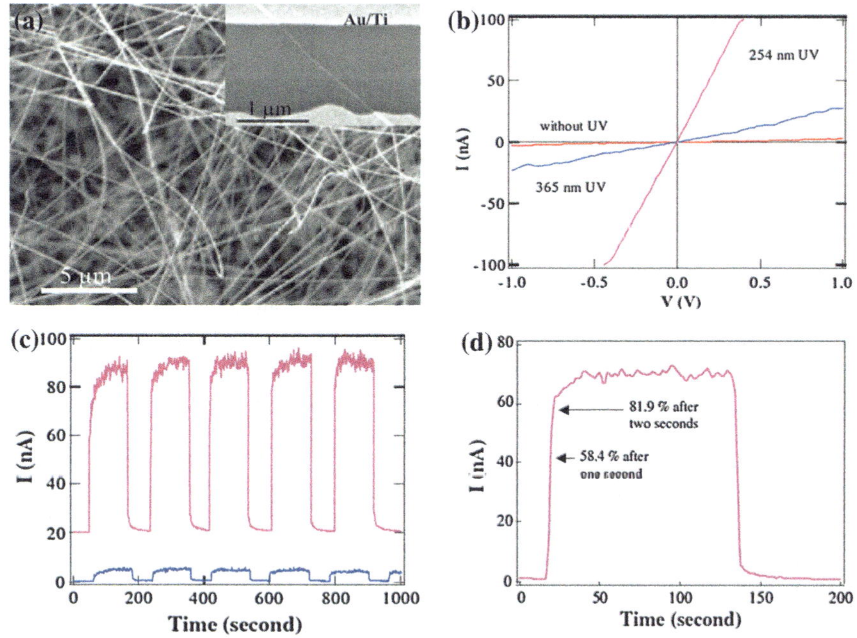

Fig. 8.14 **a** SEM image of GaN nanowires. *Inset*: SEM image of the fabricated device contacted by two Au/Ti electrodes. **b** Current–voltage characteristics of a GaN nanowire without and with UV illumination. **c** Time-resolved measurement of GaN nanowires, the UV illumination of 254 nm (*red color*) and 365 nm (*blue color*). **d** One cycle of the GaN nanowire response to UV light of 254 nm [53]

impact comes from the carrier lifetime. When the nanowire diameter is below the critical diameter, the neutral region disappears; reducing the surface barrier height (ϕ_b) results in lifetime of carrier being reduced. Meanwhile, surface band bending induces longer carrier lifetime in GaN nanowires.

The photoresponse of aluminum nitride (AlN) with the highest bandgap ($\sim$6.2 eV) and indium nitride (InN) with the lowest bandgap ($\sim$0.6 eV) among III-nitride semiconductors has been investigated [54, 55]. Chen et al. [55] synthesized InN nanowires via MOCVD at 480 °C and measured the photoresponse of InN nanowire photodetector under illumination of 808 nm in the IR region. Since high thermal current is one of the frequent problems in narrow-bandgap semiconductors, they investigated the effect of temperature on the performance. They observed a reduction in thermal current and enhanced photocurrent at low temperatures, which appear to be two major advantages for using InN nanowires.

8.5.4 Group II–VI

Three main compounds in group II–VI for fabrication of photodetectors include sulfides, selenides, and tellurides due to the wide range of bandgaps and the ability to cover from UV to IR.

8.5.4.1 Sulfides

Reports on the photoconductive response of CdS nanowires are very limited [59–61]. Kouklin et al. [59] observed that the resistance of CdS nanowires increases by one to two orders of magnitude, most likely due to electron trapping from the nanowire into the surrounding alumina following photon absorption and formation of empty traps in the bandgap. Consequently, recombination between resulting holes in the valence band and electrons in the conduction band could cause a significant drop in conductance. For this reason, CdS nanowires exhibit a negative photoresponse, which may find applications in optically controlled switches and "normally-on" infrared photodetectors.

8.5.4.2 Selenides

Various types of selenide nanowires such as CdSe, ZnSe, PbSe, and InSe respond to light illumination. CdSe (1.74 eV) is considered to be a promising material for fabricating photodetectors in the visible spectrum. Gu et al. [62] investigated the photoresponse of CdSe nanowires in the visible spectrum by illumination at 632.8 nm. They reported an increase in the nanowire conductivity and an "ON/OFF" ratio of 5.5. The response and recovery times for CdSe nanowires, synthesized via electrodeposition by using self-organized single-walled carbon nanotubes as a template, were 9.1 and 3.2 ms, while the corresponding values for lithography patterned nanocrystalline CdSe nanowires were 20 and 30 μs [63].

Zinc selenide (ZnSe) with a bandgap of 2.7 eV at room temperature is another material from the selenide group useful in optoelectronics applications. Both a single ZnSe nanowire [64] and an array of ZnSe nanowires [65] were studied as visible and UV light detector [66]. The single ZnSe nanwire-based device exhibited a very high responsivity, greater than 22 A/W at 400 nm excitation wavelength [64], although the measured photocurrent had a slow decay component attributed to the surface states or deep levels in the nanowire core; the higher responsivity of ZnSe nanowires could be related to this fact that the responsivity is proportional to the excess carrier generation rate and life time, which is long enough for ZnSe [64]. Philipose et al. [65] reported an inverse dependency of photocurrent with frequency for an array of ZnSe nanowires, while by increasing frequency a reduction in measured photocurrent was observed due to deep center capture.

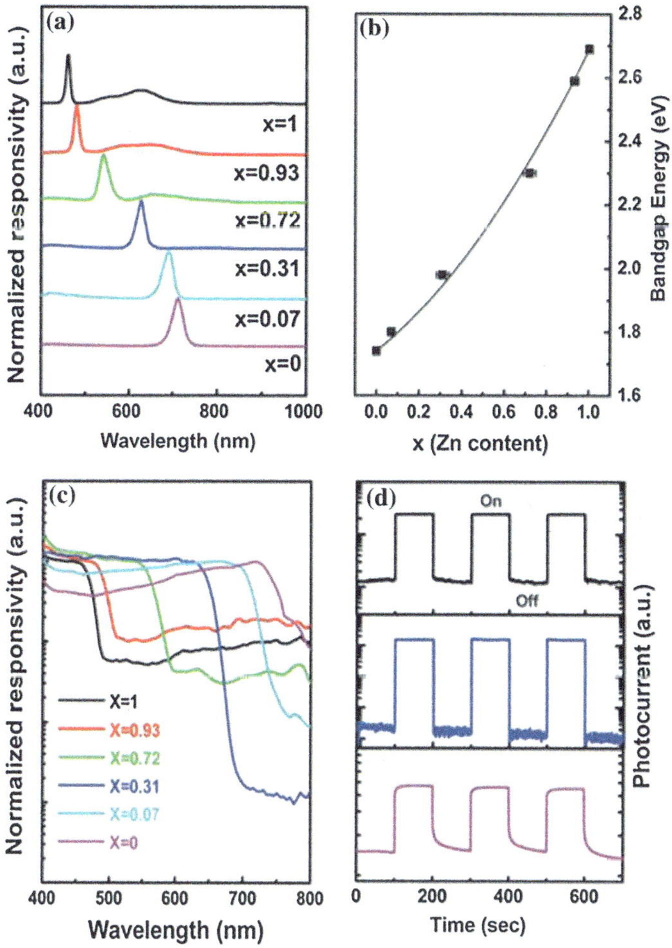

Fig. 8.15 **a** Photoluminescence spectra of $Zn_xCd_{1-x}Se$ nanowires. **b** The bandgap energies of $Zn_xCd_{1-x}Se$ nanowires versus Zn content. **c** Photocurrent of $Zn_xCd_{1-x}Se$ nanowires as a function of wavelengths. **d** Time-resolved measurement of $Zn_xCd_{1-x}Se$ nanowires for ZnSe (*black*), $Zn_{0.31}Cd_{0.69}Se$ (*blue*) and CdSe (*red*) [67]

Nanowires of a ternary compound of Zn, Cd, and Se known as alloy nanowires can be used to construct a variable-wavelength photodetector. Yoon et al. [67] synthesized $Zn_xCd_{1-x}Se$ nanowires through VLS technique by thermal evaporation of ZnSe and CdSe powders. The synthesis temperature varied from 825 to 528 °C and Si and sapphire substrates coated with gold as catalyst were used. Photoluminescence measurements were done to determine the bandgap of the nanowires by varying the amount of Zn, as seen in Fig. 8.15. Similar to other ternary compound semiconductors, bandgap bowing effect can be observed by the approximate linear dependence of bandgap energies to the Zn content. The

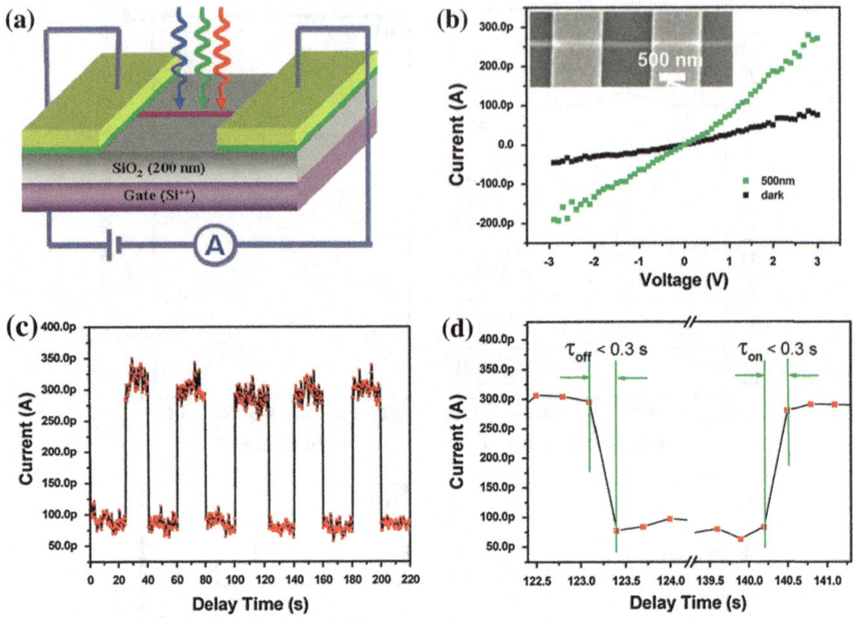

Fig. 8.16 **a** Schematic of the fabricated device for photocurrent measurement. **b** Current–voltage characteristics of In_2O_3 nanowires in dark and under illumination. *Inset*: SEM image of single-nanowire device. **c** Time-resolved measurement of In_2O_3 nanowire-based photodetector in response to light with 500 nm wavelength and a period of ~20 s at an applied voltage of 3 V. **d** The enlarged portion of 122.4–124.2 s and 139.3–141.2 s ranges, showing the fast response of the fabricated device [69]

photoresponse of $Zn_xCd_{1-x}Se$ nanowires in response to incident light from 400 to 800 nm at an applied bias voltage of 5 V is shown in Fig. 8.15c. The photocurrent of alloyed nanowires decreases when the energy of the incident light is lower than the nanowire bandgap. As mentioned before, radiation with energy higher than the bandgap has the capability of generating electron–hole pairs resulting in an enhancement in current. CdSe detectors show slower response recovery behavior, attributed to adsorption and photodesorption of O_2 on the surface of CdSe nanowires. Other detectors, for example, ZnSe and $Zn_{0.72}Cd_{0.28}Se$, show the capability of fast switching between low and high currents reversibly (Fig. 8.15d).

Individual InSe single-crystalline nanowires, upon irradiation of white light on the device, responded with an "ON/OFF" ratio as high as 50. In "ON" and "OFF" conditions, the device showed a "low" current under dark condition and upon illumination, "high" current state would be observed. Importantly, if the switching in these two states was fast and reversible, it could be concluded that the photodetector has a high performance. On the other hand, by calculating the dependence of the photocurrent of InSe nanowires on power density ($\theta = 0.67$ from Eq. 8.11), excellent photocapture in the InSe nanowires was realized [68]. Single-crystalline In_2Se_3 nanowires were synthesized by Zhai et al. [69] via thermal evaporation method and a fast, reversible and stable response to visible light was demonstrated using these nanowires. As shown in Fig. 8.16b, 500 nm-light at an intensity of 2.81 mW/cm^2 induced an enhanced photocurrent compared to dark current. A reset time of 0.3 s confirms the high photosensitivity and quick photoresponse of In_2Se_3 nanowires, attributed to superior crystal quality with lower defects and induced traps, and the large surface-to-volume ratio of the nanowires.

8.5.4.3 Tellurides

Among II-tellurides, ZnTe nanowires with a wide and direct bandgap of 2.26 eV exhibit more sensitivity to green/UV light, but only a few studies have been devoted to the photoresponse of ZnTe nanowires [70–72]. Cao et al. [70] synthesized ZnTe nanowire by a simpler and low cost vapor transport method compared to the MOCVD approach by Li et al. [72]. Figure 8.17 shows a high density of ZnTe nanowire. After dispersing them in alcohol and spreading onto a Si/SiO_2 wafer with 500 nm oxide layer, two contact pads of Ti (2 nm) and Au (60 nm) were deposited as electrodes via photolithography (Fig. 8.17b). The sensitivity of ZnTe nanowires to wavelengths longer than 600 nm is rather low, while the maximum sensitivity is observed at $\sim$500 nm green light. The photoresponse of ZnTe nanowire under dark and illumination conditions is shown in Fig. 8.17c. An enhancement of photoconductance is seen due to a hole accumulation layer which is formed near the surface of the nanowire. Oxygen adsorption by ZnTe (*p-type*) may lead to an increase in hole concentration in the nanowires by capturing electrons [70]. The high density of states due to the dangling bond at the surface of the nanowire reveals trapping mechanism via oxygen adsorption. The dependency of photocurrent on light intensity ($\theta = 0.23$) is attributed to carrier traps in the nanowire, while for other II–VI semiconductors θ varies between 0.5 and 1. The high number of trap states in the nanowire might have resulted in this small value of θ.

8.5.5 Metal Oxides

Metal oxide nanowires have been widely employed in fabrication of photodetectors due to their wide bandgap and large surface-to-volume ratio. Since their conductive and photoconductive properties are affected by surface chemistry substantially, they are considered to be ideal candidates for gas and chemical sensing [1]. These include binary metal oxides such as ZnO, SnO_2, In_2O_3, CdO, Ga_2O_3, Cu_2O, Fe_2O_3, and ternary metal oxides such as $ZnSO_3$, $ZnGa_2O_4$.

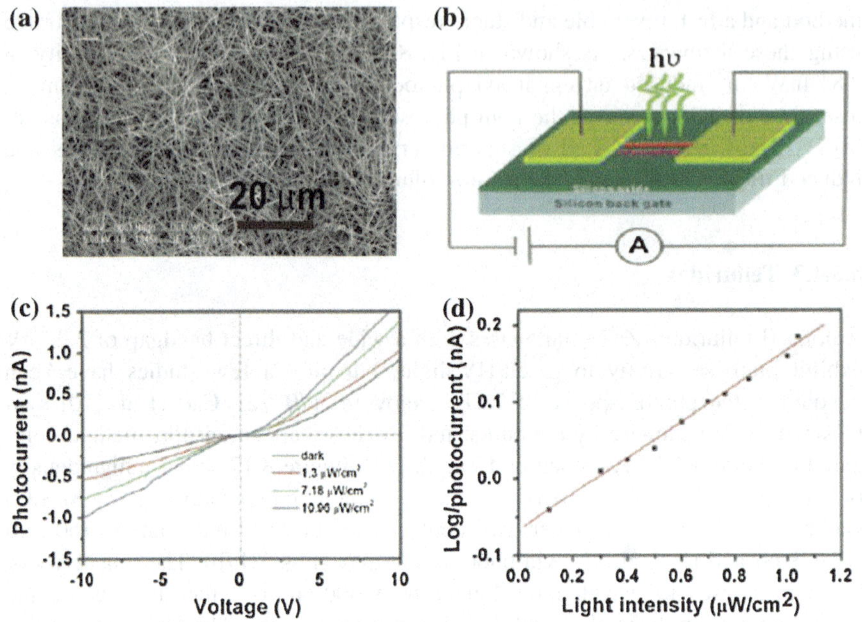

Fig. 8.17 **a** SEM image of ZnTe nanowires synthesized by VLS technique. **b** Schematic of the fabricated device for photoresponse measurement. **c** Current–voltage behavior of a single ZnTe nanowires in dark and under illumination. **d** Photocurrent of a single ZnTe nanowrie as a function of light intensity. Both the photocurrent and light intensity are in log scale [70]

8.5.5.1 Zinc Oxide

Zinc oxide, with a wide bandgap of 3.37 eV at room temperature and a large exciton binding energy of 60 meV, ensures efficient excitonic ultraviolet emission and has attracted a lot of interest for optoelectronic applications. One-dimensional ZnO nanostructures can be synthesized by either physical vapor deposition methods at high temperature (500–600 °C) or chemical approaches at low temperature ($\sim$70 °C) on diverse substrates [73]. Kind et al. [26] explored the photoconductive response of ZnO nanowires, which are sensitive to ultraviolet light. They noted that individual ZnO nanowires show insulating behavior in the dark with a resistivity more than 3.5 MΩ-cm. By illuminating with UV light, (wavelength below 380 nm and intensity of 0.3 mW/cm^2) the resistivity of nanowires was shown to decrease by 4–6 orders of magnitude (Fig. 8.18). The potential of ZnO nanowires in optical switches was evaluated with the measurement of ON and OFF states under dark and UV exposure, respectively. The sensitivity was investigated by exposure to two wavelengths of 532 and 365 nm; the ZnO nanowires responded to the UV light significantly but the green light cannot induce a photoresponse, and hence, the response exhibited a cutoff at 370 nm (Fig. 8.18b and c).

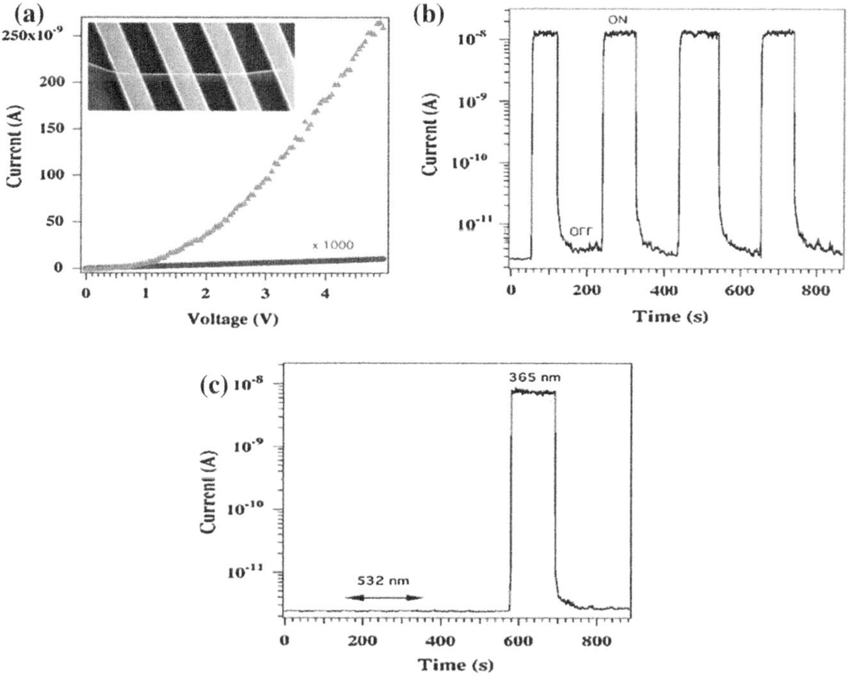

Fig. 8.18 **a** Current–voltage characteristics of a single ZnO nanowire in dark (•) *filled circle* and under illumination (▲) of UV light with 365 nm and 0.3 mW/cm^2 as the light wavelength and intensity, respectively. **b** Reversible switching of a single ZnO nanowire when the light source was on and off. **c** Sensitivity of a single ZnO nanowire to light illumination with different wavelengths of 365 and 532 [26]

Prades et al. [74] enhanced the response of the fabricated photodetector by reducing the distance between the electrical contacts, raising the width of the photoconductive area, or improving the mobility of nano material. Kim et al. [75] studied the effect of drain–source (V_{ds}) and gate–source (V_{gs}) voltages on the sensitivity of ZnO nanowire-based FET, in which they observed the highest value for on/off current ratio (photo-to dark-current ratio of $\sim 10^6$ upon UV exposure) and at the "bottom" of the subthreshold swing region. It should be emphasized that integration of individual nanowires into the photodetector is a main issue. Conventionally, many groups have employed a "pick and place" method. Collecting nanowires grown on the substrate, making a suspension, and dispersing them on the fabricated device are the main steps in this approach. Next, processes such as photolithography, electron beam lithography, and focused ion beam (FIB) have to be used to deposit metallic contacts. Therefore, fabrication of this type of photodetectors is complicated, time-consuming, and expensive. Li et al. [76] introduced a new technique to bridge a nanowire as a cost-effective and efficient alternative; in this approach, by etching a trench into a single crystal substrate, nanowires can be grown by one of the usual methods across the trench from one electrode to another (Fig. 8.19a and b). They employed CVD with gold as the catalyst. The photoresponse of their device studied by exposure to UV light (350 nm) showed a high responsivity and fast response time. The conductance of the synthesized ZnO nanowires increased by one order of magnitude at 50 nW/cm^2 and by five orders of magnitude at 48 mW/cm^2 (Fig. 8.19c). Figure 8.19d shows the dependency of photoresponse of ZnO nanowires, while the UV light is turned on and off.

8.5.5.2 Tin Oxide

Tin oxide (SnO_2) is an important *n-type* metal oxide semiconductor (3.6 eV) due to high-quantum efficiency, enhanced surface-to-volume ratio of nanowires, and conductivity change depending on surface properties such as molecular adsorption or desorption. The photoconductance of SnO_2 nanowires exhibits a strong dependency on their size. Interestingly, significant on/off ratios have been observed even for short illumination periods. A blue-shift in photoresponse follows a decrease in nanowire diameter, attributed to the confinement of charge carriers in nanowires. Kim et al. [77] reported the preparation of SnO_2 nanowires using CVD to compare the photoresponse of aligned tin oxide nanowires with that of a random network. According to their measurement, the conductance of NWs enhanced from 0.12 μS to 83.9 μS under UV exposure at $V_{ds} = 2$. Regarding the power law, the power constant of tin oxide nanowires was calculated as 0.64, which meant at the inter junction of nanowires, a number of photoexcited carriers were lost due to recombination. They attributed the slow responsivity of the grown nanowires to the slow process of adsorption and desorption as Prades et al. [78] reported before for ZnO nanowires. They found that the aligned SnO_2 nanowires show a strong polarization-dependency behavior in contrast to the random network of nanowires.

8.5.5.3 Copper Oxide

Copper oxide (Cu_2O) is a *p-type* direct bandgap semiconductor ($E_g = 2.17$ eV) with advantages such as non-toxicity, low cost, wide availability and a large exciton binding energy ($\sim$140 meV). Liao et al. [79] synthesized Cu_2O nanowires by reduction of CuO nanowires with hydrogen gas. Initially, they produced CuO by oxidation of Cu foil in the atmosphere and introducing a gas mixture (H_2/Ar 20 % atmosphere) at 200 °C for 1 h. The grown nanowires were 20–30 μm and 50–100 nm in diameter. The electrical characterization of Cu_2O nanowires under dark conditions and exposure to blue light (488 nm) clearly showed the photoresponse of the nanowires. The reversibility and reproducibility of Cu_2O nanowires from the time-resolved measurements of "ON" and "OFF" currents for six cycles remained the same, showing a fast response to blue illumination in air and at the room temperature.

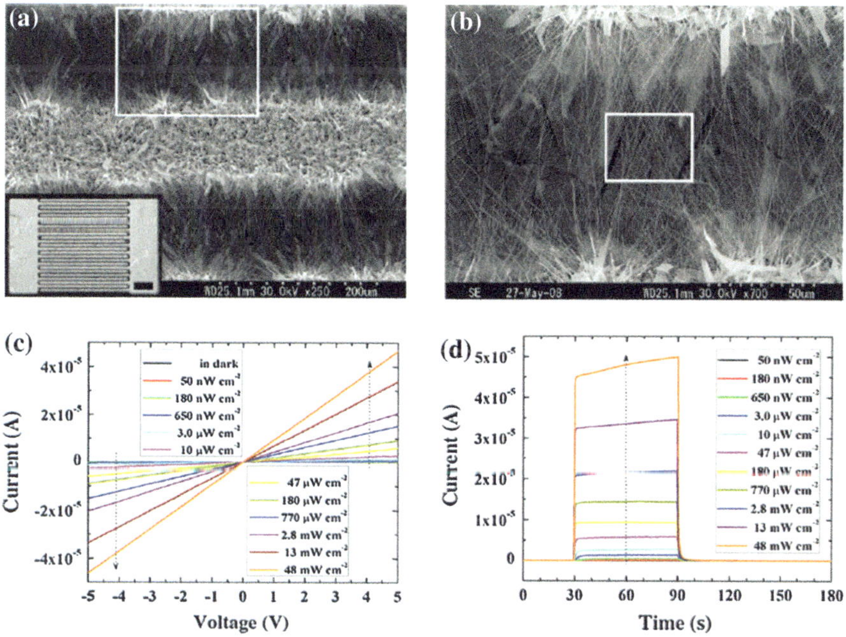

Fig. 8.19 **a** SEM images of ZnO bridging nanowires. *Inset*: an optical microscope image of the fabricated device; the scale bar is 1 mm. **b** Higher magnification of the area outlined in *white* in (**a**). **c** Current–voltage characteristics of the ZnO bridging nanowires under various intensities of light. **d** Time-resolved measurements of ZnO nanowires by switching UV light between on (at 30 s) and off (at 90 s) [76]

8.5.5.4 Gallium Oxide

Monoclinic gallium oxide (β-Ga_2O_3) is a promising *n-type* oxide semiconductor with a wide direct bandgap of 4.2–4.9 eV for use in solar-blind photodetectors. This type of photodetector, which works between 200 nm to 290 nm, is applicable for detection of weak signals under intense background radiation. Feng et al. [80] grew β-Ga_2O_3 nanowires on a silicon substrate coated with gold by evaporating gallium under controlled conditions to study their performance as a solar-blind photodetector. They observed a substantial increase, near three orders of magnitude, in the measured current upon exposure to 254 nm light. The response and recovery times over seven cycles were 0.22 s and 0.09 s, respectively.

8.5.5.5 Indium Oxide

Indium oxide (In_2O_3) has a direct bandgap of ~3.6 eV and ~2.5 eV as indirect bandgap. It has been used in solar cells and organic light emitting diodes using the bulk material, and ultra-sensitive toxic-gas detectors using thin films. Zhang et al.

[81] studied the photoconducting properties of In_2O_3 nanowires by fabricating devices based on individual nanowires. They employed a laser ablation technique via vapor–liquid–solid mechanism to produce nanowires 10 nm in diameter and 3 μm long. The HRTEM image shown in Fig. 8.20a reveals that the nanowires are single-crystalline in nature grown along [100] direction. The linear conductance of the fabricated device was around 1.7×10^3 nS, while exposure to UV light at 254 nm, raised it to 5.0×10^3 nS indicating a sensitivity of 10^4 (Fig. 8.20b). The fabricated device based on In_2O_3 nanowires exhibits a good reversible behavior while the UV light was turned on and off. The time response, defined as the time required for increasing the conductance by one order of magnitude, was estimated as 10 s. Figure 8.20d presents the photoresponse of In_2O_3 nanowires; after exposure to UV light (254 and 356 nm), significant current was observed at lower wavelength (33 nA compared with 290 nA). Considering the direct bandgap of In_2O_3 (3.6 eV), photon energy of 254 nm-light (4.9 eV) is large enough for generating photoexcitated carriers while the photon energy of 365 nm-light is smaller than the energy gap of In_2O_3 nanowire.

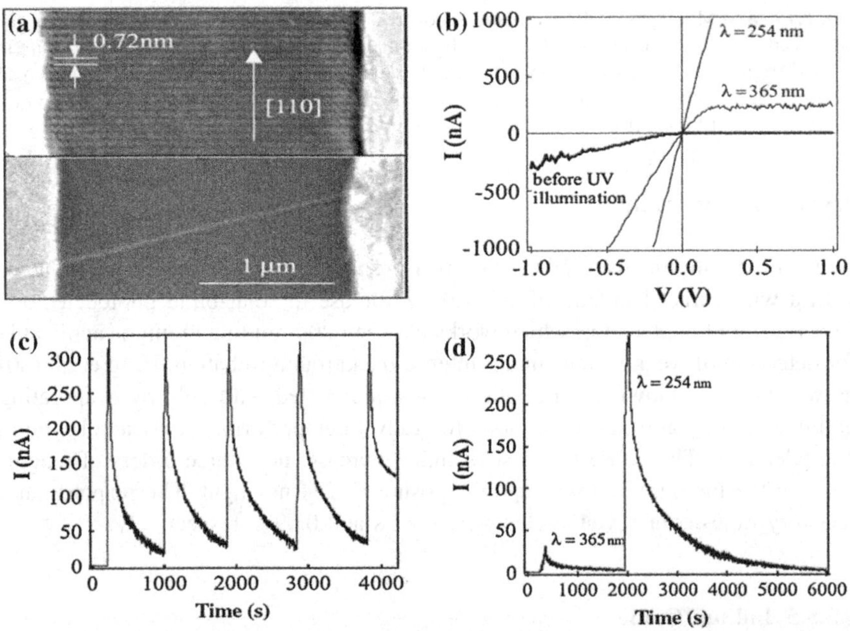

Fig. 8.20 **a** HRTEM image of an In_2O_3 nanowire (*top*) and SEM image of the fabricated device (*bottom*). **b** Current–voltage characteristics of the In_2O_3 nanowire-based device in dark and under UV illumination. **c** Time-resolved measurement of the fabricated device. **d** Photoresponse of the In_2O_3 nanowire to UV exposure with 254 and 365 nm as the wavelengths [81]

8.6 Challenges and Future Prospects

We have discussed in this chapter an overview on different concepts of photodetection, one-dimensional nanostructure-based photodetectors and diverse materials employed in device fabrication. Although construction and integration of photodetectors based on 1-D nanostructures have been advancing, several issues remain in realizing practical systems. Given the importance of material in photodetector fabrication and the complex kinetics and growth mechanisms involved, significant strides have been made toward the synthesis of high-quality nanomaterials for photodetector applications. Remaining challenges include precise control of diameter, length, density and crystallization as well as good control of nanowire quality: defects, traps, surface states, and orientation. Both direct growth on and transfer of elsewhere-grown nanowires to the substrate have been employed. Consideration must be given to the selection of the target substrate keeping in mind the lattice constant, material and thermal mismatch. The "pick and place" method is common at present in device fabrication in the photodetector field but this limits the throughput or large wafer-scale fabrication. Making contacts (Schottky or Ohmic) is a big challenge as it is critical for reliable and stable performance. Future research must focus on improvement of nanowire alignment and network assembly techniques to fabricate array of detectors on wafer-scale. Finally, sensitivity, selectivity, and stability are the anticipated features of the next generation of photodetectors. Demonstration of these attributes in statistically meaningful number of devices in a batch and system development is a must for successful commercialization.

References

1. Soci, C., Zhang, A., Bao, X.-Y., Kim, H., Lo, Y., & Wang, D. (2010). *Journal of Nanoscience and Nanotechnology, 10*, 1430–1449.
2. Smith, W. (1873). *Nature*, 303.
3. Donati, S. (2000). *Photodetector: Devices, circuits, and applications*. New Jersey: Prentice Hall PTR.
4. Sze, S. M., & Ng, K. K. (2007). *Physics of semiconductor devices*. New Jersey: Wiley.
5. Razeghi, M., & Rogalski, A. (1996). Semiconductor ultraviolet detector. *Journal of Applied Physics, 79*, 7433–7473.
6. Chai, G., Lupan, O., Chow, L., & Heinrich, H. (2009). Crossed zinc oxide nanorods for ultraviolet radiation detection. *Sensors and Actuators A, 150*, 184–187.
7. Tredwell, T. J. (1995). Visible array detectors. In M. Bass (Ed.), *Handbook of optics*. New York: McGraw-Hill.
8. Kozlowski, L. J., & Kosonocky, W. F. (1995). Infrared detector array. In M. Bass, E. W. Van Styland, P.R. Williams, & W. L. Wolfe (Eds.), *Handbook of optics*. New York: McGraw-Hill.
9. Joshi, A. M., & Olsen, G. H. (1995). Photodetection. In Optimal Society of America (Ed.), *Handbook of optics*. New York: McGraw-Hill.
10. Shen, G., & Chen, D. (2010). *Recent Patent on Nanotechnology, 4*, 20–31.

11. Tans, S. J., Verschueren, A. R. M., & Dekker, C. (1998). Room-temperature transistor based on a single carbon nanotube. *Nature, 393*, 49–52.
12. Rosenblatt, S., Yaish, Y., Park, J., Gore, J., Sazonova, V., & McEuen, P. L. (2002). High performance electrolyte gated carbon nanotube transistors. *Nano Letters, 2*, 869–872.
13. Freitag, M., Martin, Y., Misewich, J. A., Martel, R., & Avouris, PH. (2003). Photoconductivity of single carbon nanotubes. *Nano Letters, 3*(8), 1067–1071.
14. Gao, T., Li, Q. H., & Wang, T. H. (2005). Adsorption and desorption of oxygen probed from ZnO nanowire films by photocurrent measurements. *Applied Physics Letters, 86*, 173105.
15. Ye, Y., Dai, L., Wen, X., Wu, P., Pen, R., & Qin, G. (2010). Nanoporous electrodeposited versus nanoparticle ZnO porous films of similar roughness for dye sensitized solar cell application. *ACS Applied Material Interfaces, 2*, 2724–2727.
16. Fang, X., Hu, L., Huo, K., Gao, B., Zhao, L., Liao, M., et al. (2011). New ultraviolet photodetector based on individual Nb_2O_5 nanobelts. *Advanced Functional Materials, 21*, 3907–3915.
17. Lao, C. S., Park, M. C., Kuang, Q., Deng, Y., Sood, A. K., Polla, D. L., et al. (2007). Giant enhancement in UV response of ZnO nanobelts by polymer surface-functionalization. *Journal of American Chemical Society, 129*, 12096–12097.
18. He, J. H., Lin, Y. H., McConney, M. E., Tsukruk, V. V., Wang, Z. L., & Bao, G. (2007). Enhancing UV photoconductivity of ZnO nanobelt by polyacrylonitrile functionalization. *Journal of Applied Physics, 102*, 084303.
19. Fang, X., Bando, Y., Liao, M., Zhai, T., Gautam, U. K., Li, L., et al. (2010). An efficient way to assemble ZnS nanobelts as ultraviolet-light sensors with enhanced photocurrent and stability. *Advanced Functional Materials, 20*, 500–508.
20. Fan, X., Meng, X. M., Zhang, X. H., Zhang, M. L., Jie, J. S., Zhang, W. J., et al. (2009). Formation and photoelectric properties of periodically twinned ZnSe/SiO2 nanocables. *Journal of Physical Chemistry C, 113*, 834–838.
21. Li, L., Fang, X., Zhai, T., Liao, M., Gautam, U. K., Wu, X., et al. (2010). *Advanced Materials, 22*, 4151–4156.
22. Fang, X., Xiong, S., Zhai, T., Bando, Y., Liao, M., Gautam, U. K., et al. (2009). *Advanced Materials, 21*, 5016–5021.
23. Jiang, Y., Zhang, W. J., Jie, J. S., Meng, X. M., Fan, X., & Lee, S. T. (2007). *Advanced Functional Materials, 17*, 1795–1800.
24. Jie, J. S., Zhang, W. J., Jiang, Y., Meng, X. M., Li, Y. Q., & Lee, S. T. (2006). *Nano Letters, 6*, 1887–1892.
25. Yu, Z., Li, J., O'Connor, D. B., Wang, L.-W., & Barbara, P. F. (2003). *Journal of Physics and Chemistry B, 107*, 5670–5674.
26. Kind, H., Yan, H., Messer, B., Law, M., & Yang, P. (2002). *Advanced Materials, 14*, 158–160.
27. Kum, M. C., Jung, H., Chartuprayoon, N., Chen, W., Mulchandani, A., & Myung, N. V. (2012). *Journal of Physics and Chemistry C, 116*, 9202–9208.
28. Lai, W.-J., Li, S.-S., Lin, C.-C., Kuo, C.-C., Chen, C.-W., Chenb, K.-H., et al. (2010). *Scripta Materials, 63*, 653–656.
29. Chen, R. S., Chen, C. A., Tsai, H. Y., Wang, W. C., Huang, Y. S. (2012). *Applied Physics Letters, 100*, 123108.
30. Mofor, A. C., Bakin, A., Chejarla, U., Schlenker, E., El-Shaer, A., Wagner, G., et al. (2007). *Superlattices and Microstructures, 42*, 415–420.
31. Lu, Y., Dajani, I., & Knize, R. J. (2007). *Applied Surface Science, 253*, 7851–7854.
32. Chaia, G., Lupan, O., Chow, L., & Heinrich, H. (2009). *Sensors and Actuators A, 150*, 184–187.
33. Liu, N., Fang, G., Zeng, W., Zhou, H., Cheng, F., Zheng, Q., et al. (2010). *ACS Applied Material Interfaces, 2*, 1973–1979.
34. Li, Y., Dong, X., Cheng, C., Zhou, X., Zhang, P., Gao, J., et al. (2009). *Physica B, 404*, 4282–4285.

35. Soci, C., Zhang, A., Xiang, B., Dayeh, S. A., Aplin, D. P. R., Park, J., et al. (2007). *Nano Letters, 7*, 1003–1009.
36. Chen, R.-S., Chen, H. Y., Lu, C.-Y., Chen, K. H., Chen, C.-P., Chen, L. C., et al. (2007). *Applied Physics Letters, 91*, 223106.
37. Choi, H. G., Choai, Y.-S., Jo, Y. C., & Kim, H. (2004). *Japanese Journal of Applied Physics, 43*, 3916–3918.
38. Ahn, Y., Dunning, J., & Park, J. (2005). *Nano Letters, 5*, 1367–1370.
39. Bae, J., Kim, H., Zhang, X.-M., Dang, C. H., Zhang, Y., Choi, Y. J., et al. (2010). *Nanotechnology, 21*, 095502.
40. Kim, K.-H., Keem, K., Jeong, D.-Y., Min, B., Cho, K., Kim, H., et al. (2006). *Japanese Journal of Applied Physics, 45*, 4265–4269.
41. Servati, P., Colli, A., Hofmann, S., Fu, Y. Q., Beecher, P., Durrani, Z. A. K., et al. (2007). *Physica E., 38*, 64–66.
42. Yan, C., Singh, N., Cai, H., Gan, C. L., & Lee, P. S. (2010). *ACS Applied Materials Interfaces, 2*, 1794–1797.
43. Ahn, Y. H., & Park, J. (2007). *Applied Physics Letters, 91*, 162102.
44. Kim, C.-J., Lee, H.-S., Cho, Y.-J., Kang, K., & Jo, M.-H. (2010). *Nano Letters, 10*, 2043–2048.
45. Gates, B., Mayers, B., Cattle, B., & Xia, Y. (2002). *Advanced Functional Material, 12*, 219–227.
46. Liao, Z.-M., Hou, C., Zhao, Q., Liu, L.-P., & Yu, D.-P. (2009). *Applied Physics Letters, 95*, 093104.
47. Liang, F., & Qian, H. (2009). *Materials Chemistry and Physics, 113*, 523–526.
48. Liu, J.-W., Zhu, J.-H., Zhang, C.-L., Liang, H.-W., & Yu, S.-H. (2010). *Journal of American Chemical Society, 132*, 8945–8952.
49. Thunich, S., Prechtel, L., Spirkoska, D., Abstreiter, G., Morral, A. F., & Holleitner, A. W. (2009). *Applied Physics Letters, 95*, 083111.
50. Cooke, D., Wu, Z., Mei, X., Liu, J., Ruda, H. E., Kavangah, K. L., et al. (2004). *Palais des Congres de Montreal, American Physical Society* Canada.
51. Logeeswaran, V. J., Sarkar, A., Islam, M. S., Kobayashi, N. P., Straznicky, J., Li, X., et al. (2008). *Applied Physics A, 91*, 1–5.
52. Wang, J., Gudiksen, M. S., Duan, X., Cui, Y., & Lieber, C. M. (2001). *Science, 293*, 1455–1457.
53. Han, S., Jin, W., Zhang, D., Tang, T., Li, C., Liu, X., et al. (2004). *Chemistry and Physics Letters, 389*, 176–180.
54. Huang, H. M., Chen, R. S., Chen, H. Y., Liu, T. W., Kuo, C. C., Chen, C. P., et al. (2010). *Applied Physics Letters, 96*, 062104.
55. Chen, R.-S., Yang, T.-H., Chen, H.-Y., Chen, L.-C., Chen, K.-H., et al. (2009). *Applied Physics Letters, 95*, 162112.
56. Kang, M., Lee, J.-S., Sim, S.-K., Kim, H., Min, B., Cho, K., et al. (2004). *Japanese Journal of Applied Physics, 43*, 6868–6872.
57. Calarco, R., Marso, M., Richter, T., Aykanat, A. I., Meijers, R., Hart, A., et al. (2005). *Nano Letters, 5*, 981–984.
58. Chen, H.-Y., Chen, R.-S., Chang, F.-C., Chen, L.-C., Chen, K.-H., & Yang, Y.-J. (2009). *Applied Physics Letters, 95*, 143123.
59. Kouklin, N., Menon, L., Wong, A. Z., Thompson, D. W., Woollam, J. A., Williams, P. F., et al. (2001). *Applied Physics Letters, 79*, 4423–4425.
60. Li, Q., & Penner, R. M. (2005). *Nano Letters, 5*, 1720–1725.
61. Gu, Y., Romankiewicz, J. P., David, J. K., Lensch, J. L., & Lauhon, L. J. (2006). *Nano Letters, 6*, 948–952.
62. Gu, X. W., Shadmi, N., Yarden, T. S., Cohen, H., & Joselevich, E. (2012). *Journal of Physics and Chemistry C, 116*, 20121–20126.
63. Kung, S.-C., van der Veer, W. E., Yang, F., Donavan, K. C., & Penner, R. M. (2010). *Nano Letters, 10*, 1481–1485.

64. Salfi, J., Philipose, U., de Sousa, C. F., Aouba, S., & Ruda, H. E. (2006). *Applied Physics Letters, 89*, 261112.
65. Philipose, U., Ruda, H. E., Shik, A., de Souza, C. F., & Sun, P. (2006). *Journal of Applied Physics, 99*, 066106.
66. Hsiao, C. H., Chang, S. J., Wang, S. B., Chang, S. P., Li, T. C., Lin, W. J., et al. (2009). *Journal of Electrochemical Society, 156*, J73–J76.
67. Yoon, Y.-J., Park, K.-S., Heo, J.-H., Park, J.-G., Nahm, S., & Choi, K. J. (2010). *Journal of Materials Chemistry, 20*, 2386–2390.
68. Wang, J.-J., Cao, F.-F., Jiang, L., Guo, Y.-G., Hu, W.-P., & Wan, L. J. (2009). *Journal of American Chemical Society, 131*, 15602–15603.
69. Zhai, T., Fang, X., Liao, M., Xu, X. Li, L., Liu, B., et al. (2010). *ACS Nano, 4*, 1596–1602.
70. Cao, Y. L., Liu, Z. T., Chen, L. M., Tang, Y. B., Luo, L. B., Jie, J. S., et al. (2011). *Optic Express, 19*, 6100–6108.
71. Meng, Q. F., Jiang, C. B., & Mao, S. X. (2009). *Applied Physics Letters, 94*, 043111.
72. Li, Z., Salfi, J., de Souza, C., Sun, P., Nair, S. V., & Ruda, H. E. (2009). *Applied Physics Letters, 97*, 063510.
73. Wang, Z. L. (2008). *ACS Nano, 2*, 1987–1992.
74. Prades, J. D., Diaz, R. J., Ramirez, F. H., Romero, L. F., Andreu, T., Cirera, A., et al. (2008). *Journal of Physics and Chemistry C, 112*, 14639–14644.
75. Kim, W., & Chu, K. S. (2009). *Physics Status Solidi A, 206*, 179–182.
76. Li, Y., de Valle, F., Simonnet, M., Yamada, I., & Delaunay, J.-J. (2009). *Nanotechnology, 20*, 045501.
77. Kim, D., Kim, Y.-K., Park, S. C., Ha, J. S., Huh, J., Na, J., et al. (2009). *Applied Physics Letters, 95*, 043107.
78. Prades, J. D., Ramirez, F. H., Diaz, R. J., Manzanars, M., Andreu, T., Cirera, A., et al. (2008). *Nanotechnology, 19*, 465501.
79. Liao, L., Yan, B., Hao, Y. F., Xing, G. Z., Liu, J. P., Zhao, B.C., et al. (2009). *Applied Physics Letters, 94*, 113106.
80. Feng, P., Zhang, J. Y., Li, Q. H., & Wang, T. H. (2006). *Applied Physics Letters, 88*, 153107.
81. Zhang, D., Li, C., Han, S., Liu, X., Tang, T., Jin, W., et al. (2003). *Applied Physics A, 77*, 163–166.

Chapter 9
Nanowire BioFETs: An Overview

M. Meyyappan and Jeong-Soo Lee

Abstract In this chapter, the biosensing as a key element of nanotechnology and commanding a wide range of applications is discussed, e.g., fast and efficient clinical diagnostics, health care, security, environmental monitoring, etc. The operation and sensing mechanism of BioFETs and ion-sensitive FETs are elaborated on a molecular level, based upon the molecular recognition between target and probe molecules and the input gate voltage and output ON current of the conventional FETs. In particular, the extended roles of the gate electrode of BioFETs as the probing surface are highlighted, in comparison with the conventional gate electrode, together with the physical and biological processes for detecting target molecules. Moreover, the bottom-up syntheses of vertical and horizontal nanowires are presented and the ensuing nanowires are characterized. Also, the top-down and bottom-up approaches for processing nanowires are compared by taking as criteria the process complexity and quality of the nanowires produced. Finally, the future prospects of bio-sensing are presented.

Abbreviation

LED	Light-emitting diode
UV	Ultraviolet
IR	Infrared
NEP	Noise-equivalent power
SWCNT	Single-walled carbon nanotube
FET	Field effect transistor
CVD	Chemical vapor deposition
SAED	Selected-area-electron diffraction

M. Meyyappan (✉)
NASA Ames Research Center, Moffett Field, California, USA
e-mail: m.meyyappan@nasa.gov

J.-S. Lee
Department of Electrical Engineering, Pohang University of Science and Technology, Pohang, South Korea

D. M. Kim and Y.-H. Jeong (eds.), *Nanowire Field Effect Transistors: Principles and Applications*, DOI: 10.1007/978-1-4614-8124-9_9,

VLS	Vapor-liquid-solid
PECVD	Plasma enhanced chemical vapor deposition
FIB	Focused ion beam

9.1 What is the Need for Biosensing?

A biosensor is a device that can identify the presence and/or provide quantitative measurement (such as concentration) of a biochemical species in a sample. The need for biosensors is ubiquitous in modern society: clinical diagnostics, healthcare, early warning biothreat detection, water quality monitoring, food quality monitoring and related agricultural applications, environmental monitoring and many more [1–5]. The concept of a lab-on-a-chip has been around for over a couple of decades. As a system, it attempts to integrate various subsystems such as the biosensor with auxillaries including fluidic components (valves, channels etc.) to transport the samples, concentrators, separators, heaters etc. The motivation for the lab-on-a-chip is the desire to perform an entire set of analytical operations—normally done by many technicians in a laboratory—just at the chip level. The advantages are obvious: speed (minutes to an hour vs. days), smaller sample volumes, and cost savings through automation. These systems are developed typically on silicon, glass or plastic substrates and designed to be disposable, at least in the area of biomedical and clinical diagnostic applications. For example, routine health screening may need a system that can monitor pH, cholesterol, complete blood count, white blood cell count, urine analysis, biomarkers for heart disease, and the metabolic panel (Na, K, Ca, etc.). Municipal water quality monitoring typically involves analysis for about fifteen pathogens including E-coli, salmonella, and other bacteria, protozoa and viruses.

9.2 Biosensors

The sensor is the critical component of the system discussed above and may typically work based on a "lock and key" approach. Alternatively known as the affinity-based approach, it involves choosing a probe *a priori* that is designed to bind selectively to the target of interest [1]. The probe can be of any type including DNA, RNA, antibody, aptamer and others. The binding event manifests as a change in some measurable property of the surface supporting it. For example, a cantilever may show different bending levels for the bare cantilever, after functionalization with the probe, and following the probe-target hybridization. Optical techniques include monitoring the fluorescence or surface plasmon resonance. One of the most common methods is to measure current or voltage from an electrode. Aside from electrochemical measurements [2, 4, 6–10] involving cyclovoltammetry, amperometry and impedance spectroscopy, electrical transduction also includes monitoring the change in current–voltage characteristics of a field effect transistor (FET) appropriately treated with the biological probes [5, 11–24]. Fabrication and characterization of semiconducting nanowire-based biological FET or BioFET and its applications are the subject of this chapter.

9.3 What is a BioFET?

The biofield effect transistor is a special form of FET used for sensing biomolecules in a fluid system (see Fig. 9.1). The conventional field effect transistor is a three-terminal device consisting of source, gate, and drain electrodes. The gate regulates the current flow between the source and the drain just like a valve regulates fluid flow due to the pressure difference between two points in a pipe. The analogous driving force in the FET is the voltage difference between the source and the drain. The FET is a unipolar device, implying that current is conducted only by electrons or holes.

The essential part of a BioFET—also known as ion sensitive FET or ISFET—is the gate electrode replaced by the gate dielectric, electrolyte, and a reference electrode [25–30]. The electrical contact to the electrolyte is provided by a reference electrode which is placed in the electrolyte solution. The channel conductance or resistance is modulated by the electrostatic field generated by the reference electrode. Usually, the channel is covered by dielectric materials such as Al_2O_3, Si_3N_4, SiO_2, Ta_2O_5 [30]. The ion concentration in the electrolyte affects the electrostatic field and the channel potential across the electrolyte/dielectric/semiconductor system.

In pH-sensing applications, the H^+-specific binding sites on the surface can be protonated and deprotonated, and thus, a variation of pH value will change the

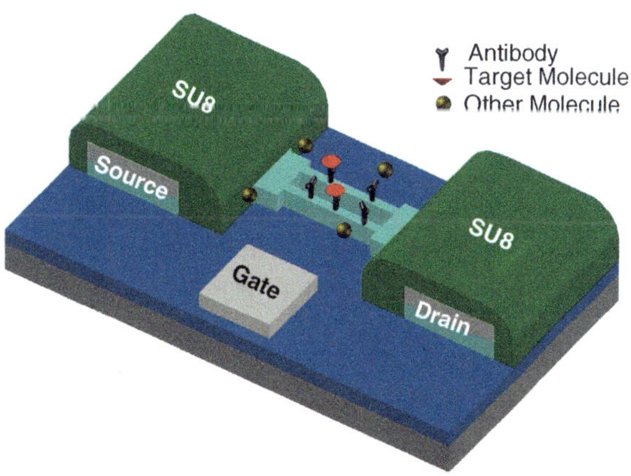

Fig. 9.1 Schematic of a BioFET

channel potential. For example, the hydroxyl group at the surface can exchange protons with the electrolyte via the following chemical reactions.

$$SiOH \Leftrightarrow SiO^- + H^+$$

$$SiOH + H^+ \Leftrightarrow SiOH_2^+$$

The typical binding sites are the hydroxyl groups on the oxide surface and the silanol (SiH_3OH) and amine ($SiNH_2$) groups on the nitride surface.

In biosensing applications, a probe molecule, which is immobilized on the dielectric surface, binds selectively with the target molecule in the electrolyte. This binding event can change the channel potential and then be detected by measuring the change of drain current or gate voltage.

9.4 Nanowire Synthesis

For the construction of nanowire-based BioFETs, a choice must be made first about the material, and then, an appropriate synthesis technique for that nanowire needs to be selected. From a conventional manufacturing point-of-view and future integration with processing chips, silicon nanowire appears to be the safe choice. Nevertheless, a variety of other materials, particularly oxide nanowires such as In_2O_3, ZnO etc., have been pursued in the literature. Regardless of the material, the nanowires can be fabricated either by a bottom-up approach or conventional top-down method involving etching to produce the desired cylindrical nanowire form out of a thin or thick film. Both approaches are described below. A detailed background on inorganic nanowires, growth methods, properties and applications including biosensors can be found in Ref. [31].

9.4.1 Bottom-Up Synthesis Techniques

The material choices for nanowires are Si, Ge, metal oxides etc. as mentioned earlier and these materials have long been grown as thin and thick films using chemical vapor deposition (CVD) and other epitaxial techniques. In the bottom-up preparation of nanowires also, the same precursors or source gases are used at more or less the same processing conditions as in the case of thin film growth. Then, what makes nanowire growth possible? There must be a guiding force to form the cylindrically shaped nanowires. An obvious choice would be to use a preformed template with the desirable pore size that can guide the growth into cylindrical nanowires. Anodized alumina is the most commonly used template for this purpose, and a wide variety of nanowires have been successfully grown using this template. The disadvantage of this method is that the template itself needs to be removed prior to the use of the nanowires in device fabrication. Often the template removal is achieved through some acid treatment which may damage the nanowire and/or introduce defects.

A successful alternative to template growth is a catalyzed-CVD approach, often known as the vapor-liquid-solid (VLS) technique [31]. A laboratory VLS reactor is a quartz tube inserted into a two-zone furnace shown in Fig. 9.2. The source vapor can be generated in any number of ways in the upstream zone depending on the desired material, as will be discussed shortly. The growth substrate consists of a catalyst, typically a thin metal layer (gold, In, Ga etc.) or mono-dispersed colloids of that metal, if readily available. The catalyst layer may be prepared either by sputtering or thermal evaporation. At the growth temperature, this thin layer breaks up into tiny droplets which remain in a molten state. The source vapor dissolves into the droplet and when supersaturation is reached, material begins to precipitate in the form of nanowires as shown in Fig. 9.3. Invariably, the catalyst particle ends up at the top of the growing nanowire in the VLS process which allows easy removal through acid wash or chemical mechanical polishing of vertical nanowires during device processing.

In the case of silicon nanowires, silane (or $SiCl_4$) diluted in H_2 may be used as feedstock as in the case of the corresponding thin film growth [31–34]. For GaAs nanowires, $(CH_3)_3Ga/AsH_3$ feedstock common in MOCVD may be used [31, 35, 36]. For most oxide nanowires, the common practice in VLS growth is sublimation of the corresponding high purity (99.999 % or higher) powder [31, 37, 38]. This sublimation approach also works well for a variety of tellurides, selenides and other materials as long as the high purity powder form of the source is available [39–42] and the sublimation temperature is reasonable without the need for high temperature furnaces (over 1,000 °C). It is possible to mix two or more source powders for growing ternary nanowires [43] (e.g., $Ge_2Sb_2Te_5$). Gold is the widely used catalyst in nanowire growth reported in the literature. Gold modulates carrier recombination in both n- and p-type silicon, since high-mobility interstitial gold atoms can transform into electrically active substitutional sites. Therefore, it is desirable to look for alternatives to gold. Fortunately, a wide variety of group III, IV, V, and VI metal elements have been found to be useful as catalyst except for Ir and Pt [44]. The only correlation appears to be the inverse dependence of nanowire growth density on the melting point of the catalyst metal. In this regard, indium (In) is a low-melting metal that is useful as a catalyst for the growth of Si and other nanowires, allowing growth zone temperature to be as low as 150 °C [45]. Gallium also is another low-melting candidate ($\sim$450 °C) suitable for nanowire growth [46].

Figure 9.4 shows a random mat of silicon nanowires grown on a silicon substrate [32]. The diameter of the nanowire correlates reasonably with the catalyst particle diameter. The length of the nanowires is several microns which can be decreased by reducing the growth time. Figure 9.5 shows vertically oriented silicon nanowires grown using a mixture of $SiCl_4/H_2$ with diameter in the range of 40–80 nm and height 1–2 μm. Only a very narrow set of conditions using this feedgas system yield vertical nanowires in the case of silicon [32]. For example,

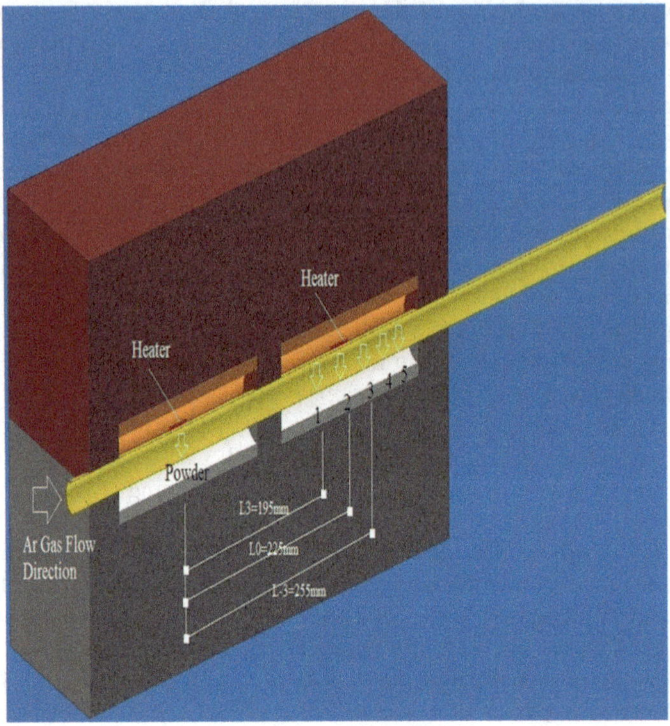

Fig. 9.2 Reactor for nanowire growth by VLS approach (figure courtesy of Keivan Davami)

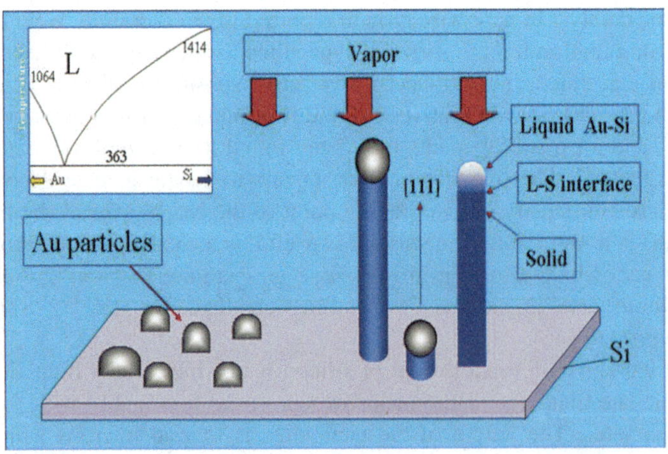

Fig. 9.3 VLS growth mechanism (figure courtesy of Xuhui Sun)

Fig. 9.4 Random mat of silicon nanowires grown by VLS method (figure courtesy of Aaron Mao)

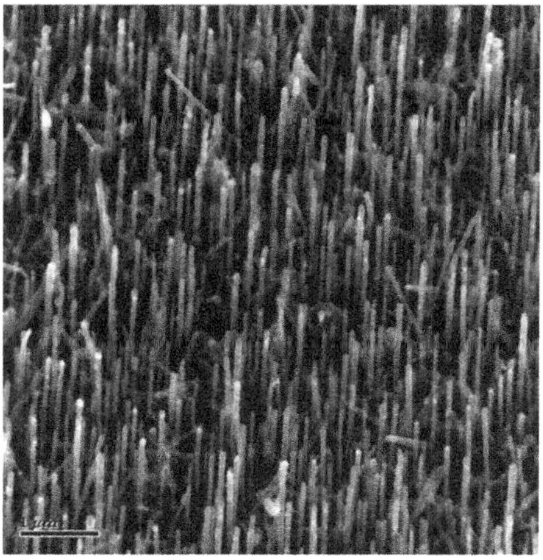

Fig. 9.5 Vertical silicon nanowires grown by VLS approach (figure courtesy of Aaron Mao)

$SiCl_4$ fractions greater than 0.25 % in H_2 and temperatures greater than 950 °C yield thin and curvy wires. Similarly, a narrow set of conditions have been identified for growing vertical silicon nanowires using the silane/H_2 system as well [33]. The silicon nanowires can be doped with boron and phosphorous to obtain p- and n-type, respectively, as in the case of thin films [31].

The as-grown nanowires typically show exceptional surface quality, ideal for device fabrication. It has been shown for a variety of different semiconductor nanowires that the bandgap increases with a decrease in diameter, particularly for sizes below the value of the corresponding Bohr radius. For example, a silicon nanowire of 1.3 nm in diameter exhibits a bandgap of 3.5 eV, substantially larger than the 1.1 eV for bulk silicon [47]. This allows bandgap engineering through control of the nanowire diameter. In spite of all the attractive features of the bottom-up technique, it is hard to use these structures in device fabrication in a straightforward manner. For example, attempting a "pick and place" approach using the random mat of nanowires seen in Fig. 9.4 over thousands of pairs of source-drain terminals on a large wafer is a difficult task. Direct growth by catalyzed-CVD (i.e. VLS) on the prefabricated contacts for thousands of devices over a large wafer is not easy either. The vertical nanowires can be used to fabricate vertical surround gate transistors. In this configuration, the source will be at the bottom with the drain contact at the top; the gate wraps around the nanowire channel connecting the source and the drain. The report on such devices using SiNWs uses many nanowires under one gate [48]. A gate surrounding only a single nanowire was reported for ZnO transistor [49]. In any case, wafer-scale fabrication of vertical nanowire transistors has not been achieved yet.

9.4.2 Top-Down Nanowires

Fabrication of nanowires using a top-down approach is straightforward with the use of patterned etching to obtain nanowires of desired diameter and length. Figure 9.6 shows an array of silicon nanowires created using dry etching [50]. Control of the nanowire diameter and length as well as the number of nanowires in the channel region is much easier compared with the bottom-up process discussed above. This is the major advantage of the top-down approach in addition to the immediate possibility of large wafer (200–300 mm) fabrication.

9.5 Device Fabrication and Performance

In recent years, a number of reports on the fabrication and performance of nanowire based BioFETs have appeared including a variety of bottom-up grown nanowires [51–64] and a few conventional top-down efforts [24, 50, 65–67]. Below a concise summary of these works is presented.

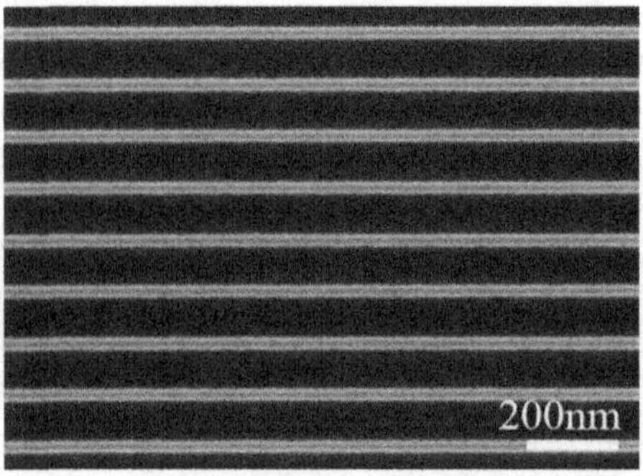

Fig. 9.6 Silicon nanowires prepared by a top-down process involving patterning followed by dry etching (figure courtesy of Taiuk Rim and Sungho Kim)

9.5.1 BioFETS Using Bottom-Up Grown Nanowires

Cui et al. [52] were one of the first to report nanowire-based devices for biosensing. They prepared silicon nanowires using a VLS approach, aligned them using a flow approach and contacted two ends of a nanowire with electron-beam lithography. The back-gated device was the first NW-BioFET to show a pH-dependent conductance. The SiNWs were also modified with biotin for the detection of streptavidin down to picomolar concentration levels. This group also demonstrated the utility of NW-BioFETs for sensitive detection of cancer markers [55]. Incorporation of surface receptors into arrays of NWs allowed selective detection of mucin-1, carcinoembryonic antigen, and prostate-specific antigen (PSA). A comprehensive analysis of their device characteristics showed that the NW-BioFETs provide the highest conductance response in the subthreshold regime along with the best charge detection limit [54].

Though silicon is the most common nanowire used in biosensor construction, oxide nanowires have been explored as well. In_2O_3 nanowires have been found to possess good conductance for NW-FET construction [56–59]. Li et al. fabricated back-gated FETs using VLS-grown In_2O_3 nanowires with a channel length of 2 μm. They attached PSA antibodies to the nanowire surface and demonstrated a sensitivity of 5 ng/mL of PSA using a single nanowire conducting channel. This group also used antibody mimic proteins which are 2–5 nm polypeptides showing strong binding capability to their targets just like antibodies to detect nucleocapsid protein [59]. The latter is a biomarker for severe acute respiratory syndrome (SARS). The In_2O_3 NW-FET using nanowires modified with a fibronectin-based binding apent detected N-protein at subnanomolar concentrations. This sensitivity is comparable to many immunological detection techniques, but the NW-FET provides results in a shorter time without the need for labeling.

Another metal oxide popular for biosensing is ZnO and the nanowire form provides a stable oxide surface and a high surface-to-volume ratio. Choi et al. [60] grew ZnO nanowires by pulsed laser deposition, which are 70 nm in diameter and about 8 μm long. The nanowires functionalized with biotin were sensitive to streptavidin binding and the device was able to detect 2.5 nM of streptavidin. Yeh et al. [61] used a ZnO nanowire device and showed that a nonsymmetrical Schottky contact under reverse bias is much more sensitive than a device with symmetric ohmic contacts. Single nanowire ZnO FET has been shown to detect uric acid as low as 1 pM concentration with a 14.7 nS conductance increase [62].

Finally, it is of interest to know the effects of nanowire diameter, number of nanowires (single vs. multiple) etc. on sensitivity and other BioFET metrics. Li et al. [63] conducted an investigation of these effects using silicon nanowire FETs. They found that the sensitivity decreased with an increase in nanowires, increase in nanowire diameter, and higher doping density. For example, devices with 4 and 7 nanowires showed 38 and 82 % lower sensitivity than nanowires 61–80 nm in diameter. An increase in doping density from 10^{17} to 10^{19} cm^{-3} results in a 69 % decrease in sensitivity.

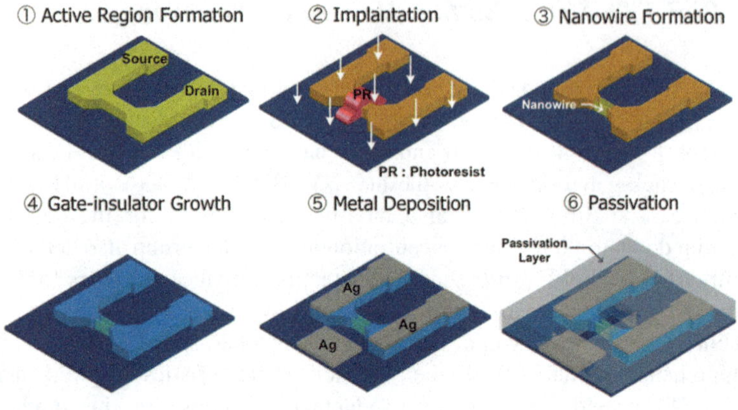

Fig. 9.7 Top-down process sequence for the fabrication of silicon nanowire ISFET (figure courtesy of Taiuk Rim and Sungho Kim)

9.5.2 Top-Down BioFETs

A schematic diagram of the SiNW ISFET fabrication process is shown in Fig. 9.7 along with a schematic of the device in Fig. 9.8. A 6-inch silicon-on-insulator (SOI) substrate with 10^{15} cm^{-3} boron doping was used. The 675-μm-thick substrate consists of 100-nm thick <100> oriented silicon layer and a 200-nm-thick buried oxide layer. The top Si layer was thinned to 40 nm using thermal oxidation at 900 °C and wet etching of SiO_2 with diluted HF (1:1,000 with deionized water) to form the rectangular-shaped nanowires. The top Si layer, except the active region, was etched using an inductively coupled plasma reactive ion etcher (ICP-RIE) with a mixture of HBr and O_2. After the device isolation process, nanowires of 50-nm width and 10 micron length were formed using electron-beam

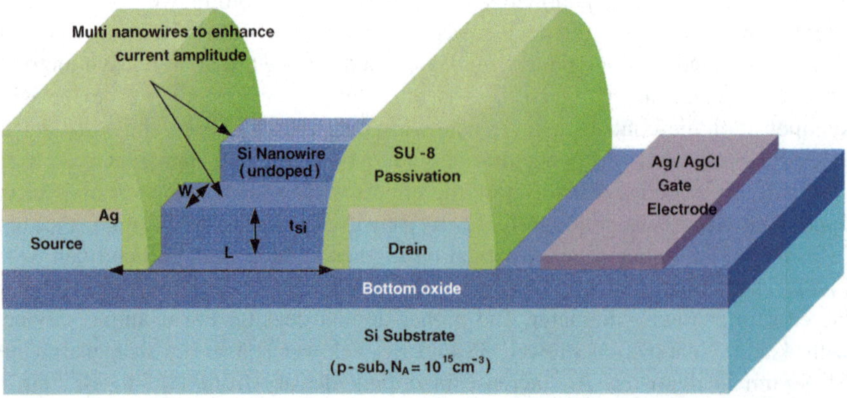

Fig. 9.8 Schematic of a Si-NW ISFET (figure courtesy of Taiuk Rim and Sungho Kim)

lithography and ICP-RIE. A 4-nm thick SiO_2 sacrificial oxide layer was grown using thermal oxidation at 900 °C to protect the interface between the nanowire channel and gate oxide from the ion implant damage. Arsenic ion implantation was then done to form the source and drain ohmic contacts, and the implanted dopants were activated using rapid thermal annealing at 1,000 °C for 25 s. The damaged sacrificial oxide was removed next by immersing in diluted HF solution and a 5-nm thick gate oxide was grown in a thermal furnace at 900 °C. The source/drain contact pads and the gate electrodes (20-nm-thick Ti and 200-nm-thick Ag layers) were deposited using an electron-beam evaporator and patterned using conventional lithography and a lift-off process. Finally, the surface of the entire device, except for the active sensing region and the contact pads, was covered with a 2-mm-thick SU-8 layer to prevent leakage current between the contacts and the solution. The width, length, and height of the well holding the solution over the devices were 1,200, 1,200, and 2 μm, respectively. For the formation of integrated Ag/AgCl pseudo reference electrode, 30 μL of 0.1 M KCl was dropped on top of the embedded Ag electrode and a voltage of 1 V was applied for 300 s. Due to the reaction between silver and chloride ions, the Ag/AgCl pseudo reference electrode (RE) was successfully fabricated for use as a gate electrode of the SiNW ISFET. The width and length of the Ag/AgCl reference electrode were 1,000 and 1,000 μm, respectively.

Figure 9.9 shows the drain current (I_D) versus gate voltage (V_G) characteristics of the above SiNW ISFET with the drain voltage (V_D) varied between 0.1 and 0.7 V. The results indicate successful modulation of the channel conductance by the embedded Ag/AgCl gate electrode. During the measurement, the gate voltage was swept from 0 to 1.5 V at a rate of 50 mV/s and the source and substrate voltages were set to ground. The SiNW ISFET shows typical n-type FET behavior with a high on–off ratio of 10^5 and a good subthreshold-slope of 100-mV/dec. Inspection of transfer curves of the SiNW ISFET under different pH conditions (not shown here) shows a lateral shift of the curves without degradation of the

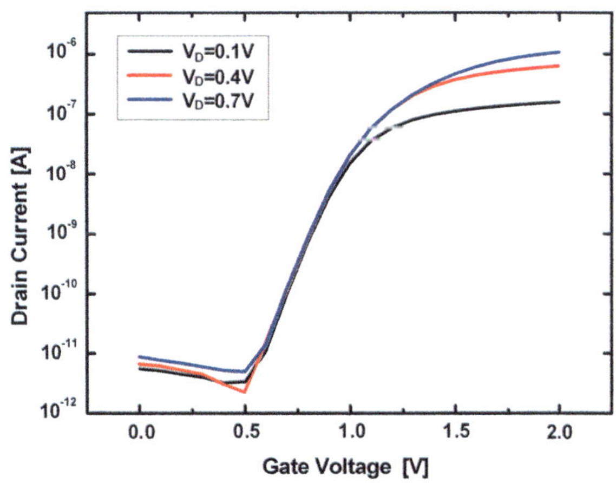

Fig. 9.9 Current-voltage characteristics of SiNW ISFET shown in Fig. 9.8 (figure courtesy of Taiuk Rim and Sungho Kim)

on-current and the sub-threshold slope. The V_{TH} of the device is found to increase linearly as the pH value increases. The total threshold voltage shift is 400 mV, and the pH sensitivity of the SiNW ISFET device is 40 mV/pH, which is a typical value for ISFETs using SiO_2 gate insulator as a sensing layer.

Stern et al. [24] also fabricated SiNW-FETs using a top-down approach and demonstrated a sensitivity of less than 10 fM specific label-free antibody detection. Tian et al. [66] fabricated a multinanowire FET and showed pH sensing with a linear response over a range of 2–9. Their NW-FET was also able to do selective detection of bovine serum albumin at concentrations of 0.1 fM.

9.6 Summary, Challenges, and Future Outlook

Biofield effect transistors (BioFETs) have been emerging as a viable candidate for biosensor systems for a wide variety of applications in health care, security, environmental monitoring etc. Silicon has been the typical material of choice for the conducting channel although other materials, such as metal oxides, have also been considered. Nanostructures for enhanced sensitivity have been explored with carbon nanotubes and graphene as leading candidates. In this chapter, a detailed account on the use of silicon nanowires in BioFETs has been presented. Both bottom-up and top-down approaches for the preparation of SiNWs have been discussed. The device fabrication details differ significantly between the two methods. The constraint today on the bottom-up SiNW synthesis techniques includes the lack of amenability for wafer scale fabrication. It is not possible to precisely grow nanowires (of chosen diameter and length) between each pair of source and drain over a 200 or 300 mm wafer. "Pick and place" approaches from bulk samples are unlikely to lead to large scale production either. As of now, there have been no credible routes to large scale device fabrication using bottom-up techniques.

In contrast, the top-down approach is readily adaptable for wafer-scale fabrication. Control of length and diameter of the nanowires is also relatively easier. Therefore, this approach seems to be a more promising route, at least for now. Regardless of the approach or material of choice, the development to date has been limited to the working of the sensor, that is, probe attachment, hybridization, and signal transduction. Fully functional systems require integration of microfluidics (channels, valves etc.) on the same chip. Microfluidics is a well evolved field, especially on silicon platforms. Therefore, integration of the sensor, signal processing, and microfluidics for sample handling along with system packaging should not pose serious challenges.

Whether it is clinical diagnostics or any other applications mentioned earlier, the early use of the sensor systems may be in the laboratory setting where the labor hours on analytical tests can be drastically reduced. From then on, migration into field use is expected to follow; this can include at-home diagnostics of viruses and infectious diseases or routine health checkup functions, municipal water quality monitoring at designated sites, pollution monitoring in rivers and lakes etc. The platforms for such field-use are likely to be different from those in the laboratory or a central lab. In this regard, smart phones may play an important role in hosting the biosensor systems. Currently, smart phone–based gas/vapor and biosensors are being investigated in many academic and industrial laboratories across the world. It is entirely possible to integrate into a smart phone a chip consisting of an array of BioFETs, or equivalent devices, along with microfluidics capabilities with the goal of performing multiplexed operations. In a health-related application, disposal cartridges will have to be employed necessarily. Even the design of microfluidics in a smart phone environment may not be challenging. The true challenge is to eliminate the need for concentrating or purifying or any other type of time-consuming sample processing, and still delivering reliable and sensitive diagnostics.

Acknowledgments The work at Postech was supported by the World Class University program through the National Research Foundation of Korea funded by the Ministry of Education, Science and Technology (Project: R31-2008-000-10100-0).

References

1. Daniels, J. S., & Pourmand, N. (2007). Label-free impedance biosensors: opportunities and challenges. *Electroanal, 19*, 1239–1257.
2. Grieshaber, D., Mackenzie, R., Voros, J., & Reimhult, E. (2008). Electrochemical biosensors—sensor principles and architecture. *Sensors, 8*, 1450–1458.
3. Cosnier, S., & Mailley, P. (2008). Recent advances in DNA sensors. *Analyst, 133*, 984–991.
4. Gooding, J. J. (2002). Electrochemical DNA hybridization sensors. *Electroanal, 14*, 1145–1156.
5. Lee, C. S., Kim, S. K., & Kim, M. (2009). Ion-sensitive field-effect transistor for biological sensing. *Sensors, 9*, 7111–7131.
6. Ronkainen, N. J., Halsall, H. B., & Heineman, W. R. (2010). Electrochemical biosensors. *Chemical Society Reviews, 39*, 1747–1763.
7. Wei, D., Bailey, M. J. A., Andrew, P., & Ryhanen, T. (2009). Electrochemical biosensors at the nanoscale. *Lab on a Chip, 9*, 2123–2131.
8. Monosik, R., Stred'ansky, M., Sturdik, E. (2012). Application of electrochemical biosensors in clinical diagnostics. *Journal of Clinical Laboratory Analysis, 26*, 22–34.
9. Yogeswaran, U., & Chen, S. M. (2008). A review on the electrochemical sensors and biosensors composed of nanowires as sensing material. *Sensors, 8*, 290–313.
10. Li, J. (2004).Biosensors. In M. Meyyappan (Ed.), *Carbon nanotubes: science and applications*. Boca Raton: CRC Press.
11. Ingerbrandt, S., & Offerhausser, A. (2006). Label-free detection of DNA using field-effect transistors. *Physica Status Solidi (a), 203*, 3395–3411.
12. Schafer, S., Eick, S., Hofmann, B., Dufaux, T., Stockmann, R., Wrobel, G., et al. (2009). Time-dependent observation of individual cellular binding events to field-effect transistors. *Biosensors & Bioelectronics, 24*, 1201–1208.
13. Poghossian, A., Ingebrandt, S., Offenhausser, A., & Schoning, M. J. (2009). Field-effect devices for detecting cellular signals. *Seminars in Cell & Developmental Biology, 20*, 41–48.
14. Sakata, T., & Miyahara, Y. (2007). Direct transduction of allele-specific former extension into electrical signal using generic field effect transistor. *Biosensors & Bioelectronics, 22*, 1311–1316.

15. Sohn, Y. S., & Kim, Y. T. (2008). Field-effect transistor type C-reactive protein sensor using cysteine-tagged protein G. *Electronics Letters, 44*, 955–956.
16. Park, K. M., Lee, S. K., Sohn, Y. S., & Choi, S. Y. (2008). BioFET sensor for detection of albumin in urine. *Electronics Letters, 44*, 185–186.
17. Park, K. Y., Sohn, Y. S., Kim, C. K., Kim, H. S., Bae, Y. S., & Choi, S. Y. (2008). Development of a FET-type albumin sensor for diagnosing nephritis. *Biosensors & Bioelectronics, 23*, 1904–1907.
18. Eteshola, E., Keener, M. T., Elias, M., Shapiro, J., Brillson, L. J., Bhushan, B., et al. (2008). Engineering functional protein interfaces for immunologically modified field effect transistor (ImmunoFET) by molecular genetic means. *Journal of the Royal Society Interface, 5*, 123–127.
19. Gupta, S., Elas, M., Wen, X., Shapiro, J., Brillson, L., Lu, W., et al. (2008). Detection of chemically relevant levels of protein analyte under physiologic buffer using planar field effect transistors. *Biosensors & Bioelectronics, 24*, 505–511.
20. Hsiao, C. Y., Lin, C. H., Hung, C. H., Su, C. J., Lo, Y. R., Lee, C. C., et al. (2009). Novel poly-silicon nanowire field effect transistors for biosensing application. *Biosensors & Bioelectronics, 24*, 1223–1229.
21. Lin, C. H., Hsiao, C. Y., Hung, C. H., Lo, Y. R., Lee, C. C., Su, C. J., et al. (2008). Ultrasensitive detection of dopamine using a polysilicon nanowire field effect transistor. *Chemical Communications, 44*, 5749–5751.
22. Freeman, R., Elbaz, J., Gill, R., Zayats, M., & Willner, I. (2007). Analysis of dopamine and tyrosinase activity on ion-sensitive field-effect transistor (ISFET) devices. *Chemistry—A European Journal, 13*, 7288–7293.
23. Zayats, M., Huang, Y., Gill, R., Ma, C. A., & Willner, I. (2006). Label-free and reagentless aptamer-based sensors for small molecules. *Journal of the American Chemical Society, 128*, 13666–13667.
24. Stern, E., Klemic, J. F., Routenberg, D. A., Wyrembak, P. N., Turner-Evans, D. B., Hamilton, A. D., et al. (2007). Label-free immunodetection with CMOS-compatible semiconducting nanowires. *Nature, 445*, 519–522.
25. Bergveld, P. (1970). Development of an ion sensitive solid state device for neurophysiological measurements. *IEEE Transactions on Biomedical Engineering, 17*, 70–71.
26. Caras, S., & Janata, J. (1980). Field effect transistor sensitive to penicillin. *Analytical Chemistry, 52*, 1935–1937.
27. Schoning, M., & Poghossian, A. (2006). BioFEDs (field effect devices): state of the art and new directions. *Electroanal, 18*, 1893–1900.
28. Bergveld, P. (2003). Thirty years of ISFETOLOGY—what happened in the past 30 years and what may happen in the next 30 years. *Sensors and Actuators, 13*(88), 1–20.
29. Schoning, M., & Poghossian, A. (2002). Recent advances in biologically sensitive field effect transistors (BioFETs). *Analyst, 127*, 1137–1151.
30. Yuqing, M., Jianguo, G., & Jianrong, C. (2003). Ion sensitive field effect transducer-based biosensors. *Biotechnology Advances, 21*, 527–534.
31. Meyyappan, M., & Sunkara, M. K. (2010). *Inorganic nanowires: applications, properties and characterization*. Boca Raton: CRC Press.
32. Mao, A., Ng, H. T., Nguyen, P., McNeil, M., & Meyyappan, M. (2005). Silicon nanowire synthesis by a vapor-liquid-solid approach. *Journal of Nanoscience and Nanotechnology, 5*, 31–835.
33. Westwater, J., Gosain, D. P., Tomiya, S., Usui, S., & Ruda, H. (1997). Growth of silicon nanowires via gold/silane vapor-liquid-solid reaction. *Journal of Vacuum Science & Technology, 13*(15), 554–557.
34. Aella, P., Ingole, S., Potuskey, W. T., & Picraux, T. (2007). Influence of plasma stimulation on Si nanowire nucleation and orientation control. *Advanced Materials, 19*, 2603–2607.
35. Johansson, J., Wacaser, B. A., Kick, K. A., & Seifert, W. (2006). Growth related aspects of epitaxial nanowires. *Nanotechnology, 17*, 5355–5361.

36. Noborisaka, J., Motohisa, J., & Fukui, T. (2005). Catalyst-free growth of GaAs nanowires by selective-area metalorganic vapor-phase epitaxy. *Applied Physics Letters, 86*, 21302.
37. Nguyen, P., Ng, H. T., Kong, J., Cassell, A. M., Quinn, R., Li, J., et al. (2003). Epitaxial directional growth of indium-doped tin oxide nanowire arrays. *Nano Letters, 3*, 925–928.
38. Nguyen, P., Vaddiraju, S., & Meyyappan, M. (2006). Indium and tin oxide nanowires by vapor-liquid-solid growth technique. *Journal of Electronic Materials, 35*, 200–206.
39. Sun, X. H., Yu, B., Ng, G., Nguyen, T.D., Meyyappan, M. (2006). III–V compound semiconductor indium selenide (In_2Se_3) nanowires: synthesis and characterization. *Applied Physics Letters, 89*, 233121.
40. Davami, K., Kang, D., Lee, J. S., & Meyyappan, M. (2011). Synthesis of ZnTe nanostructures by vapor-liquid-solid technique. *Chemical Physics Letters, 504*, 62–66.
41. Davami, K. H., Ghassemi, M., Sun, X. H., Yassar, R. S., Lee, J. S., Meyyappan, M. (2011). *In situ* observation of morphological change in CdTe nano- and submicron wires. *Nanotechnology, 22*, 35204.
42. Sun, X., Yu, B., Ng, G., & Meyyappan, M. (2007). One dimensional phase charge nanostructures: Germanium telluride nanowire. *Journal of Physical Chemistry C, 11*, 2421–2425.
43. Sun, X., Yu, B., & Meyyappan, M. (2007). Synthesis and nanoscale thermal encoding of phase charge nanowires. *Applied Physics Letters, 90*, 183116.
44. Nguyen, P., Ng, H. T., & Meyyappan, M. (2005). Catalyst metal selection for the synthesis of inorganic nanowires. *Advanced Materials, 17*, 1773–1777.
45. Sun, X., Calebotta, G., Yu, B., Seluaduray, G., & Meyyappan, M. (2007). Synthesis of germanium nanowires on insulator catalyzed by indium or antimony. *Journal of Vacuum Science & Technology, 13*(25), 415–420.
46. Sunkara, M. K., Sharma, S., Miranda, R., Lian, G., & Dickey, E. C. (2001). Bulk synthesis of silicon nanowires using a low temperature vapor-liquid-solid method. *Applied Physics Letters, 79*, 1546–1548.
47. Ma, D. D. D., Lee, C. S., Au, F. C. K., Tong, S. Y., & Lee, S. T. (2003). Small-diameter silicon nanowire surfaces. *Science, 299*, 1874–1877.
48. Schmidt, V., Riel, H., Snez, S., Karg, S., Riess, W., & Gosele, U. (2006). Realization of a silicon nanowire vertical surround-gate field-effect transistor. *Small, 2*, 85–88.
49. Ng, H. T., Han, J., Yamada, T., Nguyen, P., Chen, P., & Meyyappan, M. (2004). Single crystal nanowire vertical surround gate field effect transistor. *Nano Letters, 4*, 1247–1252.
50. Kim, S., Rim, T., Lee, U., Baek, E. H., Lee, H., Baek, C. H., et al. (2011). Silicon nanowire ion sensitive field effect transistor with integrated Ag/AgCl electrode: pH sensing and noise characteristics. *Analyst, 136*, 5012–5016.
51. Timko, B. P., Karni-Cohen, T., Quan, Q., Bozhi, T., & Lieber, C. M. (2010). Design and implementation of functional nanoelectric interfaces with biomolecules, cells and tissues using nanowire device arrays. *IEEE Transactions on Nanotechnology, 9*, 269–280.
52. Cui, Y., Wei, Q., Park, H., & Lieber, C. M. (2001). Nanowire nanosensors for highly sensitive and selective detection of biological and chemical species. *Science, 293*, 1289–1292.
53. Patolsky, F., Zheng, G., & Lieber, C. M. (2006). Nanowire-based biosensors. *Analytical Chemistry, 78*, 4260–4269.
54. Gao, X. P. A., Zheng, G., & Lieber, C. M. (2010). Subthreshold regime has the optimal sensitivity for nanowire FET sensors. *Nano Letters, 10*, 547.
55. Zheng, G., Patolsky, F., Cui, Y., Wang, W. U., & Lieber, C. M. (2005). Multiplexed electrical detection of cancer markers with nanowire sensor arrays. *Nature Biotechnology, 23*, 1294–1301.
56. Curreli, M., Li, C., Sun, Y., Lei, B., Gunderson, M., Thompson, M. E., et al. (2005). Selective functionalization of In_2O_3 nanowire mat devices for biosensing application. *Journal of the American Chemical Society, 127*, 6922–6923.
57. Curreli, M., Zhang, R., Ishikawa, F. N., Chang, K. K., Cote, R. J., Zhou, C., et al. (2008). Real-time, label-free detection of biological entities using nanowire-based FETs. *IEEE Transactions on Nanotechnology, 7*, 651–667.

58. Li, C., Curreli, M., Lin, H., Lei, B., Ishikawa, F. N., Datar, R., et al. (2005). Complimentary detection of prostate specific antigen using In_2O_3 nanowires and carbon nanotubes. *Journal of the American Chemical Society, 127*, 12484–12485.
59. Ishikawa, F. N., Chang, H. K., Curreli, M., Liao, H., Olson, A. C., Chen, P. C., et al. (2009). Label-free electrical detection of the SARS virus N-protein with nanowire biosensors utilizing antibody mimics as capture probes. *ACS Nano, 3*, 1219–1224.
60. Choi, A., Kim, K., Jung, H. I., & Lee, S. Y. (2010). ZnO nanowire biosensors for detection of biomolecular interation in enhancement mode. *Sensors and Actuators, 13*(148), 577–582.
61. Yeh, P. H., Li, Z. L., & Wang, Z. L. (2009). Schottky-gated probe-free ZnO nanowire biosensor. *Advanced Materials, 21*, 4975–4978.
62. Liu, X., Lin, P., Yan, X., Kars, Z., Zhao, Y., Lei, Y., Li, C., Du, H., Zhang, Y. (2012). Enzyme-coated single ZnO nanowire FET biosensor for detection of uric acid. *Sensors and Actuators 13*.
63. Li, J., Zhang, Y., To, S., You, L., & Sun, Y. (2011). Effect of nanowire number, diameter and doping density on nano-FET biosensor sensitivity. *ACS Nano, 5*, 6661–6668.
64. Liu, Y. C., Rieben, N., Iversen, L., Sorensen, B. C., Park, J. W., Nygard, J., et al. (2010). Specific and reversible immobilization of histidine-tagged proteins on functionalized silicon nanowires. *Nanotechnology, 21*, 245105.
65. Vu, X. I., Stockmenn, R., Wolfrum, B., Offenhausser, A., & Ingebrandt, S. (2010). Fabrication and application of microfluidic embedded silicon nanowire biochip. *Physica Status Solidi A, 207*, 850–857.
66. Tian, R., Regonda, S., Gao, J., Liu, Y., & Hu, W. (2011). Ultrasensitive protein detection using lithographically defined Si multi-nanowire field effect-transistor. *Lab on a Chip, 11*, 1952–1962.
67. Duan, X., Li, Y., Rajan, N. K., Routenberg, D. A., Modis, Y., & Reed, M. A. (2012). Quantification of the affinities and kinetics of protein interactions using silicon nanowire biosensors. *Nature Nanotechnology, 7*, 401–407.

Chapter 10
Lab on a Wire: Application of Silicon Nanowires for Nanoscience and Biotechnology

Larysa Baraban, Felix Zörgiebel, Claudia Pahlke, Eunhye Baek, Lotta Römhildt and Gianaurelio Cuniberti

Abstract Synergy between biochemistry, medicine, and material science during last decade has led to a tremendous scientific progress in the fields of biodetection and nanomedicine. This tight interaction led to the emergence of a new class of bioinspired systems. These systems are based upon utilizing nanomaterials such as nanoparticles, carbon nanotubes, or nanowires as transducers for producing novel sensor devices, or sophisticated drug delivery agents. This chapter focuses on the developments made in the area of silicon nanowire-based devices and their applications in the diverse areas of nano- and biotechnologies. Firstly, the incorporation of silicon nanowires into the electrical circuits is discussed, together with the sensing mechanism of the devices. In particular, the discussion is directed toward the most important aspects of the fabrication and functioning of the sensors, as well as the issues regarding the organic molecules interfacing with the silicon surface. Moreover, the complex interactions of organic species with nanoscale matter are addressed to as well as the need for sophisticated integration and packaging of the subsystems on a single chip. Finally, the perspectives of the potential applications of the silicon nanowires for biodetection and drug delivery are presented. Thus, the concept of "*lab on a wire*" is introduced as a set of approaches to *engineer* the nanowires and to *enrich* their functionality and potential applications in nanoscience and biotechnology.

L. Baraban (✉) · F. Zörgiebel · C. Pahlke · E. Baek · L. Römhildt · G. Cuniberti
Institute for Materials Science and Max Bergmann Center of Biomaterials, Technische Universität Dresden, 01062 Dresden, Germany
e-mail: larysa.baraban@nano.tu-dresden.de

G. Cuniberti
e-mail: g.cuniberti@nano.tu-dresden.de

F. Zörgiebel · G. Cuniberti
Center for Advancing Electronics Dresden, Technische Universität Dresden, 01062 Dresden, Germany

D. M. Kim and Y.-H. Jeong (eds.), *Nanowire Field Effect Transistors: Principles and Applications*, DOI: 10.1007/978-1-4614-8124-9_10,

Abbreviation

SiNW	Silicon nanowires
FET	Field-effect transistors
VLS	Vapor–liquid–solid
SB	Schottky barrier
VL	Vacuum level
HOMO	Highest occupied molecular orbital
LUMO	Lowest unoccupied molecular orbital
XPS	X-ray photoelectron spectroscopy
ATR-IR	Attenuated total reflectance-Infrared
SAM	Self-assembled monolayer
SEB	*Staphylococcus aureus* enterotoxin B
ssDNA	Single-stranded DNA
DNA	Deoxyribonucleic acids
PNA	Peptide nucleic acids
SELEX	Systematic evolution of ligands by exponential enrichment
VEGF	Vascular endothelial growth factor

10.1 Nanowire-Based Sensor Devices

After the decades of intense investigations in the 1980s and 1990s of the last century, nanoscale devices have finally entered the phase of diverse commercial applications, fulfilling needs of the society in even more multifunctional, faster, and smaller electronics. In recent years, the one-dimensional nanostructures, in particular semiconductor nanowires have attracted attention as highly efficient sensor elements due to their high surface-to-volume ratio and one-dimensional structure [1–3] which enables the detection of biochemical species down to single molecules [4–6]. The physical reason for the high sensitivity is to be found in the small diameter of the nanowire, allowing even single molecules to effect the conductivity in the channel by penetrating it with an electric field. This novel nanowire-based technique is capable of not only a real-time and label-free sensing of the very small quantities of the biomolecules, but also provides information on the conformation of the molecules and the strength of the biomolecular interactions. In particular, the binding efficiency of the receptor and analytes, as well as hybridization in the solution, can be tested [7]. These developments in the area of bionanosensorics can potentially become a powerful competitor to the conventional biochemical or optical detection techniques, which currently dominate the market.

Although a single nanowire can detect a single molecule, a binding event is not likely to happen on the tiny wire surface. Furthermore, integration of the nanowire elements into conventional electronic circuits, expected on the longtime perspective, requires the substantially increased current output of the sensor device. These shortcomings can be overcome with the use of parallel arrays of nanowire sensors. In so doing, the large surface areas can be covered without sacrificing efficiency of the switching behavior. Additionally, by incorporating several hundred nanowires into the parallel array, the yield of functioning transistors can be dramatically increased and device-to-device variability can be reduced compared to that of individual devices.

10.1.1 Design of Sensor Devices: Top-Down Versus Bottom-Up

The essential element of the sensor system is the silicon nanowires (SiNW) assembled as field-effect transistors (FET) and a multitude of techniques has been developed for manufacturing SiNW-based FETs (see Chap. 5). The methods of fabricating field-effect transistors are generally divided into the *top-down* and *bottom-up* processes. Figure 10.1 demonstrates the examples of the silicon nanowire devices, processed by both top-down and bottom-up techniques for sensing applications. The top-down manufacturing techniques have already been developed extensively for industrial FET production of microelectronic devices. In this approach, devices are fabricated in a lithographically assisted manner by several consecutive etching and material deposition steps. The approach enables the high-density integration of functional elements on a very small scale, a key point in the development of modern computer technology. As an example, Fig. 10.1a shows an array of ten perfectly aligned nanowires with diameters of about 50 nm, fabricated by means of the top-down nanofabrication processes and integrated with interconnects via monolithic patterning [8].

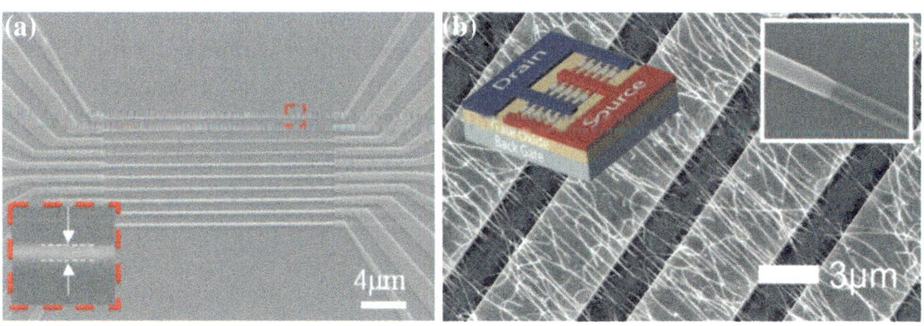

Fig. 10.1 **a** *Top-down* manufacturing: Electron-beam lithography and reactive ion etching produced silicon nanowire FETs with 50 nm diameter [8]; **b** *Bottom-up* manufacturing: Parallel array of Schottky barrier silicon nanowire FETs. Inset: Schottky barrier junction, created by thermal diffusion of the metal into semiconductor nanowire [9]

Bottom-up processes, on the other hand, rely mainly on the self-assembly and self-organization of functional elements, such as atoms, molecules, and clusters. A prominent example is the growth of carbon nanotubes from fullerenes or chemical vapor deposition assisted growth of silicon nanowires starting with gold catalyst. The nanowires thus grown can be efficiently integrated into sensor devices by means of much simpler and cost-effective processes. For instance, a contact printing approach for vapor-liquid-solid (VLS)-synthesized SiNWs can be used for transferring etched nanowires to chip substrates [9]. As shown in Fig. 10.1b, the large and high-density arrays of the silicon nanowires can be brought to the substrate and contacted by using ultraviolet lithography to form FET devices. Note that the combination of the bottom-up growth technique of nanowires and the ability to print nanowires at virtually any surface opens a great possibility for bottom-up FETs to be incorporated into a cost-efficient printable and flexible electronics.

10.1.2 Design of Sensor Devices: Single Wire Versus Parallel Array

High current densities and high transconductance, obtained with the silicon nanowire-based FETs, are the crucial parameters, which have yet to be attained to compete with state-of-the-art CMOS technology. For instance, raising the transconductance in nanowires FET to 1 mA/μm would be sufficient to apply the device as a driver for organic light emitting diodes or as transducer with stable outputs for sensors. Thus, the application of the parallel array of nanowire FETs for sensing has a great advantage, since it enables substantial increase in the drain current I_{ds}, while preserving the high subthreshold slope in the transfer characteristics.

In order to demonstrate this tremendous tendency, Fig. 10.2 reflects a comparison of transfer characteristics of a single-wire device (top-down fabricated) and a parallel array of Schottky barrier silicon nanowires (bottom-up). The top-down manufactured device shown is a *p*-doped silicon-on-insulator single nanowire device with 20 nm channel height, 400 nm channel width, and 5 μm channel length.[1] In contrast, the bottom-up manufactured device is a parallel array of 100–1,000 Schottky barrier silicon nanowire FETs with single wire diameter of 20 nm and silicon channel length of approximately 6 μm. The inverse subthreshold slope of both devices is approximately −120 mV per decade of current change for both devices, corresponding to twice of the physically achievable minimum of −59.9 mV per decade for hole-conducting FETs.

Moreover, nanowire-based FET devices exhibit excellent switching behavior that is critical for sensing applications (see Chap. 9) at relatively high current even for parallel arrays of hundreds of nanowires.

[1] Chips are fabricated and provided by IM Health, South Korea.

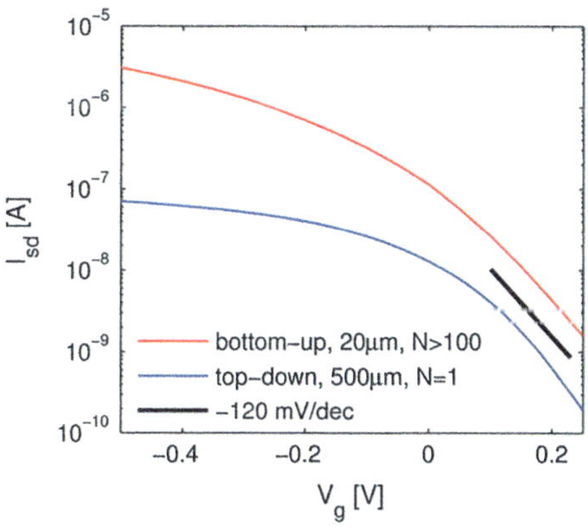

Fig. 10.2 Comparison of transfer characteristics of multiwire- (*bottom-up*) and single-wire (*top-down*) manufactured silicon nanowire FETs. *Bottom-up* device is based on Schottky barriers and doped silicon-on-insulator channels, respectively. For comparison, a slope of −120 mV/dec is depicted. The physical limit at room temperature is 59.5 mV/dec

10.1.3 Sensing Principle

The basic parameter in FET-based sensors is the gate voltage-induced surface potential, which modulates the channel conduction. The field-effect devices are very efficient transducers of the changes in the surface potential, making it superior to direct potentiometric measurements. The maximum threshold shift is determined by the Nernst equation. Ideally, an FET operating at room temperature should generate the current change of one decade with the gate voltage change of 59.5 mV. However, the imperfect interface at the metal contacts and roughness of the nanowire surface leads to an increase in this number [10, 11]. Any binding event that changes the surface charge or the surface potential due to a chemical reaction or electrostatic interaction will, therefore, be detectable with an FET-based sensor device.

In contrast to the electronic applications, where the gate electrode consists of a metal or semiconductor and is separated from the channel by an insulating dielectric layer, sensing in liquid is best performed with the liquid gate (as described previously in Chap. 9). In this case, the gate voltage is applied to the measurement solution and dielectric layer can be made of the native oxide of the silicon nanowire or an additional dielectric layer, e.g., hafnium oxide, aluminum oxide, or silicon nitride. In such measurement configuration, an additional back gate can be used to enhance the nominal sensitivity of the devices, [11, 12] without increasing the signal-to-noise ratio.

10.1.4 Example: Schottky-Barrier-Based Bottom-Up Device

Let us consider the operation of Schottky-barrier (SB)-based sensor device, produced by the bottom-up nanofabrication process. The current modulation in Schottky barrier SiNW field-effect transistors arises via the tunneling of the

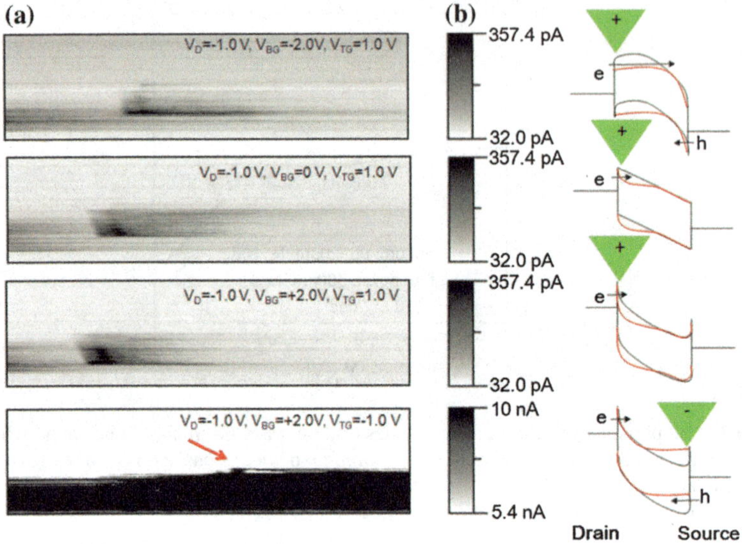

Fig. 10.3 **a** Schematics of the band diagram of Schottky junction. From top to down: Electron-tunneling modulation in hole conduction state, OFF state, and electron conduction state, respectively. **b** Scanning gate measurements at a single Schottky barrier silicon nanowire FET, using atomic force microscopy (AFM). Current map of the nanowire region, containing Schottky junction, with voltage applied to the AFM tip. Effects are seen at the Schottky barrier. Images are scanned from *Top left* to *bottom right* [15]

voltage-controlled charge through the metal–semiconductor junction. Figure 10.3 shows the corresponding band diagrams, displaying the charge transfer through Schottky contacts. This configuration was first demonstrated by Weber et al. in 2008 [13] to control the polarity of the switching behavior of these FETs and further extended by Heinzig et al. in 2012 to build a converter circuit with a single silicon nanowire [14].

It has been recently shown that Schottky junctions play crucial role in the sensitivity of the silicon nanowire field-effect transistors. An evidence that the maximum field sensitivity is localized at the Schottky barriers is demonstrated by Martin et al. in [15] using scanning gate microscopy. An electrical scanning probe technique is applied to examine the charge transport effects of a nanometer-scale local top gate during operation. The results prove experimentally that Schottky barriers control the charge carrier transport in these devices.

Schottky-barrier-based FETs are fabricated using CVD-grown intrinsic silicon nanowires, which are contact printed on a receiver substrate and contacted by interdigitated Ni electrodes in the photolithography step. Schottky barriers are then formed in a bottom-up process by diffusing nickel into the SiNWs at 500 °C in forming gas. The lattice constant of the resulting $NiSi_2$ phase deviates from the lattice constant by only 0.4 % [16, 17], leading to atomically sharp Schottky barriers between the silicon channel and $NiSi_2$ leads. An electron micrograph of silicon nanowires contacted by nickel electrodes with intruded $NiSi_2$ phases is shown in Fig. 10.1b. The semiconductor–metal interface is also shown in the inset above.

In order to use such devices for liquid-sensing applications, an isolating layer, e.g., Al_2O_3, was deposited on the entire chip using atomic layer deposition technique. Microfluidic channels should be further placed on the chip and a Ag/AgCl electrode should be attached to the channel in order to apply a gating voltage to the liquid.

The quality of the metal-semiconductor interface is therefore crucial for the function and satisfactory performance of the FET devices. A sharp interface with very low mixing of metal and semiconductor phases must be realized to attain the well-defined interface barrier for tunneling and to obtain the field enhancement effect.

10.1.5 Example: pH Sensor

Sensing of the pH value is a benchmark for the quality of surface potential sensors processed in silicon nanowires. During the device operation, the surface potential is controlled according to the Nernst equation (see Chap. 9). The change in the surface potential adds to the gating potential of the liquid electrode modulating the drain current logarithmically in the subthreshold regime of the FET. The maximum possible change of the drain current per gate voltage was discussed in the preceding chapters. The sensitivity for measurements of the pH value is given by

$$S = \partial \log_{10} I_{ds} / \partial pH \tag{10.1}$$

and is limited to $|S| \leq 1$. The sensing of the pH values by an FET is generally based on a linear dependence of current and surface charge. However, the sensitivity under discussion depends exponentially on the gate voltage because of the steep current change from OFF to ON states in the subthreshold regime. The low values of the numerator in Eq. (10.1) are due to low subthreshold current level, which in turn leads to higher noise level in electronic measurements. Accordingly, the sensitivity is not only dependent on device quality, but also on the measurement scheme. This point was first investigated by Gao et al. in 2010 [10] (Fig. 10.4).

The authors showed that the pH sensing can be optimally done by operating the sensing FET in the subthreshold regime. Because of the efficient sensing capability, a multitude of silicon nanowire-based biological sensors has been reported in recent years, which suggests viable commercial applications in the near future [5, 6, 18, 19].

Quantitative pH sensing requires more than the measurement of the changes in drain current. Rather the surface potential, which is linearly dependent on the pH value for ideal surfaces, should also be monitored. This is done by monitoring

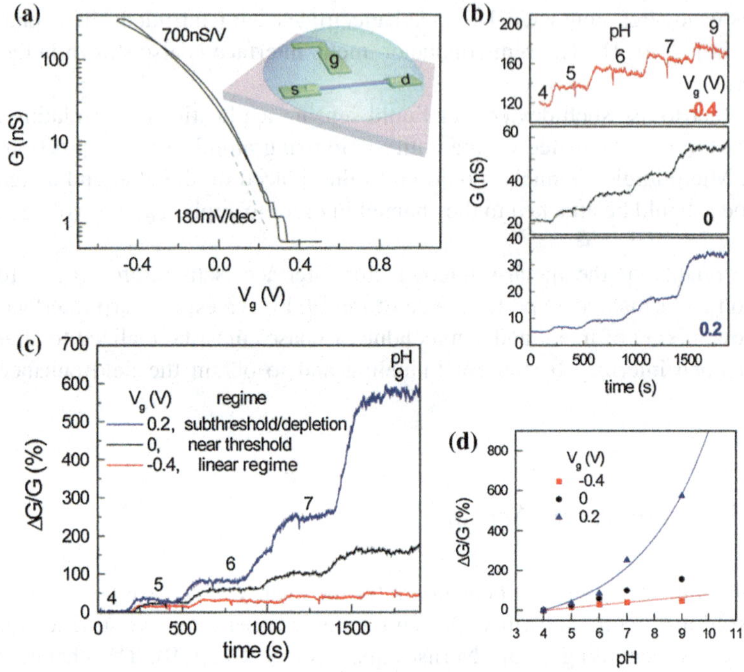

Fig. 10.4 pH sensor based on silicon nanowire FET. **a** Electric switching characteristics of a doped silicon nanowire sensor device. **b** pH sensitivity for different gate voltages. **c** Normalized data from (**b**) for comparison. **d** Relative signal change versus pH value for all gate voltages [10]

continuously the changes in gate voltage to keep the drain current pinned at a given fixed level. For an ideal liquid electrode, the measured change in gate voltage reflects the negative change of surface potential. This is because any change in the surface potential must be compensated by the potential in the liquid-gate electrode in order to keep the gating potential in the wire constant.

10.2 Nanowires as Element of Hybrid Circuits

The development and commercialization of novel hybrid intelligent systems with rich functionalities, providing strong benefit in the area of health care, environmental monitoring, and electronic applications are among the most crucial topics of the scientific community and industrial players. Recent developments in synthetic chemistry, physics, bioengineering, and nanosciences allow us to envision the appearance of novel hybrid nanoelectronic devices on a market already in the near future. Nanowires can provide excellent building blocks for hybrid nanoelectronics, due to their efficient charge transport characteristics and good compatibility with molecules. Hybrid nanowire-based devices rely on inorganic circuitry, combined with organic molecules. The molecular conjugation with nanowires has been demonstrated and recent studies focus on employing outstanding features of molecules onto the nanoelectronic devices. These hybrid devices are applied for memory or logic circuits driven by light. Light-induced switching of molecules acts as electrical gating of nanowire transistors to drive conductance changes in nanowires to induce conductance changes from OFF to ON states or vice versa.

10.2.1 Organic Molecules in Conjunction with Silicon Nanowires

Recently, there have been huge interests in studies of electrical conjunction between organic molecules and metal or semiconductor surface. For the case of silicon-based devices, a variety of well-established fabrication processes are available for use and the energy-band structures are also well known. Therefore, investigations of molecular functionalization of silicon and, in particular, SiNW surface have been extensively carried out for practical optoelectronic applications. Thus, understanding of interfacial interactions between molecular layer and silicon surface is rather important for analyzing and fabricating molecular hybrid devices.

Photochromic molecules are transformed by light between two or more stable isomers possessing different absorption spectra and geometry. The isomers are converted from one form to another under light irradiation with proper wavelength and revert to original state thermally or by light with different wavelength [20–22]. There is wide variety of the photochromic dyes, i.e., azobenzenes, diarylethenes, spiropyranes, fulgides, which are actively investigated during last decades. Chemical structures of these molecules are presented in Fig. 10.5. Chemical conformations of these molecules are changed by absorption of UV light and are converted back to original states by absorption of visible light or by thermally activated process.

In addition to the above-mentioned photochromic molecules, natural organic complex *porphyrin* has emerged as the most popular molecule for hybrid application due to its well-known and interesting characteristics. Porphyrin is easily found from living organism such as chlorophylls and Hemes. Single- and double-bond array forming porphyrin ring absorbs broad wavelength range of visible light (Fig. 10.6). Once irradiated by light of specific wavelengths in optical range, π-electrons participating in the bond array are released and move easily through molecular bonding [23].

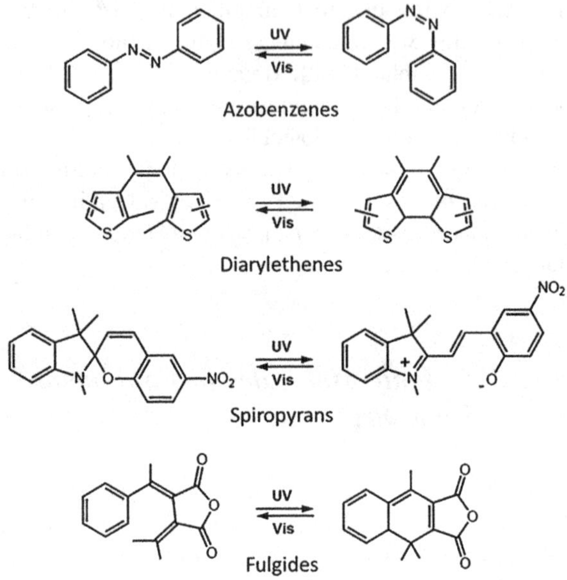

Fig. 10.5 Reversible activation and deactivation of photoswitchable molecular systems

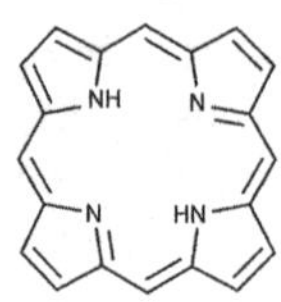

Fig. 10.6 Molecular structure of Porphyrin ring

10.2.2 Conjunction of Molecules with SiNW: Interfacial Electronic Structure

To understand electronic structure between molecules and solid surface, consider the molecular energy-band diagram [24]. Figure 10.7 a shows electronic structure of hydrogen atom and naturally the electron occupies the lowest 1 s orbital. The electron can escape from the atom to the vacuum level (VL). In a molecule, deep atomic orbitals are localized in atomic potential well, but higher lying orbitals interact with each other to form molecular orbitals, that is, the highest occupied molecular orbital (HOMO) and the lowest unoccupied molecular orbital (LUMO) (Fig. 10.7b). When molecules are located in close vicinity to each other, they interact via the weak van der Waals interactions, so that LUMO and HOMO are localized in each molecule (Fig. 10.7c). Therefore, the application of typical band theory is limited to molecular solid that has existing single-molecular electronic structure. Figure 10.7c can be simplified to Fig. 10.7d and e.

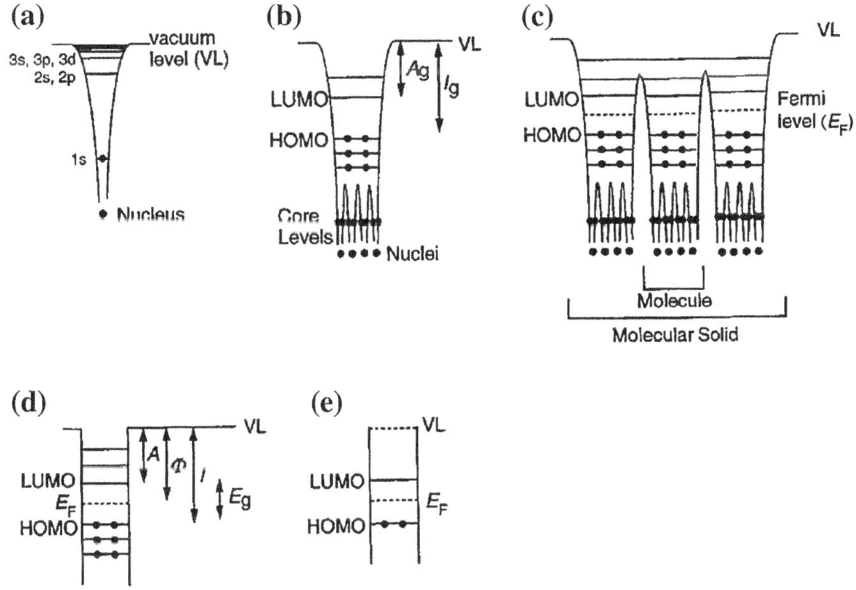

Fig. 10.7 Electronic structure with potential wells. **a** Hydrogen atom. **b** Molecule. **c** Molecular solid. **d**, **e** Simplification of (**c**). (*A*: electron affinity, *I*: ionization energy (Ag, Ig for gas phase), *Φ*: work function, E_g: HOMO–LUMO band gap) [24]

The band diagram of an interface between silicon and organic molecules is shown in Fig. 10.8a. The work function of silicon ranges from 4.6 to 4.8 eV and the energy gap E_g is about 11 eV [24]. The HOMO–LUMO gap of organic molecules that absorb visible and ultraviolet light is around 1.5–3 eV. The ionization energy of photochromic molecules from HOMO to the vacuum level is in the range of 4–7 eV. Silicon surface states are important for charge distribution, because they act as acceptor or donor states and surface states are coupled to the molecular layer via the charge exchange to align the Fermi level at equilibrium. (Fig. 10.8b).

10.2.3 Conjunction of Molecules with SiNW: Chemical Functionalization Aspects

Azobenzene is one of the most frequently used light-switching molecules, which reveals simple *cis-trans* isomerization, exhibiting a big difference in structure and electric dipole moment between isomers induced by the irradiation of ultraviolet light (Fig. 10.9). Recently, the behavior and electrical properties of azobenzene monolayer on silicon surface have been investigated by several groups [25, 26].

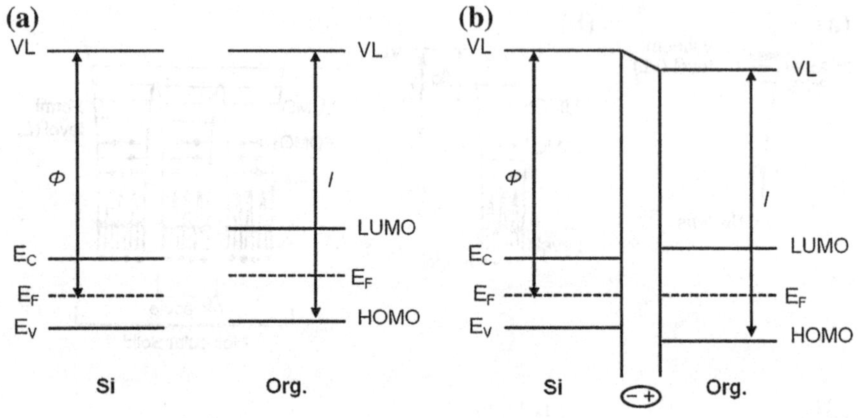

Fig. 10.8 Interfacial energy-band diagram: **a** isolated silicon and organic molecules; **b** Junction, aligned Fermi level and interfacial dipole

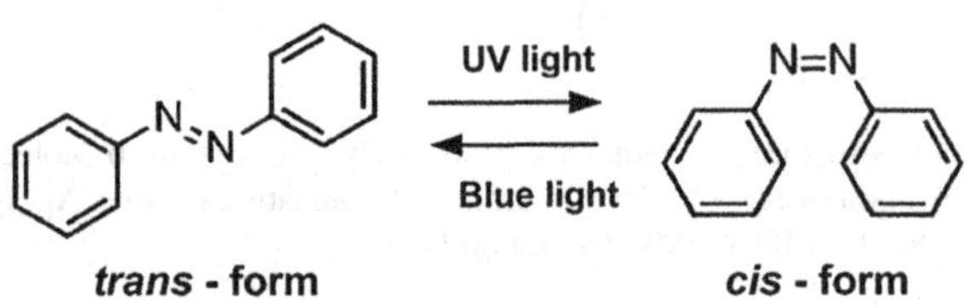

Fig. 10.9 Simple isomerization of an azobenzene molecule

Covalent bonds can be formed between specially functionalized azobenzene molecule and hydrogen-terminated silicon (Si–H) (Fig. 10.10). Authors demonstrate that the chemical bonds can be characterized by various surface analyzing methods such as contact angle, X-ray photoelectron spectroscopy (XPS), or attenuated total reflectance-Infrared (ATR-IR) spectroscopy measurement [25, 26]. Although a variety of investigations of molecular monolayer on silicon surface has been carried out, applications of Si nanowires have not been reported except for the photosensitive TiO_2 nanowires in conjunction with azobenzene [27].

For the case of porphyrin, molecules can be simply attached on silicon surface by both *chemisorption* and *physisorption*. Self-assembled monolayer (SAM) of porphyrin on silicon oxide (Fig. 10.11) has been studied by several groups [28–30] and is well investigated. In the case of established chemical binding between molecule and surface, electrons can migrate from porphyrin to silicon through covalent bond.

On the other hand, physisorption dominates as a result of the drop-casting, which is used in many applications for attaching porphyrin on silicon surface [31, 32]. When the porphyrin solution is dropped on silicon nanowires and the solvent is evaporated, porphyrin aggregates on nanowires and forms multilayer stack amorphously. This can result in distinct interaction of the molecule with the

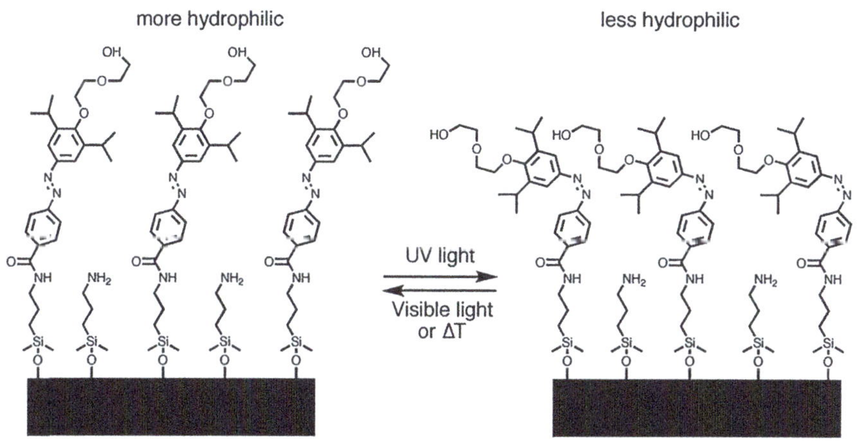

Fig. 10.10 Surface functionalization of the silicon surface by azobenzene [26]

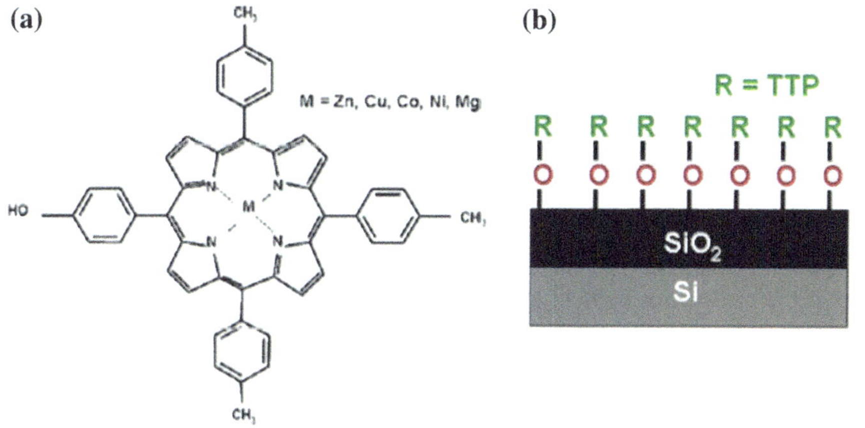

Fig. 10.11 **a** 5-(4-hydroxyphenyl)-10,15,20-tri(p-tolyl) porphyrin (TTP) with different central metal ions. **b** TTP-OH SAM on SiO_2 [30]

surface, i.e., silicon nanowire, and leads to specific influence of the light-sensitive porphyrin layer on electric characteristic of SiNW-based FET devices.

10.2.4 Optoelectronic Switching with the Use of Nanowires

Optical switching of hybrid nanowire system depends on both switching property of molecule and optical property of nanowires. Photochromic molecules switch between two or more energy states, which leads to structural and electronic change

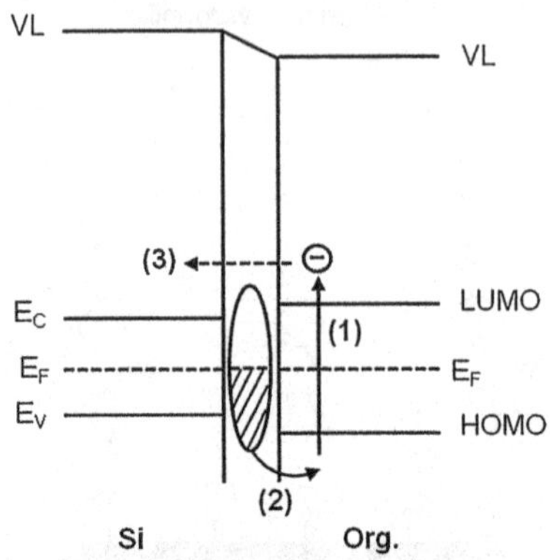

Fig. 10.12 Schematic band diagram of silicon and organic molecule interface during light illumination: (*1*) photoexcitation; (*2*) recombination with surface states; (*3*) charge carrier movement

such as geometry, dipole moment, HOMO–LUMO gap and redox potential on molecular level [21]. The individual molecular change affects macro-molecular conjugation such as wettability, charge, or conductivity. Process of charge generation by means of light irradiation is demonstrated in Fig. 10.12. During illumination, electron hole pairs are generated within organic molecules. Depending on surface states, negative or positive charge is induced at the interface by (1) the recombination processes of holes or electrons (2) [33]. Photo-generated electrons are directly injected into silicon (3) or are able to change the potential on nanowire surface.

Molecular-level switching induces charge in nanowires contacting with photochromic molecule. The conductivity of nanowires made of material with direct band gap can change by light irradiation, when the energy of photon is equal to or greater than the band gap of the material (e.g., TiO_2), as in the case of optoelectronic devices such as laser diode or LED. However, silicon does not absorb light efficiently due to its indirect band gap, so that switching of hybrid Si nanowire devices mainly reflects molecular characteristics.

Molecular dipole moment or potential change around nanowires acts as gating of nanowire FET and controls free electron or hole population in nanowires. On the other hands, molecules that release electrons with light absorption (e.g., porphyrin) exchange electrons with nanowires directly, as donor or acceptor. The difference of Fermi levels between molecules and nanowires dictates whether the molecules behave as donors or acceptors. Moreover, since the gate bias also shifts the Fermi level, gate voltage plays a key role for switching direction.

Naturally, the amplitude of switching current depends on molecular absorption spectrum, as shown in Fig. 10.13. This is because given a molecule, its HOMO–LUMO band gap determines the resonant wavelength. Thus, the absorption spectra are key factors for generating free electrons from the molecule.

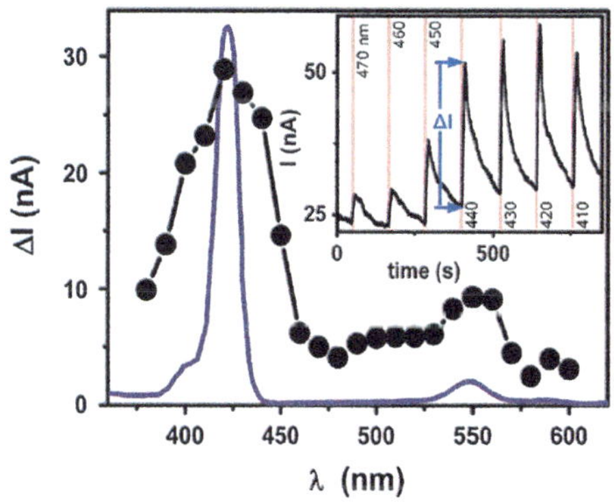

Fig. 10.13 Variation of the photoinduced drain-source current ΔI_{ds} of Porphyrin-coated nanowire FETs upon illumination as a function of light wavelength [34]

10.2.5 Example: Applications of Nanowires-Based Hybrid Devices

10.2.5.1 Memory Applications

First hybrid devices have been fabricated and applied for memory cells, employing porphyrin complexes. The memory application has been demonstrated with the use of In_2O_3 nanowires grown with self-assembled monolayers of porphyrins. Co-chelated porphyrin showed memory effect by the redox states of metal ion in the porphyrin (Fig. 10.14). They provided the potential of porphyrin that can be used for memory devices by charge storing in molecule.

Another interesting application is hybrid nanogap FETs. In these devices, porphyrin is embedded into the nanofabricated gap in the oxide layer of conventional MOS structure (Fig. 10.15). Hybrid nanogap FETs can be utilized as nonvolatile memory cell by optical charging and electrical discharging of porphyrin. Under light irradiation, Porphyrin absorbs electrons from the silicon

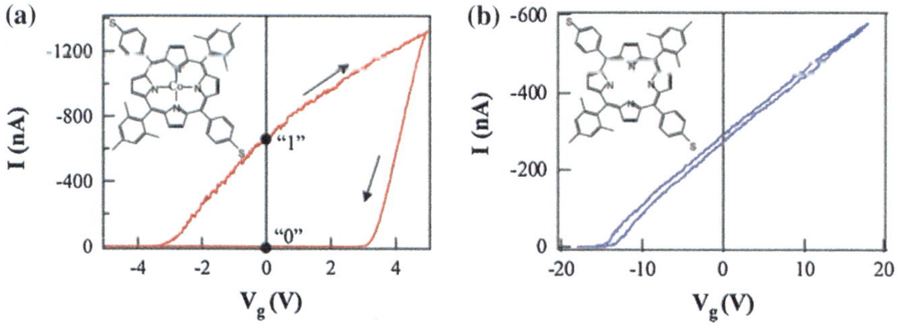

Fig. 10.14 I-V_g characteristics of In_2O_3 nanowire devices with self-assembled Co-porphyrin (**a**) and protio porphyrin (**b**), respectively [28]

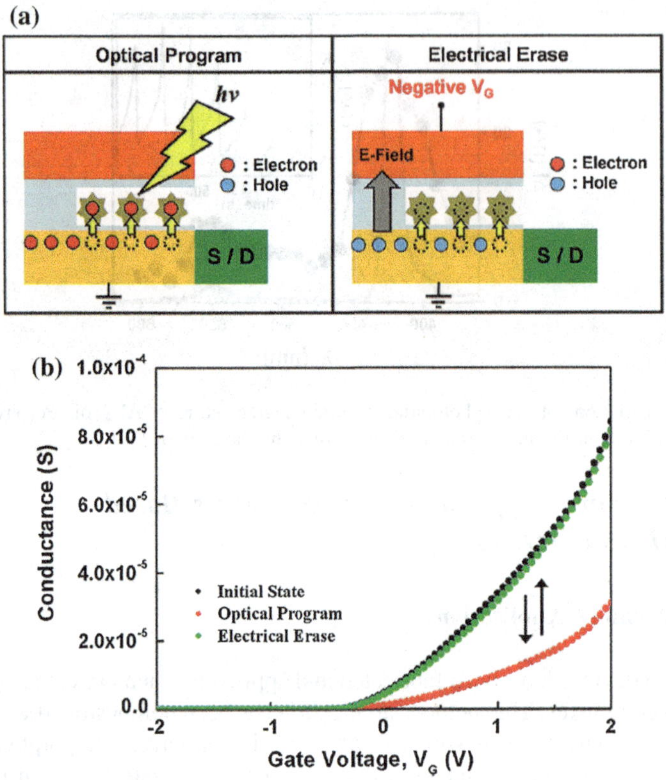

Fig. 10.15 **a** Schematic diagram of the mechanisms of the optical program and electrical erase characteristics. The porphyrin charging process is based on photoinduced charge transfer (PCT), and the discharge process is based on electron–hole pair recombination. **b** Optical program and electrical erase characteristics of the porphyrin-embedded FET [32]

substrate and stores the electrons in the ring for writing. The gate voltage V_g is used for the substrate to pull back the stored electron for erase.

10.2.5.2 Hybrid CMOS Applications

By integrating p- and n-type silicon nanowires, processed by top-down approach, hybrid CMOS has been fabricated, in which the properties of porphyrin is exploited for the incident light to control the device behavior (Fig. 10.16). The characteristics of porphyrin are undergoing changes depending on the types of nanowires in conjunction, and consequently, the conductance of silicon NW also changes in response to incident light. The switching direction is opposite between p- and n-type silicon nanowire FETs, so that the initial OFF state is switched to ON state, while the opposite switching occurs in n-type FETs upon irradiation.

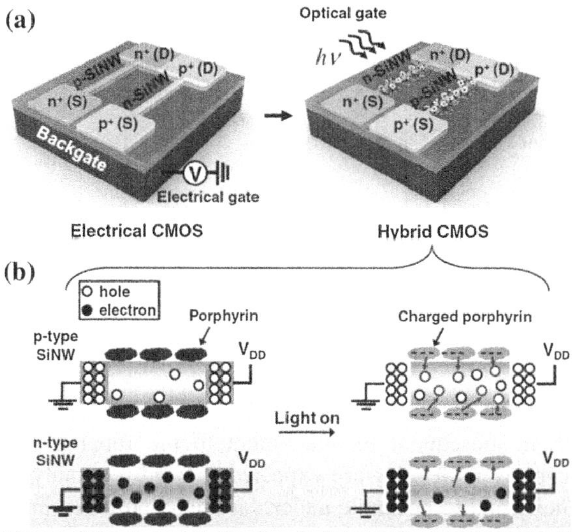

Fig. 10.16 Hybrid integrated electronic and photonic device using SiNW arrays. **a** Schematic illustration of the conversion of an electrical CMOS structure for an electrical input system into a hybrid CMOS structure for an optical input system. **b** Schematic illustration of the observed characteristics (PCT) in SiNWs coated with porphyrins. When light illuminates, the PCT and subsequent charge molecules can gate the SiNWs and control the potential inside the SiNWs [31]

Once the devices are illuminated by visible light, nanowire surface charge affects the currents in n- and p- type FETs in opposite manner, and as a consequence, the hybrid CMOS acts as logic gate with optical inputs.

10.2.6 Future Prospects and Challenges

The applications of many photochromic molecules have yet to be demonstrated aside from porphyrin. Photochromic molecules can be utilized for biological application as well by attaching biomolecules at the end of photochromic molecule to be controlled by light illumination. Also, logic circuits driven by different wavelength of light is a possibility. However, the effective functionalization of hybrid device application is still a challenge. For practical applications, stable and reproducible functionalization is required. Also, analysis of surface state has not been carried out sufficiently and the electrical transfer mechanism in interface has yet to be examined clearly

10.3 Biodetection and Diagnostics

Combination of nanowire FETs with biological recognition elements boosts up SiNWs to the level of biosensing devices (see conceptual image in Fig. 10.17), enabling label-free detection of the analyte of interest in real time [4–7], and [35]. Crucial for all kinds of biosensors is the choice of specific receptor molecules and

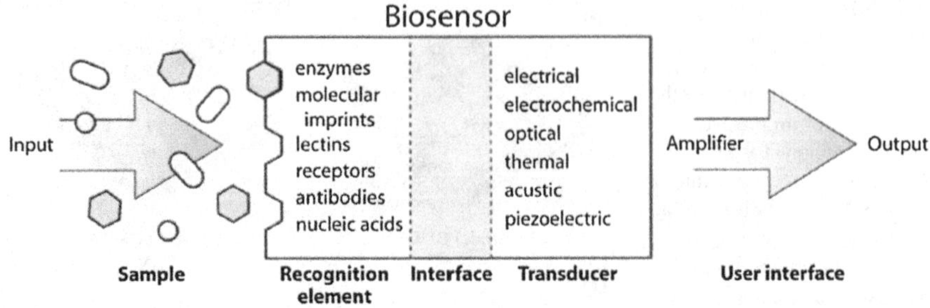

Fig. 10.17 Principle design of a biosensor, consisting of recognition elements directly associated with a transducer and combined with a signal readout unit [36]

their subsequent bioattachment to the functionalized sensor surface [36]. The capture molecules are expected to bind the analyte with high affinity and specificity and to be stable under varying conditions. For a nanowire-based biosensing, recognition elements should be small enough to enable a change of conductance of the nanowire FET once the analyte is bound to its receptor. In this section, functionalization of silicon nanowires will be described and on this basis, several options for biorecognition elements will be presented.

10.3.1 Functionalization of Silicon Nanowires

The aim of surface modification for biosensing is to place the biological receptor molecules close to the transducer surface to achieve high sensitivity. The design of each functionalization setup is crucial for the performance of the sensor. Oriented immobilization or receptor densities may increase the signal whereas steric hindrance and unspecific adsorption should be avoided for high signal quality. Electrochemical biosensors like silicon nanowire-based FETs require in addition a short distance between the biorecognition event of receptor and target molecule and the sensor surface [37]. These events are only detectable within the Debye screening length which is defined by the ionic strength of a solution (approximately 3 nm in 10 mM ionic solution [18]). A low ionic strength increases the Debye layer thickness; it is, however, competing with a reduced biosensing ability as ions are needed to stabilize the structure formation of biomolecules. Otherwise, the receptors might lose their specificity against the target.

The choice of the appropriate linker molecules for the surface functionalization strategy depends on both, the sensor surface and the possible functional groups for attachment of the receptor. In the case of Si nanowire devices, either SiO_2 or Si–H is present. Silicon nanowires are oxidized in ambient conditions resulting in a native oxide shell. Thermal oxidation leads to a more homogeneous surface which enhances reproducibility of the electrical properties. The oxide can only be removed for a short time by etching which results in hydrogen-terminated silicon nanowires. Before developing a functionalization scheme, it is necessary to decide for oxidized nanowire surfaces or hydrogen-terminated silicon. Biological receptors commonly expose free primary amines, carboxy, *thiol,* or *aldehyde* groups or can be synthesized with these functionalities. Consequently, the linker molecules should provide reactive groups able to bind the receptor in a reliable way.

10.3.1.1 Hydrogen Termination

After removing the oxide in strong acid (HF or NH4F, see Fig. 10.18), the resulting hydrogen-terminated silicon is able to react with ω-alkenes. The C = C double bond at one chain end is catalyzed in UV light and covalently attaches to Si–H forming stable Si–C bonds. The long alkyl chains tend to form well-ordered self-assembled monolayers (SAM). By using carboxy-terminated molecules, receptors with functional primary amines can be easily immobilized on the modified nanowire surface. Bunimovich et al. demonstrated that DNA probes which were surface bound via electrostatic interaction to an amino-terminated oxide-free Si nanowire show a higher sensitivity for hybridization with the complementary strand than on an equally amino functionalized oxidized nanowire [38]. This advantage of hydrogen termination has to be counterbalanced with the etching step which is aggressive toward metallic materials on the chips.

10.3.1.2 Silanization

Organosilanes are the standard class of organic molecules for covalent functionalization of SiO_2. The core atom is a Si atom with four bonds, and the principal structure can be summarized in the formula $\mathbf{R}_n\text{Si}\mathbf{X}_{(4\text{-}n)}$, see Fig. 10.19. Whereas, **R** is the organic terminal group usually chosen for further receptor attachment, **X** represents a hydroxyl or hydrolyzable group consisting of -Cl or (m)ethanol. Hydroxylated silicon dioxide can react with these silanol groups forming stable siloxane bonds (Si–O–Si). Silane layers can be stabilized by postbaking which is said to lead to a cross-linking of unbound silanol side chains. The disadvantage is the possibility of polymerization in solution and undirected attachment of the silanes on the surface leading to inhomogeneous surface layers. This can be

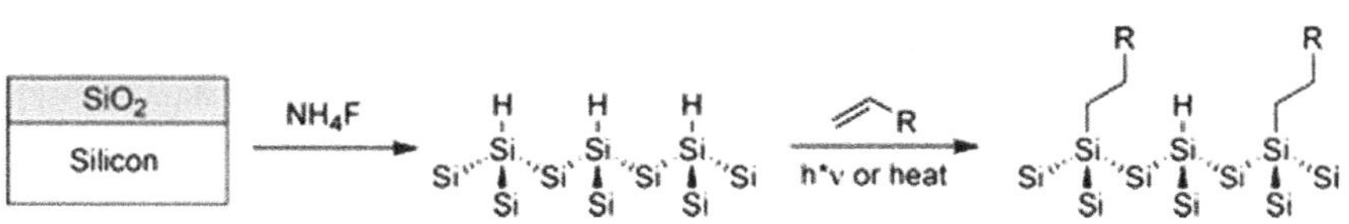

Fig. 10.18 After removing the dioxide layer in HF or NH4F, the hydrogen-terminated surface can react with alkenes by activation with temperature or UV light. **R** is the chosen terminal group for further attachment like carboxyl groups [39]

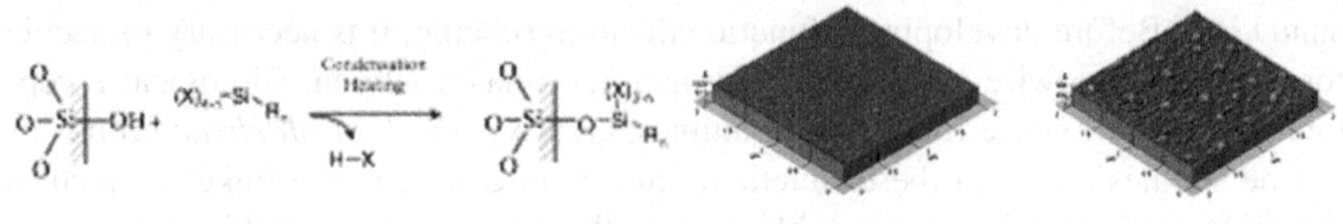

Fig. 10.19 Scheme for covalent attachment of organosilanes (RnSiX(4-n)) to a hydroxylated silicon dioxide surface via condensation(*left*). The bare Si wafer appears flat in AFM topography images (*middle*) whereas silane molecules lead to an increased roughness (*right*) [40]

improved by replacing two **X** with nonhydrolyzable methyl groups. Another option is the electric field alignment of silanes which increase the device sensitivity, see Fig. 10.20.

To these, functionalized surface receptors can be attached covalently with different functional groups. As amino, carboxy, aldehyde, or thiol groups are the most abundant groups targets display, the surfaces are usually provided with the complementary chemical group: carboxy or aldehyde for amines and vice versa and thiolated surface for thiols. For some reactions, cross-linkers or activation steps are needed [42].

To give an example, Lee et al. used an aminosilane for modification of silicon dioxide, see Fig. 10.21 [43]. The cross-linker succinic anhydride attaches to the primary amine in a ring-opening reaction. After activation using zero-length cross-linkers, the functional amino group of the DNA strand can covalently bind to the surface carboxy group. The success of the modification steps can be analyzed with techniques like AFM or fluorescence, as Lee et al. published for DNA functionalized nanowires, see Fig. 10.21. Here, the height and roughness increase significantly due to the molecular layers.

10.3.2 Role of Debye Screening Length

As mentioned in the beginning of previous paragraph, the distance between the biorecognition event and nanowire is important for the FET sensitivity. By cutting the antibody and using only *Fab* fragments (the detecting region of antibody), the biorecognition takes place closer to the transducer surface and leads to an improved device performance, see Fig. 10.22. This effect can be enhanced by tuning the receptor density to a value of mean surface coverage (see Figure E, right scheme). Avoiding steric hindrance of the receptors, they can lay down within in sensitive region, higher densities force them to stand upright sticking out of the sensitive layer, see Fig. 10.22, left scheme. This technique also allows for higher ionic strength of the buffer solution which again stabilizes the biorecognition itself [18].

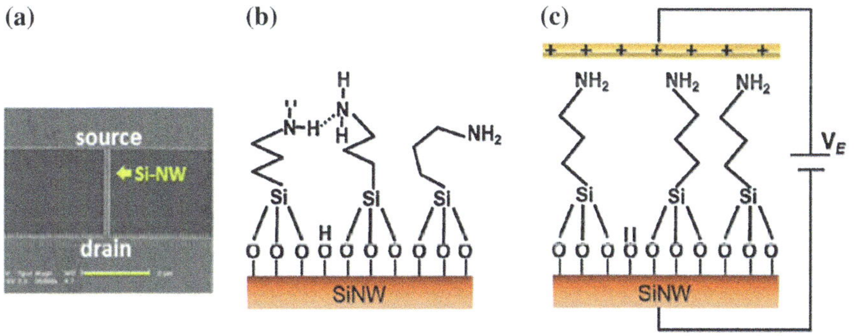

Fig. 10.20 Tuning the sensitivity of top-down fabricated SiNW FET by electric field alignment of the surface-bound organic molecules on a single NW device (*left*). Surface functionalization leads to irregular arrangement of organosilanes (*middle*: amino silane). Applying an electric field (*right*) leads to a higher degree of parallel arrangement which improves the device performance [41]

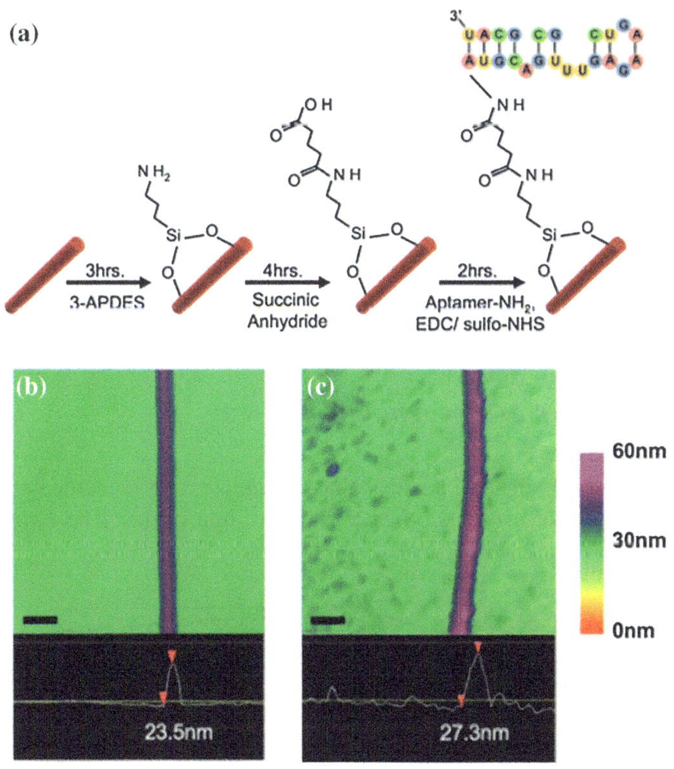

Fig. 10.21 (**a**) Functionalization scheme for a silicon nanowire with oxide shell for attachment of DNA as a capture layer. Topography analysis with AFM shows that after all immobilization steps, height and roughness of the sample have increased in (**c**) compared to the blank (**b**) [43]

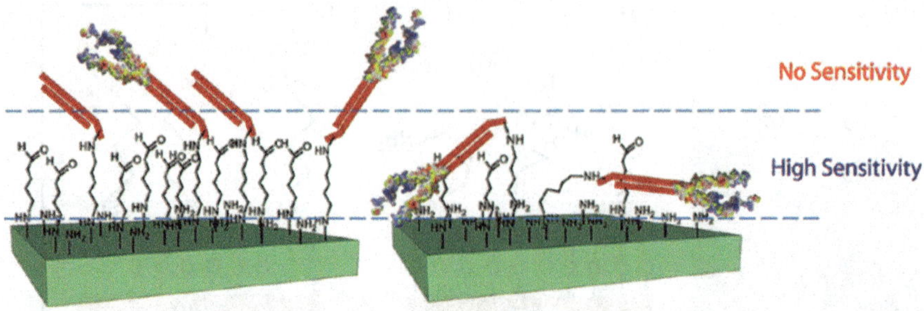

Fig. 10.22 Attachment of Fab fragments to an aldehyde surface (aminosilane linker and glutaraldehyde cross-linker). Biorecognition takes place above the sensitive layer for a high receptor density (*left*) whereas a mean density allows for molecule bending (*right*) [18]

10.3.3 Noncovalent Binding of the Molecules

Noncovalent adsorption of molecules on SiNWs can also be employed for sensing purposes. Let us consider an example of lipid bilayers that have shown a great potential for multiple biotechnological applications. As an application SiNWs, they can be used to shield the nanowire surface (see Fig. 10.23, blue reference curve in the inset) and to prevent it from unspecific adsorption of the analyte molecules, like proteins. Due to the lipid mobility in bilayers, defects can heal easily increasing the device lifetime. By integrating pores in the membrane, it is possible to sense specific ions or molecules as the pores open upon their presence which results in a conductance change. Also pH changes initiate pore openings and can be detected [44]. Kinked nanowires have been coated with lipid bilayers to increase biocompatibility for measurements of single-cell potentials [45].

10.3.4 Antibodies as Recognition Element

Antibodies are biological molecules that recognize very specifically a certain target. Antibodies consist of two heavy and two light chains. The *Fab* domain is

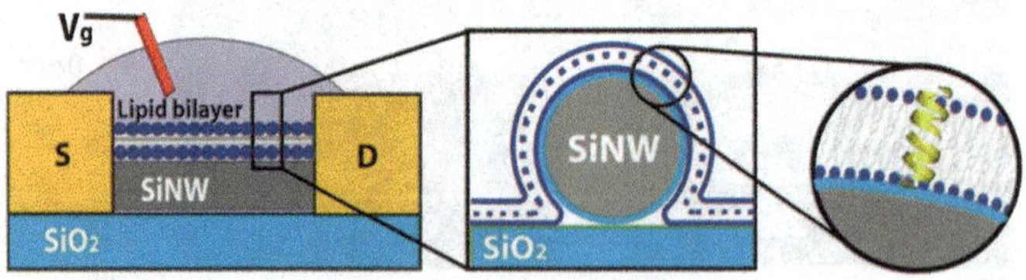

Fig. 10.23 Scheme of lipid bilayer with pore protein on SiNW FET and pH sensitivity (*upper graph*) compared to pore-free bilayer (*lower graph*) which shows no change in the electrical behavior due to pH change from 6 to 9 as indicated [44]

responsible for target recognition and each antibody is thus in principle able to bind two target molecules. It is the variable region that adapts to new pathogens and thus helps the immune system to fight the disease. The *Fc* end of a class of antibodies is fixed on the contrary. There are two groups of antibodies—polyclonal and monoclonal, depending on their production process. Polyclonal antibodies are a pool of different antibodies that can detect an analyte whereas monoclonal antibodies are all identical. Antibodies are highly specific, but they are produced in animals and are relatively sensitive to varying environmental conditions which challenges the application in biosensing. In addition, the target variety is limited as antibodies have to occur naturally [46]. Several examples employing antibodies on Si nanowire-based FETs have been demonstrated so far showing their potential. Cancer markers like PSA (prostate specific antigen) could be detected down to a low concentration of 90 fg/ml with antibodies covalently attached to the SiO_2 surface via primary amines to the aldehyde terminated surface, see Fig. 10.24 [6].

Silicon nanowire FETs are able to detect single biorecognition events (single molecule sensing) as Patolsky et al. demonstrated and confirmed optically detecting influenza A virus using surface-attached antibodies as illustrated in Fig. 10.25 [5]. Another application was demonstrated by Mishra et al. for detection of bacterial toxin SEB (*Staphylococcus aureus* enterotoxin B) [47]. The antibodies were immobilized as a capture layer on the transistor via covalent bonding to the carboxy-terminated surface. By analyzing the impedance upon SEB recognition, the sensor response down to 1 fM of SEB was monitored.

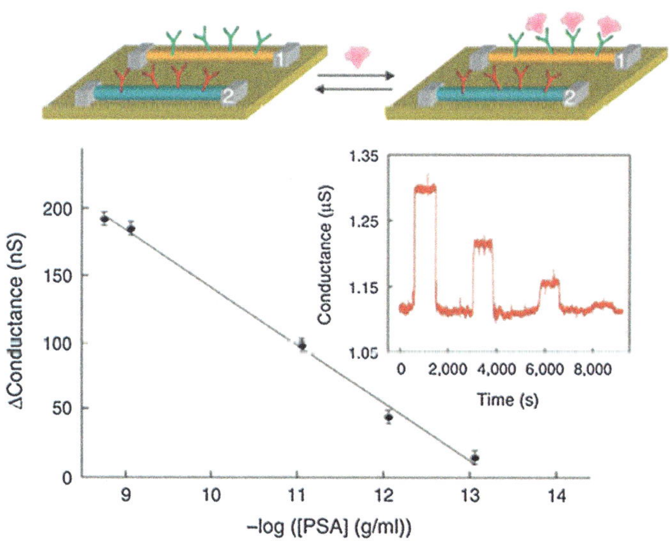

Fig. 10.24 Multiplexed detection of cancer markers on p-type SiNW array FETs by using different antibodies attached to the NWs (*top scheme*). PSA biorecognition leads to conductance changes with 90 fg/ml as a detection limit. The inset shows the conductance development with time [6]

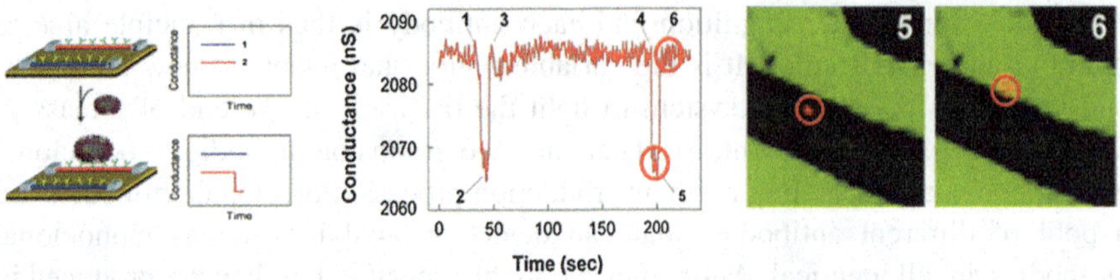

Fig. 10.25 Multiplexed detection of viruses by using different antibodies attached to the NWs. Upon biorecognition, the conductance changes with time (*left*). Parallel acquisition of conductance (*middle*) and an optical images (*right*) show single virus attachment and detachment and corresponding signal change

10.3.5 Peptides as Recognition Element

10.3.5.1 Structure

In addition to antibodies, peptides can also represent specific capture molecules enabling for nanowire-based sensing of biochemical species. Oligopeptides consist of a small number of amino acids linked by peptide bonds (see Fig. 10.26). Due to their little size they are more stable than antibodies and can reveal high specificity that make them perfect recognition elements for biosensors at the nanoscale.

A size of the peptides allows the binding of the target molecules in close proximity to the nanowire surface, which potentially can increase the sensitivity of the biodetection. This argument together with the fact that such receptors can be artificially developed for a large variety of the targets represents a great advantage of using peptides for future biotechnology.

10.3.5.2 Peptides Development

Peptides for sensing can be chosen taking nature as a model [49]. Furthermore, peptides can be identified by a selection process similar to SELEX for aptamers against almost every imaginable target. This screening technology, called *bio-panning*, is based on phage display—the presentation of peptides on the surface of bacteriophage particles. Phage display was firstly introduced in 1985 by George P. Smith [50] and is now a widely expanding research field [51–53]. The marvelous idea behind is the linkage between the genotype and the phenotype of the phage: a foreign DNA with the information for the peptide is inserted into the genome of the phage and the peptide is displayed as a fusion to one of the coat proteins of the phage [52]. If the foreign DNA insert has got a randomized sequence, each phage will present a different peptide. Such a wide diversity of phages is called a phage display library and can be used to select those peptides by biopanning which are able to bind to a certain target molecule with high affinity and specificity. Peptides

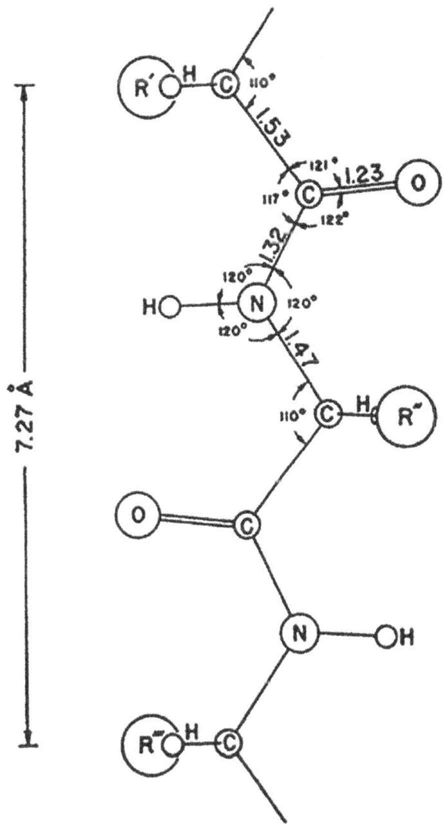

Fig. 10.26 Peptide structure. Each residue R represents an amino acid side chain [48]

identified by phage display have become part of many biosensing applications [54] and would be a promising option for nanowire-based biosensors, too.

10.3.5.3 Applications in Nanowires Sensing

Regarding nanowire-based sensing, peptides have been used as biorecognition elements for metal ion detection. Two different peptides—specific for ions Cu2+ and Pb2+ —were immobilized on independently addressable clusters of silicon nanowires. The metal ions could be simultaneously detected and quantified. Although high amounts of additional Cu2+ ions influenced Pb2+ concentration measurements at low concentration, almost no interaction was observed for clinical relevant concentrations (see Fig. 10.27) [49].

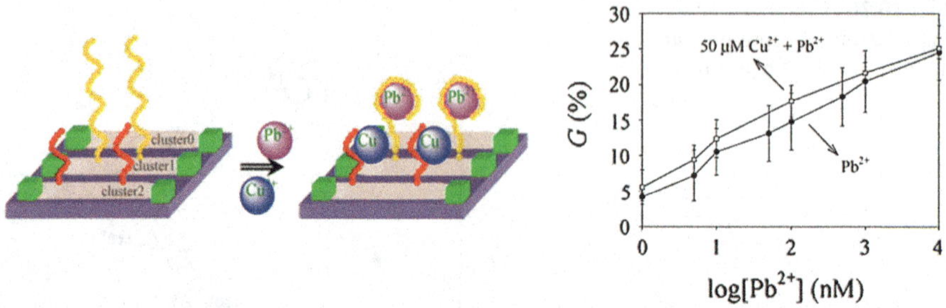

Fig. 10.27 Detection of Pb2+ ions (with or without additional Cu2+ in solution) and effect of Pb2+ concentration on conductance of SiNWs modified with Pb2+—specific peptide [49]

Fig. 10.28 Comparison of peptide nucleic acid (PNA) and deoxyribonucleic acids (DNA) [36]

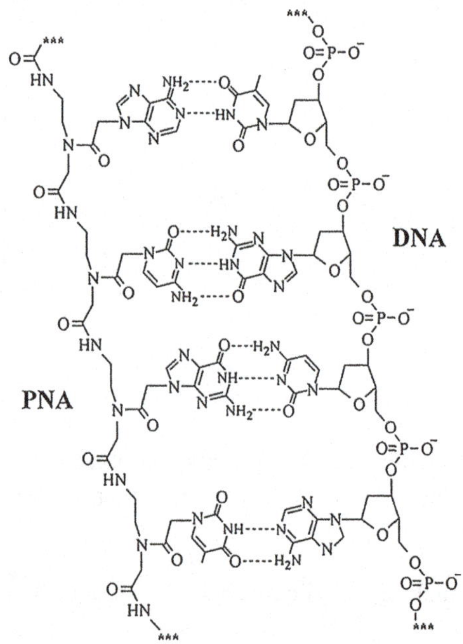

10.3.6 DNA and PNA as Recognition Elements

10.3.6.1 Structure

Single-stranded DNA (ssDNA) can be detected by hybridization (see Fig. 10.28, right-hand side) using complementary single-stranded deoxyribonucleic acids (DNA) or artificial peptide nucleic acids (PNA). DNA is a polymer of nucleotides, which are made up of nucleobases and a negatively charged sugar phosphate backbone. In contrast to DNA, PNA has neutral backbone and consists of

N-(2-aminoethyl)glyine units linked by peptide bonds (see Fig. 10.28, left-hand side) [55].

Compared to DNA, PNA is more stable and shows a higher affinity [5].

10.3.6.2 Applications in Nanowires Sensing

Li et al. immobilized ssDNA oligonucleotides on SiNWs and could show a strong conductance change when adding a 25 pM solution of complementary DNA, but no signal change for mismatch DNA. They could achieve a signal-to-noise ratio of 8 and 6 for p- and n-type wires, respectively [7]. SiNWs modified with PNA were used to detect wild-type DNA down to fM-level and to discriminate from mismatch DNA. Therefore, sensors were functionalized with PNA specific for a sequence from cystic fibrosis transmembrane receptor and it could be shown that DNA containing the ΔF508 mutation, an indicator for cystic fibrosis, gives a much smaller signal than wild-type DNA (Fig. 10.29) [56].

10.3.7 Aptamers as Recognition Elements

Aptamers are a new class of receptors based on DNA and were first reported 1990 [57]. Aptamers are artificial ligands consisting of short oligonucleotides (single-stranded DNA or RNA) of a length of usually 15 up to less than 100 bases. Depending on their sequence, they possess a characteristic structure (quadruplex, pinhole or others). The structure is stabilized by ions, (see Fig. 10.30 for a *thrombin* binding aptamer). This enables the detection of target molecules ranging from small organic molecules (e.g., TNT [58]) up to large proteins like thrombin [59].

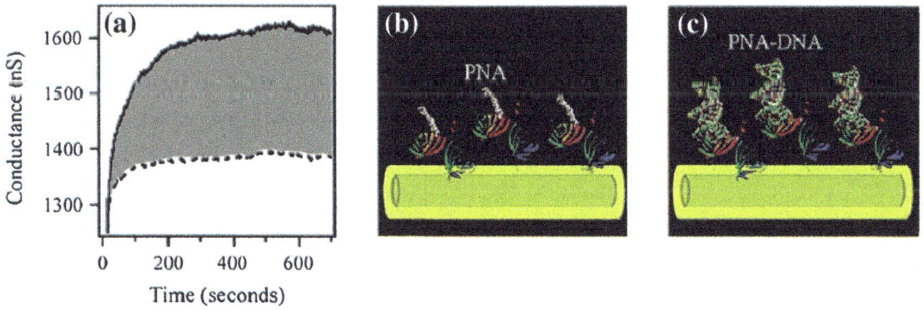

Fig. 10.29 Sensing of DNA using PNA as recognition element. Effect of 100 fM wild-type DNA (*solid line*) and 100 fm mutated DNA (*dashed line*) on SiNW conductance (**a**), Schematic binding of DNA to PNA on SiNW surface (**b** and **c**) [56]

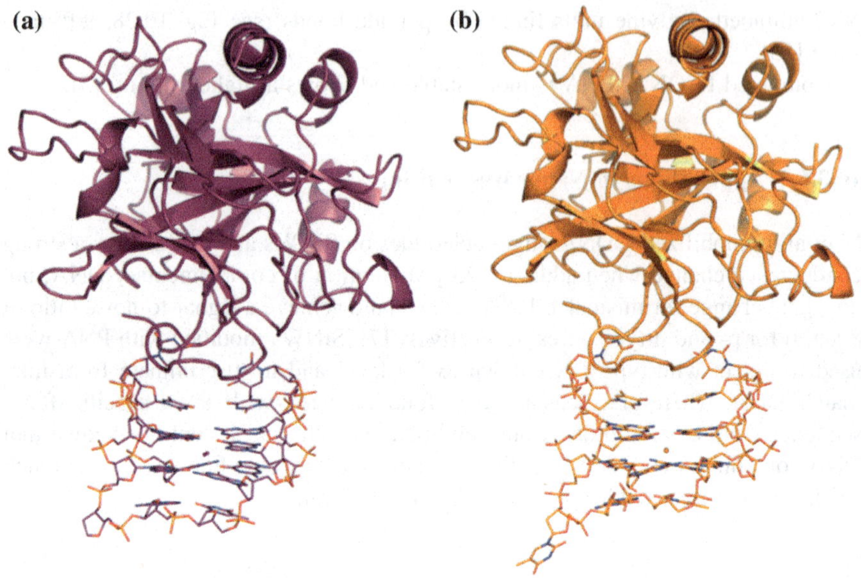

Fig. 10.30 Model of aptamer–thrombin complex formation in the presence of sodium (**a**) and potassium (**b**) [60]

Compared to antibodies, the advantages are their low costs due to in vitro production and easy labeling with functional groups or fluorophores, the little batch-to-batch variations, their relatively high stability under nonphysiological conditions, and the larger variety of targets such as toxins [46]. However, developing a new aptamer is relatively complicated and requires multiple processing steps.

Aptamers are artificially designed receptors and have to be identified for each new target. This is carried out in a so-called SELEX process (systematic evolution of ligands by exponential enrichment) out of a library of ssDNA. Each oligonucleotide has the same length consisting of a randomized sequence of the predetermined number of bases. On both ends, primers with a fixed sequence are attached. They are used, e.g., for immobilizing the DNA. This library of typically 1,013–1,018 different combinations is incubated with the immobilized target, unbound DNA is washed away [46]. By changing the solution, bound molecules are eluted, followed by a further enrichment and are again incubated with the target. By repeating several rounds, one can possibly identify unique sequences for detection. Depending on the target characteristics, various SELEX procedures have been developed. In order to identify aptamers for small target molecules, Stoltenburg et al. developed a method called Capture-SELEX to attach the potential aptamer strand onto magnetic beads via hybridization with a short complementary strand [61]. If a very strong affinity occurs between the target and the aptamer, dehybridization takes place. By this, only excellent binders (=aptamer) are identified.

10.3.7.1 Applications in Nanowires Sensing

Since reported, aptamers have gained increasing interest for application in Si nanowire-based sensing due to their advantages for FET-based sensors. Because of the smaller size of aptamers compared to antibodies, biorecognition can take place closer to the transducer surface within the Debye layer [62]. Sensing of thrombin which plays a major role in the blood coagulation cascade was carried out with a 15-bases-long aptamer [63]. Thrombin was detected in model solutions (thrombin in buffer) as well as in blood samples (see Fig. 10.31). Devices with a control aptamer of a randomized sequence on the contrary did not show a signal change upon sample injection (see inset in Fig. 10.31). Tumor growth or the level of angiogenesis is often indicated by the presence and concentration of the vascular endothelial growth factor (VEGF) and thus its detection can be interesting in clinical applications. Lee et al. could measure VEGF down to the subnanomolar range with aptamer-functionalized Si nanowire FETs [43].

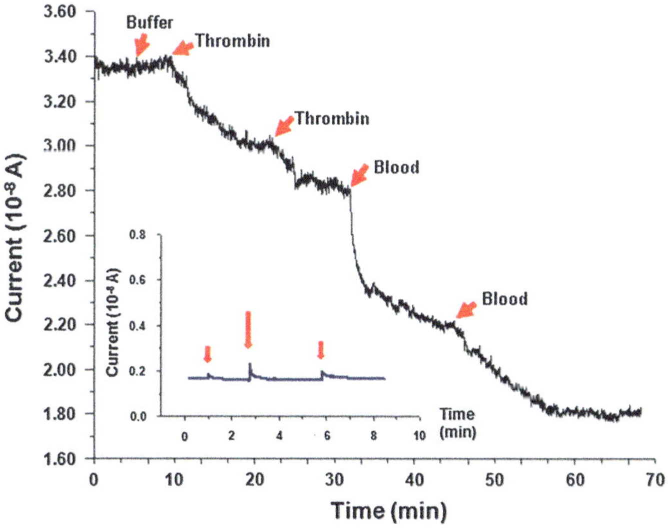

Fig. 10.31 Real-time detection of thrombin in different samples with an aptamer-functionalized Si nanowire FET. The current changes due to biorecognition. The inset shows a control aptamer without conductance change after injection [43]

10.4 Drug Delivery Applications

As it is reported in numerous publications and reviews, nanomaterials and in particular silicon nanowires can be considered not only as a backbone in diverse in vitro detection tests, but also as a highly selective drug carrier for in vivo applications [64–66]. Some of the prerequisites, which make nanowires to be superior for a number of applications in nanomedicine are related to their size properties, i.e., high surface area and elongated shape (high size aspect ratio), allowing easy penetration through the biological tissue [67, 68]. Furthermore, biochemical modification of the nanowires, usually covered by the shell of amorphous silica, is relatively easy [40]. Another important argument to explore 1D nanostructures for spying at the scale of single proteins or cellular machines is a simple size domain comparison of the nanowires and typical biomolecules. Because dimensions of the single molecules and nanoparticles nearly coincide, investigation of biological processes using nanotechnological tools become possible and can be performed without too much interference [69, 70].

Thus, after the decade of intense investigations of physical and chemical properties, nanowires entered the phase of application relevant research and even numerous commercial realizations [71]. Nowadays, the nanowires are in the scope of intense investigations for development of novel fluorescent biological labels, [72] drug and gene delivery, [73], tissue engineering, tumor destruction, to name just a few.

10.4.1 Construction of Nanowire-Based Biologically Active Drug Carrier

In order to use the nanowire as biologically active unit, i.e., vehicle for drug delivery, or contrast agent for in vivo imaging, it has to acquire necessary functionality. A biological or molecular coating, acting as bioorganic interface should be linked to the nanowires surface. Important role of the coating is to reduce the *toxicity* and to provide the *biocompatibility* of the nanowires with the surrounding environment. Some of the prominent examples of biological layers include monolayers of small molecules (e.g., fibronectin coating for better attachment of Fibroblasts), polymers like collagen [74, 75], DNA brushes, or antibodies. Furthermore, the functionalized "nano vehicles" has to exhibit ability to be detected inside of the organism or testing tube using optical, electrical, electromagnetic, etc. techniques. Set of standard approaches used to construct the biologically active nanowires are schematically summarized in Fig. 10.32.

Silicon NW typically represents a core of the bioactive object. Since the core can carry important properties (e.g., luminescence or radio-frequency response), it should be covered by inert protective layer. In the case of silicon nanowires, amorphous native silicon oxide SiO_x represents a natural protection of the core.

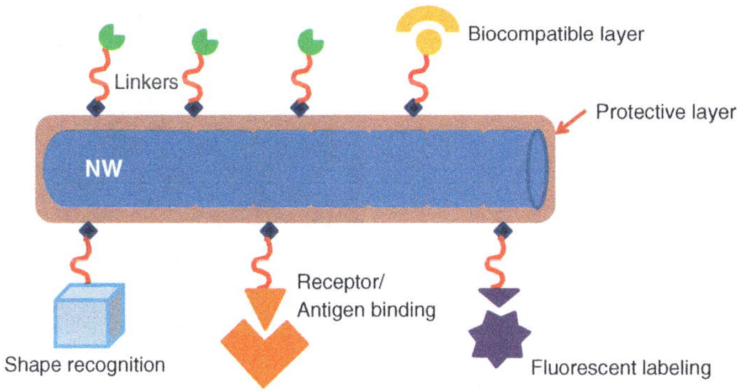

Fig. 10.32 Schematic description of the typical design, applied for development of the bionanomaterials

Alternatively, additional organic layers, consisting of, i.e., lipid shells can be formed around the nanowire to play a protective role [76]. In order to make nanowires suitable for biological tagging and labeling, linkers with diverse functional groups can be further immobilized at the surface of the covering shell. End group of the linkers is always aimed to attach various species, like antibodies, magnetic tags, fluorophores, etc.

10.4.2 Internalization of the Nanowires: In Vitro Tests

One of the strategies to deliver the relevant medications and contrast agents, or to probe and manipulate biological processes occurring inside the cells [67, 75], is *internalization* of the nanowires by the living cells. Internalization is defined by the process when the nanomaterials (i.e., nanoparticles, nanotubes, nanowires) are engulfed by the cells through their membrane [64, 77]. After numerous investigations in this field, it is assumed that so-called *endocytosis* (or *phagocytosis* for uptaking of the objects larger than 0.5 μm) describes most precisely the processes of the intracellular injection of the biochemical species and nanoparticles [78]. This process typically consists of few stages: (1) approach of the object to the lipid membrane; (2) formation of the lipid-based vesicle, wrapping the transported species; (3) fusion of the vesicles inside of the cells; (4) unwrapping of the vesicle for sorting and targeting of the delivered species. To facilitate the internalization, surface properties of the nanowires must be predesigned and specifically functionalized to meet the requirements of cellular delivery and targeting (see Sect. 10.4.1).

In order to demonstrate the interfacing of the nanomaterials and living cells, let us consider an example of the mammalian cells such as mouse embryonic stem (mES) cells placed into direct contact with vertically grown silicon nanowires

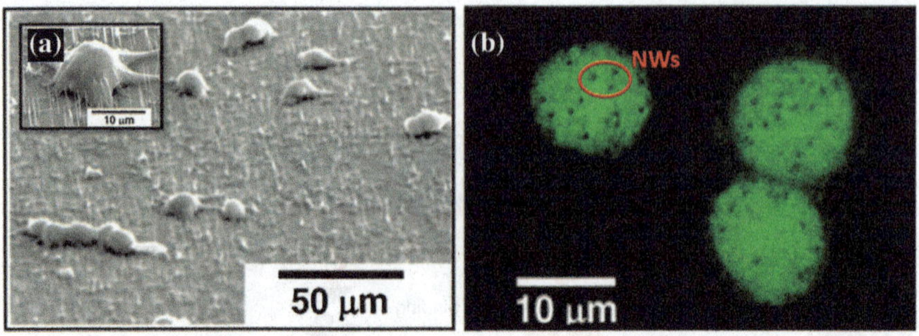

Fig. 10.33 Penetration of the silicon nanowires inside of the stem cells mES. **a** incubation of the stem cells on the silicon substrate in order to internalize nanowires; **b** Confocal microscopy image to visualize mES with penetrated SiNWs [67]

(Fig. 10.33). For these experiments, stem cells were cultured directly on a silicon wafer with as grown nanowires [67], as shown in Fig. 10.33a. SiNW array penetrated into the cell naturally, without applying any external force. It has been shown that nanowires internalization did not affect dramatically cellular metabolism, since the cells survived up to several days on the nanowire substrates. Confocal microscopy was used to visualize the cells with nanowires penetrated inside (Fig. 10.33b).

Whereas first studies have demonstrated the possibility of vertical nanowires to penetrate *spontaneously* cells' membrane, further work was dedicated to generalize the use of silicon nanowires and to represent nanowires as a universal platform to deliver a large variety of the biological species, which can affect cell activity [79]. In particular, it is known that large variety of the complex cellular processes can be probed and analyzed, by integrating surface-modified SiNWs, delivering diverse range of biomolecules into living cells. To achieve this rich output, an approach, introduced in Fig. 10.33, is further developed to internalize a broad range of bioeffectors, i.e., DNA, RNA, peptides, and proteins (see Sect. 10.3) into almost any cell type. This ability is summarized in Fig. 10.34.

Note that silicon nanowires can be covered by aminosilane groups, which provide noncovalent binding of the molecules to the NWs. Thus, molecules can be released from the wires surface, once internalization happened [64, 66, 67, 79]. Thus, in vitro internalization of silicon nanowires carrying biomolecules might be considered as an excellent tool to assist in discovery new aspects of fundamental cell behaviors, such as motility, proliferation and differentiation, adhesion, etc.

10.4.3 Nanowires-Based Drug Delivery: In Vivo Tests

After the number of successful in vitro tests where the uptake of the silicon nanowires by numerous cell types were studied, the research of SiNW cells

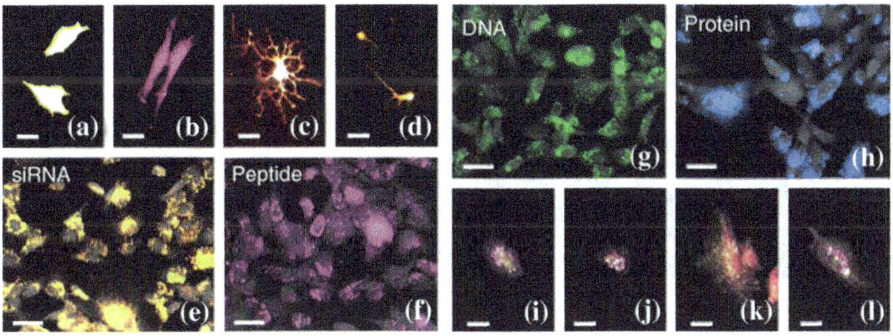

Fig. 10.34 Diverse biochemical species can be delivered by means of SiNWs vertically grown [79]. **a–d** Transfection of fluorescent proteins into **a** HeLa S3 cells, **b** human fibroblasts, **c** NCPs, **d** hippocampan neuron cells. Transport of biomolecules such as **e** siRNA, **f** peptides, **g** plasmid DNA, **h** recombinant TurboRFP-mito protein. Co-transfection of two different molecules: **i** two different siRNAs, **j** peptide and siRNA, **k** plasmid DNA and siRNA, **l** a recombinant protein and siRNA

conjugates reached the phase of in vivo assays. Traditionally, one of the most common areas where the nanomaterials meet biotechnology is closely related to anticancer therapy [66, 73]. The intense investigation of the drug delivery processes is motivated by certain deficiencies of the traditional treatment. Among them are the facts that (1) only a small amount of conventionally delivered anticancer drug can penetrate into tumors; (2) strong distribution of drugs into non-target tissues; (3) short circulation time in the blood. Use of the nanomaterials for novel diseases (i.e., cancer) treatment is considered as a new paradigm, which enables the dramatic increase in the drug concentration inside of target tissue due to its high loading capacity

In order to demonstrate the capability of the silicon nanowires to internalize and deliver concentrations of the drugs, sufficient to inhibit growth of the tumor, let us focus on in vivo therapeutic examinations of the mice bearing epidermoid carcinoma tumor (KB) on their back [66]. For these purposes, the anticancer antibiotic doxorubicin (DOX) is associated with SiNWs by physisorption. An ultrahigh drug-loading capacity of 20800 mg/g has been shown for these particular experiments. A summary of the effect of SINW-DOX complexes on mice KB cells is depicted in Fig. 10.35.

Mice population with KB tumors was divided into four groups, treated with (a) physiological saline, (b) pure SiNWs, (c) free DOX, and (d) SiNW-DOX complexes. In the case of doxirubicine therapy (c and d), two concentrations of 5 and 25 μg/kg were administered, respectively. Afterward, the mouse autofluorescence analysis was performed to estimate the efficacy of the treatment. Obviously, SiNW-DOX complexes revealed high killing rate of the tumor on the one hand, and extremely longtime accumulation of the drug in the tumor region.

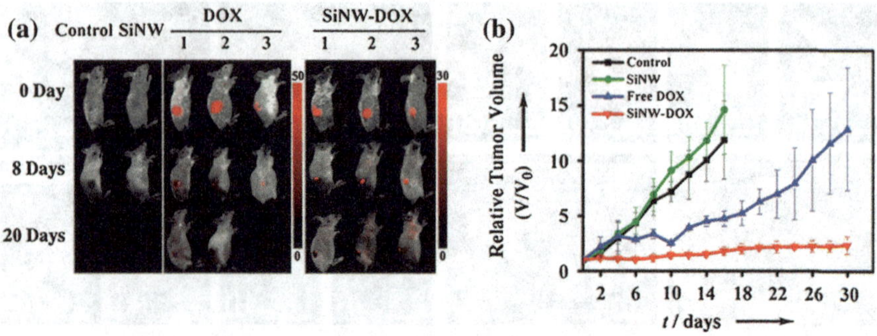

Fig. 10.35 *In-vivo* anticancer assay of the mice, carrying KB tumor. **a** in vivo fluorescence images of four groups mice with labeled KB cells, tested with saline, SiNW, DOX, and SINW-DOX; **b** Tumor growth inhibition at different time points

10.5 Challenges and Perspectives

SiNW-based sensors hold up the potential as efficient and powerful vehicles for biodetection and biomedical diagnostics. These sensor devices offer the capability of high-sensitivity sensing and easy signal readout in real time without time- and money-consuming labeling steps. However, the detection sensitivity depends on the ionic strength of the sample, and therefore, high ionic strength solutions like blood as a desalting step is advisable. Additional challenges consist of checking the reproducibility of biosensing results for a given target molecule and producing the market-ready biosensing devices. A vision would be to implement parallel detection of a huge number of different analytes to get a high-throughput screening method, comparable and even better than DNA or protein microarrays, but on a biosensing level.

SiNWs conjugates with organic and biomolecules represent excellent platform for studying fundamental behaviors of the cells such as adhesion or proliferation. From the point of view of practical applications, such as drug delivery, there are ongoing research on specificity, toxicity, and biodegradability of the nanowires in the living organism. Nevertheless, nanowires are considered as very promising candidates for the future therapy, since they are able to provide high drug-load capacity and targeted transport of the medicines.

Therefore in the future, it might be possible to simultaneously screen hundreds of biomarkers for medical or environmental purposes, and to use nanowires with specific bioactive layers for disease treatment, making thereby “lab on a wire” a reality.

References

1. Iijima, S. (1991). Helical microtubules of graphitic carbon. *Nature, 354*(6348), 56–58.
2. Wu, Y., Xiang, J., Chen, Y., Wei, L., & Lieber, C. M. (2004). Single-crystal metallic nanowires and metal/semiconductor nanowire heterostructures. *Nature, 430*(6995), 61–65.
3. Nerowski, A., Poetschke, M., Bobeth, M., Opitz, J., & Cuniberti, G. (2012). Dielectrophoretic growth of platinum nanowires: Concentration and temperature dependence of the growth velocity. *Langmuir, 28*(19), 7498–7504.
4. Cui, Y., Wei, Q., Park, H., & Lieber, C. M. (2001). Nanowire nanosensors for highly sensitive and selective detection of biological and chemical species. *Science, 293*, 1289–1292.
5. Patolsky, F., Zheng, G., Hayden, O., Lakadamyali, M., Zhuang, X., & Lieber, C. M. (2004). Electrical detection of single viruses. *Proceedings of the National Academy of Sciences of the United States of America, 101*(39), 14017–14022.
6. Zheng, G., Patolsky, F., Cui, Y., Wang, W. U., & Lieber, C. M. (2005). Multiplexed electrical detection of cancer markers with nanowire sensor arrays. *Nature Biotechnology, 23*(10), 1294–1301.
7. Li, Z., Chen, Y., Li, X., Kamins, T. I., Nauka, K., & Williams, R. S. (2004). Sequence specific label-free DNA sensor on silicon nanowires. *Nano Letters, 4*, 245–247.
8. Park, I., Li, Z., Pisano, A. P., & Williams, R. S. (2010). Top-down fabricated silicon nanowire sensors for real-time chemical detection. *Nanotechnology, 21*(1), 9. 015501.
9. Pregl, S., Weber, W.M., Nozaki, D., Kunstmann, J., Baraban, L., Opitz, J., Mikolajick, T., Cuniberti, G. (2013). Parallel arrays of Schottky barrier nanowire field effect transistors: Nanoscopic effects for macroscopic current output. *Nano Research*, DOI: 10.1007/s12274-013-0315-9
10. Gao, X. P. A., Zheng, G., & Lieber, C. M. (2010). Subthreshold regime has the optimal sensitivity for nanowire FET biosensors. *Nano Letters, 10*(2), 547–552.
11. Knopfmacher, O., Tarasov, A., Fu, W., Wipf, M., Niesen, B., Calame, M., et al. (2010). Nernst Limit in Dual-Gated Si-Nanowire FET Sensors. *Nano Letters, 10*(6), 2268–2274.
12. Spijkman, M., Smits, E. P. C., Cillessen, J. F. M., Biscarini, F., Blom, P. W. M., & de Leeuw, D. M. (2011). Beyond the Nernst-limit with dual-gate ZnO ion-sensitive field-effect transistors. *Applied Physics Letters, 98*(4), 043502.
13. Weber, W., Graham, A.P., Duesberg, G.S., Liebau, M., Cheze, C., Geelhaar, L., Unger, E., Pamler, W., Hoenlein, W., Riechert, H., Kreupl, F., & Lugli, P. (2008) Tuning the Polarity of Si-Nanowire Transistors Without the Use of Doping. In *Nanotechnology, 2008. NANO'08. 8th IEEE Conference on* (pp. 580–581). IEEE Xplore.
14. Heinzig, A., Slesazeck, S., Kreupl, F., Mikolajick, T., & Weber, W. M. (2012). Reconfigurable silicon nanowire transistors. *Nano Letters, 12*(1), 119–124.
15. Martin, D., Heinzig, A., Grube, M., Geelhaar, L., Mikolajick, T., Riechert, H., et al. (2011). Direct Probing of Schottky Barriers in Si Nanowire Schottky Barrier Field Effect Transistors. *Physical Review Letters, 107*(21), 216807.
16. Cherns, D., Anstis, G. R., Hutchison, L., & Spence, J. C. H. (1982). Atomic structure of the NiSi 2/(111) Si interface. *Philosophical Magazine A, 46*(5), 849–862.
17. Mi, S. B., Jia, C. L., Urban, K., Zhao, Q. T., & Mantl, S. (2009). NiSi2/Si interface chemistry and epitaxial growth mode. *Acta Materialia, 57*(1), 232–236.
18. Elnathan, R., et al. (2012). Biorecognition layer engineering: Overcoming screening limitations of nanowire-based FET devices. *Nano Letters, 12*(10), 5245–5254.
19. Patolsky et al. (2006). Detection, stimulation, and inhibition of neuronal signals with high-density nanowire transistor arrays. *Science, 313*, 1100–1104. 5790.
20. Tian, H., & Feng, Y. (2008). Next step of photochromic switches? *Journal of Materials Chemistry, 18*, 1617–1622.
21. Russew, M., & Hecht, S. (2010). Photoswitches: from molecules to materials. *Advanced Materials, 22*, 3348–3360.

22. Browne, W. R., & Feringa, B. L. (2009). Light switching of molecules on surfaces. *Annual Review of Physical Chemistry, 60*, 407–428.
23. Kadish, K. M., Smith, K. M., & Guilar, R. (2003). *The porphyrin handbook: Inorganic, organometallic and coordination chemistry* (Vol. 3). New York: Academic Press.
24. Lshii, H., Sugiyama, K., Ito, E., & Seki, K. (1999). Energy level alignment and interfacial electronic structures at organic/metal and organic/organic interfaces. *Advanced Materials, 11*(8), 505–625.
25. Dietrich, P., Michalik, F., Schmidt, R., Gahl, C., Mao, G., Breusing, M., et al. (2008). An anchoring strategy for photoswitchable biosensor technology: azobenzene-modified SAMs on Si(111). *Applied Physics A, 93*, 285–292.
26. Pei, X., Fernandes, A., Mathy, B., Laloyaux, X., Nysten, B., Riant, O., et al. (2011). Correlation between the structure and wettability of photoswitchable hydrophilic azobenzene monolayers on silicon. *Langmuir, 27*, 9403–9412.
27. Lv, Xiaojun, Changc, Haixin, & Zhang, Hao. (2011). Photoelectrochemical switch based on cis-Azobenzene chromophore modified TiO_2 nanowires. *Optics Communications, 284*(20), 4991–4995.
28. Li, C., Ly, J., Lei, B., Fan, W., Zhang, D., Han, J., et al. (2004). Data storage studies on nanowire transistors with self-assembled porphyrin molecules. *The Journal of Physical Chemistry B, 108*, 9646–9649.
29. Li, Q., Surthi, S., Mathur, G., Gowda, S., Zhao, Q., Sorenson, T. A., et al. (2004). Multiple-bit storage properties of porphyrin monolayers on SiO_2. *Applied Physics Letters, 85*(10), 1829–1831.
30. Khaderbad, M. A., Roy, U., Yedukondalu, M., Rajesh, M., Ravikanth, M., & Ramgopal Rao, V. (2010). Variable interface dipoles of metallated porphyrin self-assembled monolayers for metal-gate work function tuning in advanced CMOS technologies. *IEEE Transactions on Nanotechnology, 9*(3), 335–337.
31. Choi, S.-J., Lee, Y.-C., Seol, M.-L., Ahn, J.-H., Kim, S., Moon, D.-I., et al. (2011). Bio-inspired complementary photoconductor by porphyrin-coated silicon nanowires. *Advanced Materials, 23*, 3979–3983.
32. Seol, M.-L., Choi, S.-J., Kim, C.-H., Moon, D.-I., & Choi, Y.-K. (2012). Porphyrin silicon hybrid field-effect transistor with individually addressable top-gate structure. *ACS Nano, 6*(1), 183–189.
33. Zidon, Y., Shapira, Y., & Dittrich, Th. (2007). Illumination induced charge separation at tetraphenyl-porphyrin/metal oxide interfaces. *Journal of Applied Physic, 102*, 053705.
34. Winkelmann, C. B., Ionica, I., Chevalier, X., Royal, G., Bucher, C., & Bouchiat, V. (2007). Optical switching of porphyrin-coated silicon nanowire field effect transistors. *Nano Letters, 7*(6), 1454–1458.
35. Bi, X., Agarwal, A., & Yang, K. L. (2009). Oligopeptide-modified silicon nanowire arrays as multichannel metal ion sensors. *Biosensors & Bioelectronics, 24*(11), 3248–3251.
36. Chambers, J. P., Arulanandam, B. P., Matta, L. L., & Valdes, J. J. (2008). Biosensor recognition elements. *Curr Issues Mol Biol, 10*(12), 1–12.
37. Stern, E., Wagner, R., Sigworth, F. J., Breaker, R., Fahmy, T. M., & Reed, M. A. (2007). Importance of the debye screening length on nanowire field effect transistor sensors. *Nano Letters, 7*, 3405–3409.
38. Bunimovich, Y., Shin, Y., Yeo, W., & Amori, M. (2006). Quantitative real-time measurements of DNA hybridization with alkylated nonoxidized silicon nanowires in electrolyte solution. *Journal of the American Chemical Society, 128*(50), 16323–16331.
39. Bocking, T., James, M., Coster, H., Chilcott, T., & Barrow, K. (2004). Structural characterization of organic multilayers on silicon (111) formed by immobilization of molecular films on functionalized Si-C linked monolayers. *Langmuir, 20*, 9227–9235.
40. Dugas, V., Elaissari, A., & Chevalier, Y. (2010). Surface sensitization techniques and recognition receptors immobilization on biosensors and microarrays. In *Recognition Receptors in Biosensors* (pp. 47–134), Zourob, M., Ed., Springer, New York.

41. Chu, C.-J., Yeh, C.-S., Liao, C.-K., Tsai, L.-C., Huang, C.-M., Lin, H.-Y., et al. (2013). Improving nanowire sensing capability by electrical field alignment of surface probing molecules. *Nano Letters,*. doi:10.1021/nl400645j.
42. Hermanson, G. T. (2008). Bioconjugated Techniques. Academic Press pp. 1–1233 .
43. Lee, H., Kim, K., Kim, C., Hahn, S., & Jo, M. (2009). Electrical detection of VEGFs for cancer diagnoses using anti-vascular endothelial growth factor aptamer-modified Si nanowire FETs. *Biosensors & Bioelectronics, 24*, 1801–1805.
44. Misra, N., Martinez, J., Huang, S., Wang, Y., Stroeve, P., Grigoropoulos, C., et al. (2009). Bioelectronic silicon nanowire devices using functional membrane proteins. *Proceedings of the National Academy of Sciences of the United States of America, 106*, 13780.
45. Gao, R., Strehle, S., Tian, B., Cohen-Karni, T., Xie, P., Duan, X., et al. (2012). Outside looking in: Nanotube transistor intracellular sensors. *Nano Letters, 12*, 3329–3333.
46. O'Sullivan, C. K. (2002). Aptasensors–the future of biosensing? *Anal Bioanal Chem, 372*, 44–48.
47. Mishra, N. N., Maki, W. C., Cameron, E., Nelson, R., Winterrowd, P., Rastogi, S. K., et al. (2008). Ultra-sensitive detection of bacterial toxin with silicon nanowire transistor. *Lab on a Chip, 8*, 868–871.
48. Pauling, L., Corey, R. B., & Branson, H. R. (1951). The structure of proteins: Two hydrogen-bonded helical configurations of the polypeptide chain. *Proceedings of the National Academy of Sciences of the United States of America, 37*(4), 205–211.
49. Bi, X., Agarwal, A., & Yang, K. L. (2009). Oligopeptide-modified silicon nanowire arrays as multichannel metal ion sensors. *Biosensors & Bioelectronics, 24*(11), 3248–3251.
50. Smith, G. P. (1985). Filamentous fusion phage: Novel expression vectors that display cloned antigens on the virion surface. *Science, 228*, 1315–1317.
51. Pande, J., Szewczyk, M. M., & Grover, A. K. (2010). Phage display: Concept, innovations, applications and future. *Biotechnology Advances, 28*(6), 849–858.
52. Paschke, M. (2006). Phage display systems and their applications. *Appl Microbiol Biotechnol, 70*(1), 2–11.
53. Kehoe, J. V., & Kay, B. K. (2005). Filamentous phage display in the new millennium. *Chemical Reviews, 105*(11), 4056–4072.
54. Mao, C., Liu, A., & Cao, B. (2009). Virus-based chemical and biological sensing. *Angewandte Chemie International Edition, 48*(37), 6790–6810.
55. Sassolas, A., Leca-Bouvier, B. D., & Blum, L. J. (2008). DNA biosensors and microarrays. *Chemical Reviews, 108*(1), 109–139.
56. Hahm, J., & Lieber, C. M. (2004). Direct ultrasensitive electrical detection of DNA and DNA sequence variations using nanowire nanosensors. *Nano Letters, 1904*(1), 51–54.
57. Ellington, A. D., & Szostak, J. W. (1990). In vitro selection of RNA molecules that bind specific ligands. *Nature, 346*, 818–822.
58. Menger, M., Glökler, J., & Rimmele, M. (2006). Application of aptamers in therapeutics and for small-molecule detection. *Handbook of Experimental Neuropharmacology 894*(173), 359–373.
59. Bock, L. C., Griffin, L. C., Latham, J. A., Vermaas, E. H., & Toole, J. J. (1992). Selection of single-stranded DNA molecules that bind and inhibit human thrombin. *Nature, 355*, 564–566.
60. Russo Krauss, I., Merlino, A., Randazzo, A., Novellino, E., Mazzarella, L., & Sica, F. (2012). High-resolution structures of two complexes between thrombin and thrombin-binding aptamer shed light on the role of cations in the aptamer inhibitory activity. *Nucleic Acids Research, 40*, 8119–8128.
61. Stoltenburg, R., Nikolaus, N., & Strehlitz, B. (2012) Capture-SELEX: Selection of DNA Aptamers for Aminoglycoside Antibiotics. *Journal of Analytical Methods in Chemistry,902* 415697.
62. Chiu, T. (2009). Aptamer-functionalized nano-biosensors. *Sensors, 9*, 10356–10388.
63. Kim, K., Lee, H., Yang, J., Jo, M., & Hahn, S. (2009). The fabrication, characterization and application of aptamer-functionalized Si-nanowire FET biosensors. *Nanotechnology, 20*, 235501.

64. Chen, H.-H., Chien, C.-C., Petibois, C., Wang, C.-L., Chu, Y. S., Lai, S.-F., et al. (2011). Quantitative analysis of nanoparticle internalization in mammalian cells by high resolution X-ray microscopy. *Journal of Nanobiotechnology, 9*, 14.
65. Rim, H. P., Min, K. H., Lee, H. J., Jeong, S. Y., & Lee, S. C. (2011). pH-Tunable calcium phosphate covered mesoporous silica nanocontainers for intracellular controlled release of guest drugs. *Angewandte Chemie International Edition, 123*(38), 9015–9019.
66. Peng, F., Su, Y., Wei, X., Lu, Y., Zhou, Y., Zhong, Y., et al. (2013). Silicon-nanowire-based nanocarriers with ultrahigh drug-loading capacity for in-vitro and in-vivo cancer therapy. *Angewandte Chemie International Edition, 52*(5), 1457–1461.
67. Kim, W., Ng, J. K., Kunitake, M. E., Conklin, B. R., & Yang, P. (2007). Interfacing silicon nanowires with mammalian cells. *Journal of the American Chemical Society, 129*, 7228–7229.
68. Salata, O. (2004). Applications of nanoparticles in biology and medicine. *Journal of Nanobiotechnology, 2*, 3.
69. Taton, T. A. (2002). Nanostructures as tailored biological probes. *Trends Biotechnol, 20*, 277–279.
70. Whitesides, G. M. (2003). The "right" size in nanotechnology. *Nature Biotechnology, 21*, 1161–1165.
71. Mazzola, L. (2003). Commercializing nanotechnology. *Nature Biotechnology, 21*, 1137–1143.
72. Bruchez, M., Moronne, M., Gin, P., Weiss, S., & Alivisatos, A. P. (1998). Semiconductor nanocrystals as fluorescent biological labels. *Science, 281*, 2013–2016.
73. Panatarotto, D., Prtidos, C. D., Hoebeke, J., Brown, F., Kramer, E., Briand, J. P., et al. (2003). Immunization with peptide-functionalized carbon nanotubes enhances virus-specific neutralizing antibody response. *Chemistry & Biology, 10*, 961–966.
74. Sinani, V. A., Koktysh, D. S., Yun, B. G., Matts, R. L., Pappas, T. C., Motamedi, M., et al. (2003). Collagen coating promotes biocompatibility of semiconductor nanoparticles in stratified LBL films. *Nano Letters, 3*, 1177–1182.
75. Zhang, Y., Kohler, N., & Zhang, M. (2002). Surface modification of superparamagnetic magnetite nanoparticles and their intracellular uptake. *Biomaterials, 23*, 1553–1561.
76. Jiang, Z., Qing, Q., Xie, P., Gao, R., & Lieber, C. M. (2012). Kinked p–n Junction Nanowire Probes for High Spatial Resolution Sensing and Intracellular Recording. *Nano Letters, 12*, 1711–1716.
77. Zhang, W., Tong, L., & Yang, C. (2012). Cellular Binding and Internalization of Functionalized Silicon Nanowires. *Nano Letters, 12*, 1002–1006.
78. Marsh, M., & McMahon, H. T. (1999). The Structural Era of Endocytosis. *Science, 285*(5425), 215–220.
79. Shalek, A. K., Robinson, J. T., Karp, E. S., Lee, J. S., Ahn, D. R., Yoon, M. H., et al. (2010). Vertical silicon nanowires as a universal platform for delivering biomolecules into living cells. *Proceedings of the National Academy of Sciences of the United States of America, 107*(5), 1870–1875.

Chapter 11
Nanowire FET Circuit Design: An Overview

Jinyong Chung

Abstract Thanks to the short-channel effect suppression of nanowire FET (NWFET), it has become a candidate for the next-generation VLSI device. Different types of NWFETs are introduced to meet the system requirements, with pros and cons. The bottom-up-processed transistor provides the full performance in transistor level, but the integration capability is bounded by the single type of devices per layer and logic design for NMOS and PMOS combined structure is severely restricted. Besides, gate length is relatively large sized, and to overcome this, crossed nanowire transistor-based nanoscale integrated circuit (NASIC) is proposed. Even though this design achieves the minimum area and promises for the large-scale integration so far, the transistor performance is far from the nanotransistor, because of its non-closed gate structure. Top-down process allows flexible design, but the performance and integration are in its early stage. CMOS interface is required to activate the NW logic and to provide signals from/to conventional system.

Abbreviation

NWFET	Nanowire FET
NASIC	Nanotransistor application-specific integrated circuits
SCE	Short-channel effect
FPGA	Field-programmable gate array
GAA	Gate-all-around
SBD	Schottky barrier diodes
xNW	Crossing nanowire
LB	Langmuir–Blodgett
SNAP	Superlattice nanowire pattern transfer
MW	Microwire

J. Chung (✉)
Inha University, Incheon, Korea
e-mail: andychung@inha.ac.kr

D. M. Kim and Y.-H. Jeong (eds.), *Nanowire Field Effect Transistors: Principles and Applications*, DOI: 10.1007/978-1-4614-8124-9_11,

PLA	Programmable logic array
PCM	Phase change memory
TMR	Triple modular redundancy

11.1 Introduction

"Why nanoscale circuits," instead of unlimitedly integrating CMOS VLSI, will be the first question to every person who is starting in circuit design. As minute power-consuming individual transistors are being integrated in tens of billions into smaller silicon chips, heat dissipation and power management is a prime concern in the system design. Simple solution is the power supply lowering, but CMOS technology has met the limit due to relatively high off-leakage and V_T, and alternative FINFET CMOS is sought after to improve the on-to-off-state current ratio. The sub-100-nm planar transistor structure is vulnerable to losing control of the gate over the channel, and the device on or off states. NWFET exhibits the immunity of short-channel effect (SCE) due to reduced source and drain depletion field by the surrounding gate structure, and bulk inversion provides higher conductance. As it provides low V_T—one third or less than that of CMOS and low off-leakage, it is a strong candidate for the next-generation circuit technology enabler.

The early NW transistor dimensions were 50–100 nm in diameter and 1 μm channel length, to demonstrate the nanoscale transistor operation. Narrow diameter NW is advantageous in SCE suppression with high i_{ON}/i_{OFF} ratio, but the resulting narrow gate width reduces the driving capability. The optimum diameter is calculated to be $\sim$3 nm.

The bottom-up process can take advantage of externally grown high-quality NWs. There are multitude of NW types by the material of single or compound, and the growth direction of radial and axial. The gates can be placed only along the transferred parallel wires. CMOS interface circuits are required to provide the signals, and the microwire supplies the power. The gate length defined by the lithography is well over nanoscale, and for the nanoscale gate length, the second layer wires are placed on top of the first layer orthogonally. Meanwhile, top-down process provides structural direction of vertical or horizontal NWs.

The externally grown wire importing and gate processing on top of the wires accompany wide variability in parameter and high defect rate. The device may fail or show poor performance, and yield can be very low. The circuit deteriorates as the operation continues, and therefore the defect and fault detection circuit and the correction schemes are necessary.

This chapter reviews the SNW transistor for circuit design and the integrated circuits implementation. The single nanowire transistor based on bottom-up process and exploiting the 1D carrier flow is introduced. This is followed by the discussions on the 2D array for designing the nanotransistor application-specific integrated circuit (NASIC) field-programmable gate array (FPGA), and top-down device, which is compatible with CMOS processing. After covering the topics on the error-checking and correcting topic, the chapter closes.

11.2 NWFET I–V Characteristics

The electrostatics can be analogously described as fully depleted gate-all-around (GAA) SOI MOSFETs, and the surface potential can be found by solving [1, 2],

$$\frac{\partial^2 \phi_f(x)}{\partial x^2} - \frac{\phi_f(x) - \phi_{gs}}{\lambda^2} = -\frac{e\rho(x)}{\varepsilon_{NW}} \tag{11.1}$$

where

ϕ_f surface potential
ϕ_{fs} gate potential

$$\lambda = \sqrt{\frac{2\varepsilon_{NW} d_{NW}^2 \ln(1 + 2d_{ox}/d_{NW}) + \varepsilon_{ox} d_{NW}^2}{16\varepsilon_{ox}}}, \text{ (the natural length)} \tag{11.2}$$

$\rho(x)$ carrier density
ε_{NW} the nanowire dielectric constant
ε_{ox} the dielectric constant of oxide
d_{ox} gate oxide thickness
d_{NW} NW diameter.

The natural length λ, also called as characteristic length, represents the channel region controlled by the drain, and the device is free of SCE, when the channel length is larger than 5–10λ.

Equation (11.2) indicates that the further scaling can be achieved by reducing the NW diameter, instead of thinning the gate oxide of 1.5 nm or below which can cause gate tunneling current. Avoiding the thin gate dielectric prevents trapped charges caused by strong electric field across the thin dielectric, and therefore more reliable circuit operation life is expected.

The threshold voltage (V_{TH}) can be calculated by [1],

$$V_{TH} \approx \frac{qN_A}{\varepsilon_{NW}} \lambda^2. \tag{11.3}$$

Where V_{TH} is determined by work function, channel, and dielectric thickness and is not affected by the short-channel length.

The drain current formula for circuit simulation, valid for ballistic and drift–diffusion transport has been proposed [3], and the drain current due to ballistic transport is given by

$$I_{ds} = \frac{qk_BT}{\pi\hbar}\sum\nolimits_n \sum\nolimits_v g_v \times \left[\ln\left(1 + e^{(q\psi_s - E_v^n)/k_BT}\right) - \ln\left(1 + e^{(q\psi_s - E_v^n - qV_{ds})/k_BT}\right)\right] \tag{11.4}$$

where,

g_v valley degeneracy
E_v^n conduction band minimum for the n-th sub-band

$$\psi_s = V_{gs} - \phi_{ms} - \frac{Q_{NW}}{C_{INS}} \tag{11.5}$$

Q_{NW} total charge per unit length in the nanowire

The drift diffusion current is given by

$$I_{ds} = \frac{\mu}{L_{\text{eff}}}\int_0^{V_{ds}} Q_{NW}(V_{gs}, V_{ds})dv \tag{11.6}$$

where $\mu = \frac{\mu_0}{\left(1 + \frac{\mu_0}{v_{\text{sat}}L_{\text{eff}}}V_{ds}\right)}$

The metal contacts to source and drain junctions are Schottky contacts, and the Schottky barrier diodes (SBD) are included for rigorous modeling (Fig. 11.1) [4]. SBD drop of 0.3–0.5 V causes the output signal swing not rail to rail, and this drop should be lower than the threshold voltage not to turn on next stage off transistors in CMOS configuration. The annealing process after metal lithography makes the source–drain contacts close to ohmic contacts, and simple resistor replacements are accepted.

11.3 NWFET as Circuit Element

11.3.1 Transistor Level

Depending on the gate shape over the NW, GAA, Ω, Π type or two crossing nanowire (xNW) gates can be formed. The electrical channel width depends on the

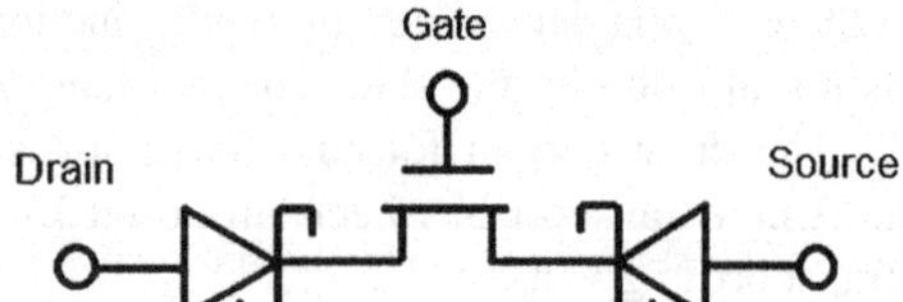

Fig. 11.1 Equivalent circuit of n-type NWFET

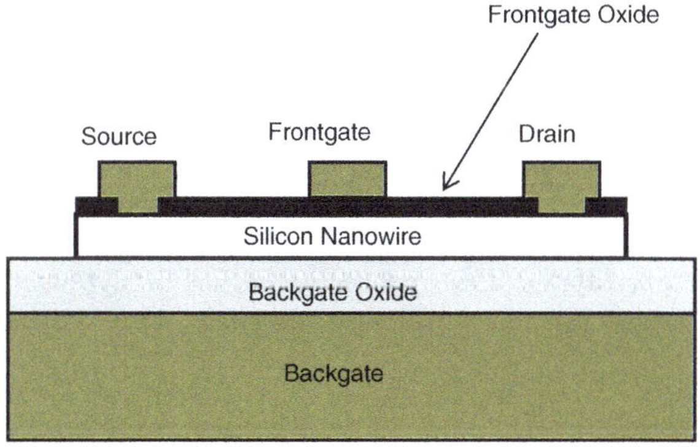

Fig. 11.2 Double-gate NWFET

wire cross-section, cylindrical or non-cylindrical, and gate voltage, where the stronger gate voltage brings the channel closer to the dielectric interface, hence wider channel in n-type device. While the GAA type can take full advantage of nanowire transistors, the xNW transistor has small layout area with nanoscale gate length, and the performance can be very poor due to small effective gate area and thick effective gate dielectric formed by two orthogonal wires. The source/drain junction is normally underlapped by the gate, and this affects the channel property and parasitic capacitance values in a manner different from the conventional overlapped gate CMOS. Due to narrow channel diameter of the source/drain region near the metal, the interconnect resistance is high, and the resistance is reduced by the wider extension region in the top-down process.

The depletion body does not require the bias connection which is necessary for the CMOS VLSI. A double-gate junctionless NWFET (Fig. 11.2), where source–drain junction implantation is not required, can provide ambipolar characteristics. The front gate determines the transistor switching, while the back gate controls the n-type or p–type transistor. The positive voltage, back-gate bias will make the transistor function as n type, while the negative bias makes the device p type, and the convenience of type change, even on-the-fly flip, can provide a flexible circuit design potential.

The continuous flow of DC current will raise the channel temperature higher than the AC operating case due to self-heating effect, and there is drain current degradation [5].

11.3.2 Inverter and Logic Gate Configuration

In bottom-up process, NFET and PFET can be built side by side on one wire, or on separate wires, which may result in complex routing. Single-wire inverter may require large area between p- and n-type transistor registrations. So, for the dense layout, clocked pseudo-NMOS (or PMOS) type circuit is advantageous, even though it burns higher power dissipation. To build the CMOS configuration it is easier to use the top-down process technology for implementation.

PFET and NFET conductance matching by width, like CMOS design, is not possible for single-diameter NW, and two p-type wires for one n type [6] or different diameters for p-type and n-type NWFETs are proposed [7].

The vertical NWT is good for inverter design, but multi-input gate design is complex, for its serial or parallel connection involves large layout space, and especially, the series connection results in high interconnect resistance with poor conductance. Two stack-gate structures were implemented [8].

11.4 Bottom-Up Process Technology VLSI Design

In the bottom-up process, the prefabricated nanowires are placed on substrate by pattern transfer method, Langmuir–Blodgett (LB) or superlattice nanowire pattern transfer (SNAP), with controlled pitch, followed by the dielectrics growth and gate material deposition (GAA NWT) or NW gate stacking (cross-NWT) to process the NW transistors. By the number of nanowire layers transferred, single-layer, dual-layer, and multi-layer (3D) devices have been demonstrated [9, 10]. This technology application to design is adopted earlier than the top-down approach, due to the better dimension and performance-controlled single-transistor fabrication. The gate characteristics are determined by on–off drain current ratio which is affected by the neighboring NW.

11.4.1 Gate-All-Around NW Transistor Technology

The transferred NWs on substrate require gate dielectric and gate material deposition. The gate terminal is made of polysilicon or metal, and its geometry is defined by lithography [9]. To drive the NW transistors, CMOS circuit is required to provide power and signals through microwire (MW). Since this technology can achieve the best performance NWFETs, the terahertz signal-handling Ge-core/Si-shell NW transistor has been reported [11]. Combined with the low-temperature CMOS process, the consistent layer-to-layer device performance has been obtained for multi-stack device. Up to 10 layers of stacking has been achieved, and the circuit can be manufactured on flexible substrate, paper, plastic, or glass [10].

The transistor length control with lithography has limited the minimum size, and the one-directional NW alignment is severely restricting the flexible logic gate configuration. As only p- or n-type NWs are transferred per layer, nano-CMOS circuit design is impractical, since the two different layer p- and n-type NWFETs can be connected through less compatible MW interconnection.

11.4.2 Cross-Point NW Transistor Technology

To improve the channel length of GAA transistor, nanosize transistor length has been achieved by nanowire gate placement. When the parallel horizontal layer (row pattern) and vertical layer (column pattern) NWs are transferred consecutively, then the cross-point NWFET array structure results (Fig. 11.3). The elimination of gate geometry delineation by lithography results in deteriorated transistor performance, where nanowire transistor advantage of GAA structure is replaced by rather inferior planar-type cross-point gate structure at this size. This process limits the transistor orientation as horizontal and vertical, and the routing of interconnection is not flexible—just along the NW path with nanometal wire.

As only one type of NWs is transferred per layer, n-type input transistors can form series input gates, together with precharging/predischarging and evaluating devices at both ends. The former yields NAND function, while the latter becomes AND. If p-type NWs are used, then NOR or OR function plane is composed. The following second-level logic, where a different type of transistor logic is provided, provides nanoscale programmable NAND–NOR or AND–OR logic array (PLA) design (Fig. 11.4). The CMOS logic block microscale interface to the nanoscale NW blocks for signal mapping is a key design issue, and decoder or demultiplexer circuits are adopted. As the series-connected device strength is weak, it is not suitable for high performance; rather, it is suitable for smaller area and low power.

The nano-PLA [12] and NASIC (Fig. 11.5), [13] have drawn the attention, for its implementation potential of cross-point NWFET array into PLA or processors. Two externally grown layers of NWs, usually n type and p type, are transferred onto silicon, and FET channel and gates are selectively formed at the cross-points with gate lithography and oxidation. At the cross-points where the transistor gates are not formed, the NW is made to act as interconnect, with lowering resistivity process, typically, silicidation. The source–drain region functionalization requires

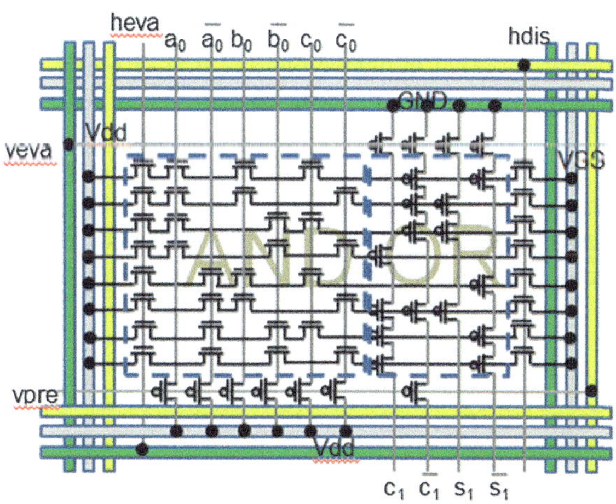

Fig. 11.3 Cross-NW AND–OR two-level logic [8]

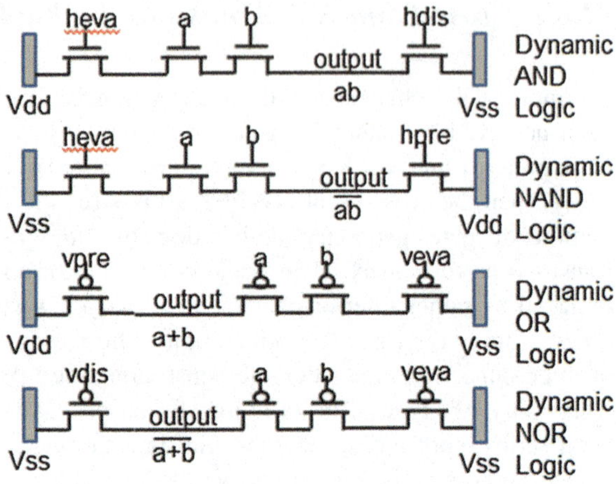

Fig. 11.4 PLA gate structure with single-type NW transistor

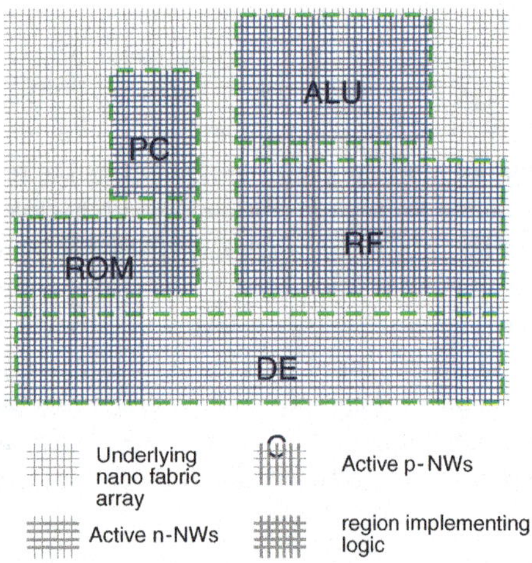

Fig. 11.5 Floor plan of WISP-0 processor

n- and p-active implantation, and non-self-align junction transistors are formed. For the two cylindrical wire crossing gate area is not of true planar shape, and a weak channel is formed due to narrow effective width. The lower and upper layer NWs can be either gate layer or source–channel–drain layer; hence, very flexible gate array design is resulted. With the multiple flexible array tiles, image processors, streaming processors are fabricated. Even single-transistor-type logic is proposed to limit the parameter variability to nNWFET only [14] (Fig. 11.6).

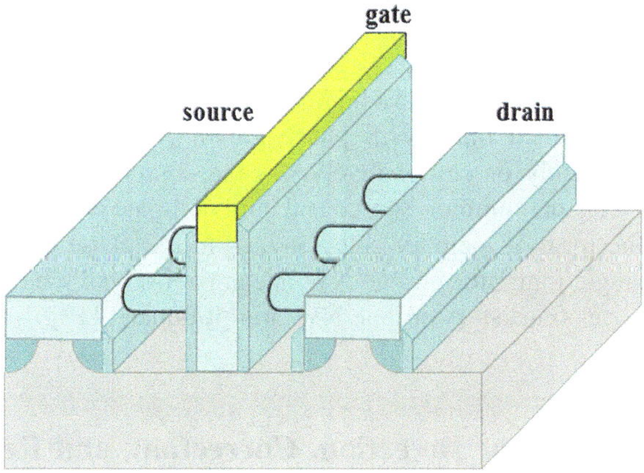

Fig. 11.6 Horizontal NW FET, multi-wire channel

11.5 Circuits for Top-Down Process

11.5.1 Horizontal NWFET Circuits

The top-down horizontal NWFET (Fig. 11.6) is the closest to the conventional CMOS structure, where the transistor channel is replaced by one or multiple NWs. On the contrary to the bottom-up approach where the externally grown NWs are transferred to VLSI circuit substrate, the top-down processing builds the NWs on the device substrate directly. The NW transistor is a natural advancement of Ω or Π gate FINFET, whose gate is modified to extend to the bottom side, hence maintains the advantages of the CMOS lithography for arbitrary transistor location, size, and complementary logic gate construction. The wire cross sectional shape is non-cylindrical, and the transistor performance varies widely in chip due to less controllability over the NW peripheral shape. The available device is limited to Si channel, and line edge roughness affects the carrier mobility more than the bottom-up one.

As the NW transistor can be configured by substituting the planar CMOS transistor channels by one or multi-NWs, the existing CMOS layout can be converted with minimum modification compared with other NWT technology. As the gates are surrounded and affected by neighboring wires in multi-NWs, depleting gate material cannot be used. The horizontal and vertical stack multi-wire channel transistors were fabricated for 25-stage ring oscillators which showed 10 pS delay per inverter stage [15]. This technology is the most suitable for designing complex gates, due to its flexible transistor connection—serial or parallel—freedom, compared with bottom-up or vertical NW. The suspended beam structure of the channel makes the wire affected by mechanical tensile pressure.

11.5.2 Vertical NWFET Circuits

Vertical device allows higher device integration for source and gate and drain that are formed on a VLS phase grown NW. P-type and n-type wires are grown separately or consecutively. As the source and gate are buried deep inside the insulator, common-source and common-gate circuit will allow best areal density in the form of memory cell array, as in the phase change memory (PCM) [16]. As single transistor on one NW accompanies heavy burden in interconnect and area, series transistors on one NW are introduced [17].

11.6 Error Detection, Correction, and Reliability

The nanoscale manufacturing accompanies wider variation in device dimensions and electrical characteristics, and results in highly unacceptable electrical performance in speed or power. And the device fail rate is more sensitive to defects or device parameter spread, compared with the conventional CMOS technology. Even the good device can experience transient faults introduced by power glitches, more internal noises due to inherently weaker signal charge and the radiation originated soft errors. The device suffers ongoing reliability, where the continued operation may degrade the electrical parameters and damage the gate dielectric materials.

Many error-checking and correcting techniques developed in CMOS can be adopted to cope with the NWFET faults. The built-in fault tolerance method detects the faults by scan path, BILBO, memory testing, or other types of error checking. The permanent fault self-repair method disables the failing function blocks, and the faulty blocks are replaced by the redundancy circuit. The transient faults can be suppressed by majority polling method, and the circuit implementation requires large area for duplicate functions. Reconfiguration of physical function that skips the faulty area can enhance the error immunity. This practice has been implemented in the FPGA's and requires the programmable devices. The application of these methods to NWFET circuits needs modification and improvements, for they require far-less process variability and defects.

For the bottom-up process, VMATCH [18] is proposed to improve the yield in nano-PLA by improving the timing performance. It identifies the physical gates strength by decreasing order of i_{ON}, through V_T measurements, and maps to the high-fanout (logically weak) gates in sequence. In NASICs, hierarchical method [19] is proposed to cover 5–15 % defect rate; it combines the redundancies for gate level and system level, and built-in error-correcting circuitry. Stuck faults are masked at gate-level redundancy, and voting-based triple modular redundancy (TMR) screens above gate-level faults. As TMR circuitry is as defective as the other parts, the insertion point is discretionary where the timing performance is minimally affected.

11.7 Conclusion and Future Prospectus

With continued scaling for the near-future system design, the NWFET advantage of performance and scale supports next-generation system following the CMOS. To interface with the nanoscale devices, microscale connection is currently indispensable for either bottom-up process or top-down process, and the NWFET only circuit in top-down technology may take time for a while. For application-specific designs, bottom-up design is suitable for small-scale and early design adoption for its easy generation of high-quality NWFET, but, diversified function integration of sensors, signal processors, logics, and memories for ultralow power beyond CMOS can be implemented with top-down approach. The wide process variability and high defect rate challenge the design for production yield and longevity of reliable system.

The functional hybrid system, with high-precision NWFET sensor and its analysis or results evaluation in CMOS logic, can be early application. Energy-generating devices and memory devices can be mounted together on various NWFET designs through 3D integration for self-operation.

As NWFET utilizes the benefits of high-performance nanodevices and mature CMOS technology, it will integrate far-diverse functions on one package, and it will take the role of next-generation system design technology.

11.8 Problems

P11.1. Please generate I–V characteristics curve with SPICE

a. Source and drain resistances of GAA transistor are 30 Ω each.
b. For GAA transistor, the source and drain series resistances of 10 KΩ.
c. For xNW transistor, the source and drain resistances are of 100 Ω. Discuss how the gate can be modeled for the simulation.

P11.2. Drain current calculation in AC and DC for self-heating effect (SHE) evaluation and dissipation through surrounding dielectric;

a. Draw equivalent circuit for SHE evaluation.
b. Calculate drain current/volume, heat/volume, for 3 × 3 stacked NWs. $V_{DD} = 1.0$ V, $V_{TH} = 0.3$ V, $V_{GS} = 0$–1.0 V with 2 V step, with continuous V_{DS}, from 0 to 1.0 V. T = room temperature. Repeat the calculation under 1 kHz gate and drain voltage signal.

P11.3. Draw two-stage double-wire inverters for top-down, vertical NW technology.

P11.4. Design 3-input NAND and NOR gate with vertical NWFET and horizontal NWFET. Discuss the pros and cons of each technology.

P11.5. Design a 4-bit adder with NASIC approach. Assume the junctionless device is used, and bottom NWs are n type, upper NWs p type.

P11.6. Discuss the advantages of each NWFET technology in circuit design.

References

1. Yan, R., et al. (1992). Scaling the Si MOSFET: From bulk to SOI to bulk. *IEEE Transactions on Electron Devices, 39*, 1704–1710.
2. Colinge, J. et al. (2007). From gate-all-around to nanowire MOSFETs. IEEE CAS International Semiconductor Conference, pp. 11–17.
3. Paul, B., et al. (2007). An analytical compact circuit model for nanowire FET. *IEEE Transactions on Electron Devices, 54*, 1637–1644.
4. Lee, S., et al. (2009). A SPICE-compatible new silicon nanowire field effect transistors (SNFETs) model. *IEEE Transactions on Nanotechnology, 9*, 643–649.
5. Huang, R. et al. (2010). Self-heating effect and characteristic variability of gate-all-around silicon nanowire transistors for highly-scaled CMOS technology (invited). IEEE International SOI Conference (SOI), pp. 1–4.
6. Liao, Y. et al. (2011). A high-density SRAM design techniques using silicon nanowire FETs. International Semiconductor Device Research Symposium.
7. Bindal, A., et al. (2007). The design of dual work function CMOS transistors and circuits using silicon nanowire technology. *IEEE Transactions on Nanotechnology, 6*, 291–302.
8. Li, X., et al. (2011). Vertically stacked and independently controlled twin-gate MOSFETs on a single Si nanowire. *IEEE Electron Device Letters, 32*, 1492–1494.
9. Cui, Y., et al. (2003). High performance silicon nanowire field effect transistors. *Nano Letters, 3*, 149–152.
10. Javey, A., et al. (2007). Layer-by-layer assembly of nanowires for three-dimensional, multifunctional electronics. *Nano Letters, 7*, 773–777.
11. Liang, G., et al. (2007). Performance analysis of a Ge/Si core/shell nanowire field-effect transistor. *Nano Letters, 7*, 642–646.
12. DeHon, A. et al. (2007). Architecture approaching the atomic scale. IEEE European Solid State Circuits Conference (ESSCIRC), pp. 11–20.
13. Wang, T. et al. (2008). NASICs: A nanoscale fabric for nanoscale microprocessors. IEEE International Nanoelectronics Conference (INEC), pp. 989–994.
14. Narayanan, P. et al. (2008). CMOS control enabled single-type FET NASIC. IEEE Computer Society International Symposium on VLSI (ISVLSI), pp. 191–196.
15. Sleight, J. et al. (2010). Gate-all-around silicon nanowire MOSFETs and circuits. IEEE Device Research Conference (DRC).
16. Sun, X., et al. (2008). Germanium antimonide phase-change nanowires for memory applications. *IEEE Transactions on Electron Devices, 55*, 3131–3135.
17. Lee, S., et al. (2007). Highly scalable non-volatile and ultra-low-power phase-change nanowire memory. *Nature Nanotechnology, 2*, 626–630.
18. Gojman, B. et al. (2009). VMATCH: Using logical variation to counteract physical variation in bottom-up, nanoscale systems. IEEE International Conference on Field-Programmable Technology.
19. Moritz, C., et al. (2007). Fault-tolerant nanoscale processors on semiconductor nanowire grids. *IEEE Transactions on Circuit and Systems, 54*, 2422–2437.

Made in the USA
Monee, IL
13 September 2026

62976347R00164